The authors
dedicate this book
to their families,
who, with great patience,
have witnessed the making
of this volume

Cover photos:
Title and spine: *Cliona delitrix*, a boring sponge—Hans A. Baensch
Back: t. l. *Alcyonium palmatum*, a soft coral—Dr. Harry Erhardt
t. r. *Latrunculia magnifica*, a sponge—Dr. Harry Erhardt
b. l. *Hexabranchus sanguineus*, a nudibranch—Norbert Wu
b.r. *Favia* sp., a stone coral—Dr. Harry Erhardt

Erhardt, Dr. Harry; Moosleitner, Dr. Horst
[Meerwasser Atlas, Band 2. English]
Marine atlas, Volume 2: the joint aquarium care of invertebrates.
Erhardt, Dr. Harry; Moosleitner, Dr. Horst:
with the collaboration of Dr. Rober Patzner ... [et. al.];
translated and revised by Gero W. Fischer, Shellie E. Borrer.
1. English softcover edition 1998, 736 p., 12x18 cm.

ISBN 1-890087-10-6 (U.S.A. only)
ISBN 3-88244-502-5 (For other countries)

Published in the United States by:
Microcosm Ltd., 2085 Shelburne Road, Shelburne, Vermont 05482

Distribution
USA: Microcosm Ltd., 2085 Shelburne Road, Shelburne, Vermont 05482
Canada: Rolf C. Hagen Inc., 3225 Sartelon Street, Montreal, Que. H4R 1E8
Great Britain: Rolf C. Hagen (U.K.) Limited, California Drive, Whitwood Industrial Estate, Castleford WF 10 50H, West Yorkshire
Australia: Pet Pacific Pty. Ltd., Unit C, 30 Skarratt Street, Auburn N.S.W 2144
P. O. Box 398, Rydalmere N.S.W. 2116

Printed in Singapore

Dr. Harry Erhardt   Dr. Horst Moosleitner

MARINE ATLAS 2

— Invertebrates —

1. Softcover edition

With the collaboration of
Dr. Robert A. Patzner

Translated and revised by:
Gero W. Fischer, Ph.D.
Shellie E. Borrer, M.S.

Publishers of Natural History and Pet Books
Hans A. Baensch – Melle – Germany

Foreword

Over the past 10 years, the interest in marine invertebrates has escalated. While Marine Atlas Volume 1 dealt with anemones (Anthozoa) and crustacea (Decapoda), Volumes 2 and 3 present us with additional invertebrates. However, due to the scope of the subject, only a representative number of invertebrates can be presented, even though these two volumes together consist of more than 1,300 pages.
Approximately 1,000 species of anemones and about 10,000 species of crustacea are known. All of them will never be photographed. The invertebrates in the following two volumes were selected on the basis of their abundance in nature, the depths at which the organisms occur (to 40 m, corresponding to the diving range of novices), and on the possibility of actually maintaining most of these species in an aquarium. We have chosen representatives from every class to give readers interested in marine biology an overview of the multitude of organisms in our seas.
No other book concerning invertebrates has dealt so exhaustively with the husbandry of these beauties, and very few, if any, books provide such expansive information. With over 700 photos (and more than 500 in Volume 3) and explicit textual information, the reader is introduced into a multitude of marine organisms.

Identifying the pictured animals to genus and species was a daunting task, one taken up by renowned specialists from all over the world. Because it was not always possible to identify the pictured animal to species (external features are not always sufficient for finer taxonomic levels) and some of the presented creatures remain unclassified or are awaiting scientific description, the designation of "species" was used. Diving and photography are still relatively new tools for scientists; hence only a fraction of the organisms of the sea's biotopes have been identified. Unfortunately, identifying species on the basis of a picture alone is often confounded by intraspecific variations, i.e., conspecifics may have different colors, designs, and shapes as influenced by geography and environmental factors.

Foreword

It is to be hoped that the sea's millions of diverse organisms will be preserved. As divers and aquarists we should do everything possible to contribute to this goal.

Animal breeding in marine installations and aquaria will continuously improve as new techniques are discovered and specific biological requirements are met. Encouraging examples thereof are giant clams (*Tridacna*), crustacea, oysters, and mussels. However, the pollution of the seas by industrial waste and river discharges is so serious, that it has been speculated that the world's coral reefs will be extinct within 20 years!
All the oceans of the Earth are interconnected, and the currents carry chemical pollutants (heavy metals, phosphates, etc.) to even the most remote areas of the world's seas. The seas are unable to cope with long term pollution, despite the magnificent natural cycles that purify the ocean's waters.

Fritzlar, Autumn 1997

Dr. Harry Erhardt

Acknowledgments

We are thankful to the many marine aquarists and underwater photographers who responded to our call and sent us their photographs and contributions for critical review, thereby aiding us towards the realization of Marine Atlas 2. Those who have made extraordinary contributions, e.g., sharing their vast experience in husbandry, are gratefully acknowledged.

Due to the great number of animal phyla involved, it was necessary to find specialists the world over. Without their knowledge and expertise it would have been impossible to identify the numerous species. Their generous help deserves our special thanks.

Nowadays it is impossible for any one person to be an expert on all the multifarious invertebrates of the globe—the information garnered over the last two decades is just too extensive.

In most cases it is very problematic, if not impossible, to identify a species solely based on photographs, particularly in the case of sponges, soft corals, and gorgonians. Positive identification is based on skeletal material, which in most cases was unavailable and unprocureable. Similarly, hydrozoans, bryozoans, tubeworms, feather and brittle stars, and tunicates can only be identified with certainty when a specimen is available to the taxonomist. This is why many marvelous underwater photographs of invertebrates could not be included in the Marine Atlas; even the most renowned specialists were unable to identify many species for the reasons mentioned.

Nevertheless, we hope that many familiar species have been presented, thereby creating a unique reference book. We are aware that only a small percentage of invertebrates have been presented in the following pages. Hundreds of volumes would be required to address all species presently classified!

This volume is not exclusively dedicated to the marine aquarist, although many of the species presented here have been successfully maintained in aquaria for years. It is our intention to appeal to a broad segment of the population, one that includes snorkelers, divers, biologists, students, underwater photographers, and scientists. From the richness of information presented, you will be able to distill from it what is of particular interest to you and gain a deeper understanding about the variety of species, growth forms, and lifestyles of invertebrates.

Acknowledgments

We would like to thank the following scientists for their assistance:

Dr. J. Godeaux (Thaliacea)
Dr. Manfred Grashoff (Octocorallia)
Dr. Csesa Hartmann-Schröder (Polychaeta)
Dr. John N. A. Hopper (Porifera)
Prof. Dr. Donald E. Keith (Octocorallia)
Dr. Loisette Marsh (Asteroidea)
Prof. Dr. Hans-Jörg Marthy (Mollusca - Cephlaopoda)
Prof. Dr. David L. Meyer (Crinoidea)
Dr. Claude Monniot (Tunicata)
Dr. Rich Mooi (Echinoidea)
Dr. Leer von Ofwegen (Octocorallia)
Dr. Frank W. E. Rowe (Holothuroidea)
Dr. Klaus Rützler (Porifera)
Dipl. Biol. Sabine Schoppe (Ophiuroidea)
Prof. Dr. Helmut Schuhmacher (Hexacorallia)
Prof. Dr. Manfred Sieger (manuscript)
Dr. Rob van Soest (Porifera)
Dr. Armin Svoboda (Hydroidea)
Dr. Jean Vacelet (Porifera)
Dr. W. Vervoort (Hydroidea)
Dr. George F. Warner (Antipatharia)
Dr. Erhard Wawra (Mollusca—in part)
Prof. Dr. A. Wöhrmann-Repenning (manuscript)
Dr. Natalie Yonow (Nudibranchia, Opistobranchia)
Dr. Sven Zea (Porifera)
Dr. Helmut Zibrovius (Stylasteridae)

We would also like to thank Helmut Debelius, who prepared the chapter on nudibranchs.

Table of Contents

Table of Contents

Additional groups such as bivalves, segmented worms, echinoderms (sea stars, sea urchins, etc.) and sea squirts are presented in Volume 3.

Symbols Used in Species Descriptions

Fam.: Family.

Common name: The most commonly used trade names have been listed. Frequently, a new common name is suggested, or none is listed. Should you have knowledge of a common name, please write to us. It will be included in the next edition.
Next to the Latin name, the name of the describer and the year of the description are listed. If the describer and year are in parenthesis, the species was initially classified under a different genus.

Hab.: Habitat. This refers to the species' natural geographic origin.

Sex.: Sexual differences.

Soc.B./Assoc.: Social Behavior/Association. An indication on how the organism interacts with other species and animal groups, and when possible, what animals it can and cannot be housed with.

M.: Conditions recommended for maintenance. Deviations from normal conditions are given. Additional information can be obtained from the data line at the bottom of every species description.
Because marine organisms (except those from the Red Sea) should always be maintained under similar conditions, indications about pH, redox, and carbonate hardness are not specifically given.

Light: Light requirement.
Sunlight zone = 30,000 to 50,000 lux;
moderate light zone = 10,000 to 20,000 lux;
dim light zone = 500 to 1,000 lux.
These values are measured at the water surface of the aquarium.

B./Rep.: Breeding/Reproduction. Indications are given on breeding the various species to the best of our knowledge. There are no breeding reports for the majority of animals, and we ask for information dealing with possible breeding successes.

F.: Feeding. The following abbrviations are used:
C = Carnivore (meat eater)
H = Herbivore (plant eater)

O = Omnivore (meat and plant eater)
More specific recommendations are given in the individual animal descriptions.
Note that FD = freeze-dried.

S.: Specialties. Deals with subject matter not corresponding to any other category.

T: Temperature.

L/Ø : Length/Diameter. For fishes and crustacea the total length, including the tail, is given. For anthozoans, we think it is more appropriate to indicate a diameter.

TL: Tank length. Although aquaria outside Europe are usually sold by volume, length gives a better feel for the size of the aquarium, since tall models generally offer less "living space" per unit volume.

WM: Water movement: w = weak, m = moderate, s = strong.

WR: Water region: b = bottom, m = middle, t = top.

AV: Availability: 1 = regularly in the trade, 2 = intermittently in the trade, 3 = rarely in the trade, 4 = not available (possibly a protected species).

D: Degree of maintenance difficulty.

1= Species for beginners, i.e., aquarists which begin the hobby without prior experience. Organisms in this group are very robust species that tolerate all but the most blatant maintenance errors.
2 = For aquarists with half a year of experience.
3 = For advanced hobbyists with approximately 2 years experience in marine aquarium maintenance.
4 = For advanced hobbyists with specialized knowledge. Poisonous animals, while not necessarily difficult to maintain, are listed under this category. Mostly, however, it identifies animals with special dietary difficulties and/or those that require optimal water conditions. A beginner should initially refrain from buying these species.

Underwater landscape at the island of Saba (Caribbean)

Aplysina sp., Caribbean.

Phylum: Porifera

Phylum Porifera

Sponges are primitive multicellular organisms, yet despite their elementary development, some specializations have evolved. Over 5000 species of sponges have been described to date, and additional species will definitely be discovered as frontiers such as underwater caves, the deep sea, and Antarctica are focused upon.

Exclusively sessile, sponges are largely goblet-shaped animals that do not have true tissues or organs. A nervous system is absent, and instead of muscles, they have contractile cells. The mesohyl layer constitutes the bulk of the sponge. Within its matrix there are about 10 "types" of multifunctional, wandering ameboid cells. Though these cells can be differentiated from each other, and all have distinct functions, differentiation is transient. It is probably correct to assume that most of these cells are in different functional states and that they are all just one type of cell. In keeping with their primitive organization and development, metabolism, osmoregulation, and nutrition of every cell is virtually independent.

An epithelial-like layer made up of pinacocytes, the pinacoderm, covers the exterior of the sponge and lines the channels that transverse the sponge's body wall. Pinacocytes are polygonal flat cells. Interspersed among the ordinary pinacocytes are special pinacocytes, the porocytes or "pore cells." Like a section of tubing, porocytes have a bore that opens a channel from the environment, through the body wall, and into the atrium, allowing the sponge to pump water into its body. As the porocyte contracts, the bore is closed. Effluent water is passed through the main opening, the osculum.

The sponge's atrium is covered by a layer of digestive cells, the choanocytes, which constitutes the extent of the digestive system. Collectively, these cells are called the choanoderm. Choanocytes have a collar of microvilli around the base of their long, swinging flagellum; hence they are often called collar cells.

The mesohyl, a gelatinous proteinaceous matrix, lies between the pinacoderm and the choanoderm.

Spicules provide support and texture and shape for the sponge. These spicules come in a variety of shapes and compositions, depending on species. One type of ameboid cell within the mesohyl is the sclerocyte; it is these cells which are responsible for the formation of spicules. Often this is a concerted effort by many ameboid cells, largely dependant on the complexity of the spicule (simple needlelike spicules may require no more than one sclerocyte). There are calcium and silica spicules, giving rise to calcareous or siliceous sponges respectively, as well as spicules made of organic compounds called spongin (horny sponges). Some spicules are made of a combination of two or more of the above substances. In light of their species specific nature, it is not surprising that most sponge taxonomy is based on spicule morphology and composition.

The epithelial layer of a sponge is transversed by numerous pores that are in contact with a richly branched three dimensional channel system. The opening and closing of the pore is regulated by the porocyte. The extensive channel system leads to a centrally located atrium, where the water is filtered by the choanocytes before it is released through the main opening (osculum).

Sponges are divided into three organizational types—asconoids, syconoids, and leuconoids.

The simplest are the asconoid sponges. Asconoid sponges have but one large chamber or atrium lined with flagellated choanocytes. Because of their simple structure and organization, these sponges grow no larger than 8 mm.

The second, more advanced type of sponge is the syconoid sponge. Folds in the body wall give rise to cup-shaped structures called radial tubes. Consequently, the surface area and the number of choanocytes is enhanced. Sponges with this structure have a maximum length of 4 cm.

The majority of sponges, including all large sponges, belong to the third and final group, the leuconoid sponges. Connected by

affluent and effluent channels, numerous chambers are distributed throughout the sponge. The millions of flagellae of the choanoderm generate the current necessary to draw fresh water in through the porocytes and expel the filtered, waste-laden water through the osculum. Waterborne nutrients and dissolved oxygen are carried into the sponge via the microvilli of the collar cells. The microvilli retain microorganisms and organic detritus from the incoming water, and the cells then assimilate the food particles. Actual nutrient transport is subsequently accomplished by free-moving cells, the amoebocytes. The discharged water is largely void of organic substances.
Oxygen is absorbed by the cells from the water without specialized respiratory organs.
Sponges are efficient filtering mechanisms that unflaggingly cleanse the water of the seas. Considering that a small sponge just 10 cm in diameter pumps over 100 l of water through its body every 24 hr, the rate of filtration is astounding. Indigestible substances as well as catabolites are expelled with the effluent water.

Sponges undergo sexual as well as asexual reproduction (budding). Gametes are not produced in particular organs, but throughout the body of the sponge. Most sponges are hermaphrodites. Sperm cells, which are released into the open water, fertilize the amoeboid egg cells of conspecific sponges. Resulting mobile larvae are expelled from the female's osculum. The spherical, flagellated larvae have a simple morphology. With the aid of flagella and the water current, they reach new sites, fall out of the water column, and adhere to a suitable substrate. *Ergo*, sexual propagation is an important method of species distribution for sessile sponges.
After the larvae have anchored themselves to a substrate, the cells commence to realign themselves. In this process the exterior layer of flagellated cells moves to the interior of the newly developing sponge and becomes the choanoderm. The cells of what was previously the interior layer of the larva move to the exterior of the animal to form the pinacoderm. Freshwater sponges and some marine sponges are capable of reproducing with en-

capsulated cysts (gemmules), particularly when unfavorable conditions (drought or cold weather) threaten the life of the parent sponge.
Sponges have an exceptional capacity of regeneration. Small pieces can grow into new sponges. Even when several species of sponges are minced, scraped, mixed, and pressed through a sieve, they regenerate into complete organisms—probably one of the most miraculous processes in nature.

Sponges assume various growth forms which reflect their adaptation to the environment, particularly in regard to substrate, available space, and neighboring organisms.
Despite the intraspecific variability of sponge growth, there are distinc growth forms which facilitate identification. Almost all imaginable shapes and colors are found in this phylum, which illustrates that their simple morphology gives them a wide latitude to adapt to their surroundings. Sites exposed to strong currents support encrusting, rindlike, or matlike growth patterns. Deep, protected areas foment polymorphism; spheres, vases, tubes, and bells are commonly found in such locals.
Sponges thrive in a variety of habitats, even in dark locations avoided by many animals, such as grottos, caves, and overhangs. Many species are cosmopolitan. Crustacea commonly use sponges to camouflage themselves. They simply place the sponge on their body and proceed with their normal routine. The sponge's body is rich in crevices and pores that safeguard many small invertebrates and bacteria. Numerous small crabs, worms, feather and brittle starfishes, snails, and bivalves live in the channel system and play an active role in the enormous filtration capacity of these organisms. Oftentimes there is an established relationship between two species, a symbiosis. One such example is the association between brittle stars and sponges. The starfish feeds on the fouling sediments it finds in the sponge's fine pores. Since both organisms benefit from this relationship, it is termed mutualistic symbiosis, or simply symbiosis.

Sponges are vulnerable in front of certain organisms. Some algae and colonial anemones are space competitors, capable of inhibiting the sponges' growth. Their natural enemies include a few snails, nudibranchs, free-living worms, crustacea, fishes, and turtles with a diet centered on sponges. In turn, some sponges are destructive to their environment, notably boring sponges. They cause considerable damage to dikes and harbor installations, ship hulls, and oyster beds.

Multiple chemical compounds have been isolated from sponges and are now synthesized worldwide. Antibiotics, hormones, and even some compounds able to halt the growth of certain types of tumors are among them. Hormones, compounds that chemically transmit information that influences growth and metabolism in higher animals, seem to play a predominant role in the regulation and communication processes of sponges, compensating for their lack of a central nervous system. It is extremely probable that chemists, pharmacologists, and immunologists will be able to glean further useful substances from sponges in the immediate future.

Sponges in the Aquarium

Current literature considers sponges to be very difficult to impossible to maintain in captivity. The difficulty probably stems from inadequate collection and transportation procedures rather than from care itself. Despite their highly developed ability to regenerate, sponges are very sensitive to environmental changes. Unfortunately, drastic environmental flux is the rule rather than the exception during collection and transport. Deficient oxygen concentrations, elevated temperatures, and other variations all take their toll on the sponge's health. When purchasing, select healthy specimens without injuries (e.g., pressure sites). The sponge should be attached to its substrate. Exposure to air will prove injurious, since they very quickly oxidize. Because of these and other reasons, sponges can only be kept for a limited period of time, even in well maintained aquaria. Sponges will not tolerate being placed

on or buried within a sand substrate. Situate them among rocks or anchor them to the decoration instead.
The majority of sponges sold in German pet stores hail from the Mediterranean, the Atlantic, and the North and Baltic Seas. Tropical sponges are rarely offered.
In aquaria, sponges must be placed on a shaded site. The heightened algae growth that often accompanies strong illumination is to the detriment of sponges, often causing an early demise. Upon death, sponges release toxins which usually kill all tankmates.
The very small particulate foodstuff sponges require add another substantial burden to their husbandry. Just their dietary requirements make them virtually impossible to maintain in community with other organisms. They generally starve if the filter and protein skimmer are not turned off during and for a while after phytoplankton and zooplankton are fed. Since sponges create their own current to suck waterborne foodstuffs into their body, it is sufficient to place the food into the surrounding water column.

Taxonomy

Phylum: PORIFERA — Sponges

Class: CALCAREA — Calcareous sponges

Order: HETEROCOELIA — Honeycomb sponges
- Family: Clathrinidae
  - Genus: *Clathrina*
- Family: Leuconiidae
  - Genus: *Leuconia*
- Family: Leucascidae
  - Genera: *Pericharax*, *Leucetta*

Class: DEMOSPONGIA — Common sponges

Subclass: TETRACTINOMORPHA

Order: ASTROPHORIDA
- Family: Epipolasidae
  - Genus: *Astreopus*

Order: HOMOSCLEROPHORIDA
- Family: Oscarellidae
  - Genus: *Oscarella*

Order: HADROMERIDA
- Family: Chondrillidae
  - Genus: *Chondrosia*
- Family: Latrunculiidae
  - Genus: *Latrunculia*
- Family: Spirastrellidae
  - Genus: *Spirastrella*
- Family: Clionidae
  - Genus: *Cliona*
- Family: Tethyidae
  - Genus: *Aaptos*

Order: AXINELLIDA
- Family: Axinellidae

Genera: *Axinella*
Acanthella
Cymbastela
Pseudoaxinella
Family: Raspailiidae
Genus: *Ectyoplasia*

Order: SPIROPHORIDA
Family: Tetillidae
Genus: *Cinachyra*

Order: HALICHONDRIDA
Family: Halichondriidae
Genera: *Axinyssa*
Liosina
Family: Hymeniacidonidae
Genera: *Hemimycale*
Scopalina

Order: POECILOSCLERIDA
Family: Esperiopsidae
Genera: *Neofibularia*
Monanchora
Family: Myxillidae
Genera: *Crambe*
Iotrochota
Iophon
Lissodendoryx
Family: Mycalidae
Genus: *Mycale*
Family: Crellidae
Genus: *Crella*
Family: Adociidae
Genus: *Aka*
Family: Clathriidae
Genera: *Clathria*
Echinoclathria
Rhaphidophlus

Family: Anchinoidae
Genus: *Anchinoe*
Family: Agelasidae
Genus: *Agelas*

Order: HAPLOSCLERIDA
Family: Haliclonidae
Genus: *Haliclona*
Family: Niphatidae
Genera: *Gelloides*
Niphates
Family: Callyspongiidae
Genus: *Callyspongia*
Family: Petrosiidae
Genera: *Petrosia*
Xestospongia

Order: DICTYOCERATIDA
Family: Spongiidae
Genera: *Cacospongia*
Carteriospongia
Spongia
Family: Thorectidae
Genera: *Ircinia*
Sarcotragus
Smenospongia

Order: VERONGIIDA
Family: Aplysinidae
Genera: *Aplysina*
Verongula
Family: Druinellidae
Genus: *Pseudoceratina*

Order: DENDROCERATIDA
Family: Haliscaridae
Genus: *Haliscara*

Clathrina clathrus (SCHMIDT, 1868)

Yellow calcareous lattice sponge

Hab.: Atlantic, Mediterranean.

Sex.: None.

M.: Unknown.

F.: O; plankton.

S.: The intense yellow color and characteristic lattice structure make this an eye-catching sponge. Young specimens are vase-shaped, while older specimens grow a multitude of 0.5–1 mm intertwining tubes. This sponge lives in seaweed lawns, crevices, and caves on stones, rocks, and deep coral substrates. *Clathrina* are cosmopolitan.

T: 10°–18°C, Ø: 8 cm, AV: 4, D: 4

Order: HETEROCOELIA
Fam.: Clathrinidae, Leuconiidae

Clathrina coriacea (MONTAGU, 1818)
White calcareous lattice sponge

Hab.: Mediterranean, Atlantic, North Sea. Probably cosmopolitan!

Sex.: None.

M.: Unknown.

F.: O; plankton, detritus.

S.: *C. coriacea* has a flat to matlike reticulate growth pattern. The growth form assumed largely depends on the amount of current the sponge is exposed to—flat and encrusting in moving waters and slightly protuberant in protected areas. It can be found in small caves and seagrass lawns, on sandy and rocky substrates, and underneath of stones. The mesh of interconnected tubes converges into one effluent opening, the osculum. The body of the sponge is contractile.

T: 10°–24°C, **Ø**: 10 cm, **AV**: 3–4, **D**: 4

Leuconia sp.
Small knobbed calcareous sponge

Hab.: Indo-Pacific, Australia.

Sex.: None.

M.: Unknown.

F.: O; plankton.

S.: This sponge forms small, thin, partially branched tubes that have a smooth surface and terminal oscula. Dense groups colonize shaded cave walls and areas underneath overhanging ledges. *Leuconia* sp. is often overlooked because of its small size and low growth form.

T: 22°–26°C, **L**: 3 cm, **AV**: 4, **D**: 4

Clathrina coriacea

Leuconia sp.

Order: HETEROCOELIA
Fam.: Leucascidae

Pericharax heterorhaphis (HAECKEL, 1872)
Yellow calcareous sponge

Hab.: Indo-Pacific, Australia.

Sex.: None.

M.: Unknown.

F.: O; plankton.

S.: Notoriously bright yellow, *P. heterorhaphis* assumes an encrusting or vase-shaped growth form. The osculum of the latter form has a broad collar, a fold of the lighter-colored inner body wall. With its small base, the yellow calcareous sponge affixes to hard substrates of caves or shaded vertical rock walls. The white figures on the sponge's surface are probably small sea cucumbers.

T: 22°–26°C, L: 30 cm, AV: 4, D: 4

Leucetta philippinensis (HAECKEL, 1872)
Yellow calcareous sponge

Hab.: Red Sea, Indo-Pacific, Australia.

Sex.: None.

M.: Unknown.

F.: O; plankton.

S.: *L. philippinensis* is found in niches, crevices, caves, and areas under overhanging ledges at a broad range of depths. The upright bright yellow growth form only occupies a few square centimeters of rocky substrate. Current-exposed overhangs are its preferred habitat.

T: 22°–26°C, L: 6–12 cm, AV: 4, D: 4

Asteropus sarassinorum (THIELE, 1903)
Starpore sponge

Hab.: Indo-Pacific, Malaysia.

Sex.: None.

M.: Unknown.

F.: O; plankton.

S.: This massive sponge has a round, almost spherical form and a small base. The surface is rough and interspersed with many relatively large pores. The excurrent openings are elevated like little volcanos. *A. sarassinorum* anchors itself to corals along the upper reef slope, mostly in waters rich in suspended matter; it is always covered by a fine layer of sediment.

T: 22°–26°C, L: 60 cm, AV: 4, D: 4

Oscarella lobularis, orange.

Oscarella lobularis SCHMIDT, 1868
Flesh sponge

Hab.: Atlantic, Mediterranean, North Sea.

Sex.: None.

M.: Unknown.

F.: O; plankton.

S.: This encrusting sponge has a soft, fleshy texture and diverse surface contours and colors, though it is generally flesh-colored. A collection of furrows, bulges, and knots with the oscula on the protuberances is the general presentation. It forms crusts up to 6 mm thick over stones, rocks, and cave walls. Some specimens inhabit seagrass beds of *Posidonia*. The flesh sponge lives down to great depths, conquering new substrates through long appendages.

T: 10°–18°C, **L:** 12 cm, **AV:** 4, **D:** 4

Order: HADROMERIDA
Fam.: Chondrillidae

Chondrosia reniformis NARDO, 1833
Kidney sponge

Hab.: Mediterranean.

Sex.: None.

M.: Unknown.

F.: O; plankton.

S.: As denoted by its name, this massive, irregular sponge forms thick kidney-shaped encrustations. Its smooth, shiny surface and slimy and rubbery to the touch. Typical skeletal structures such as calcareous spicules and spongin fibers are lacking; collagen fibers act as support instead. The kidney sponge generally inhabits dim recesses beneath ledges, within crevices, and along vertical walls. Coloration depends on light intensity—the brighter the illumination, the darker the sponge and vice versa. It is found from shallow water to 30 m depth.

T: 10°–18°C, Ø: 6–18 cm, AV: 4, D: 4

Petromica ciocalyptoides, a rare sponge species of the Caribbean.

Chondrosia reniformis

Chondrilla nucula

Order: HADROMERIDA
Fam.: Latrunculiidae

Latrunculia magnifica KELLER, 1889
Magnificent fire sponge

Hab.: Red Sea, Indo-Pacific.

Sex.: None.

M.: Unknown.

F.: O; plankton.

S.: This erect arboreal sponge has blunt-tipped branches which are round to oval in cross-section. Its smooth surface is riddled with numerous small incurrent pores and a few scattered chimneylike oscula that are chiefly round, but occasionally oval. Exposed areas along reef slopes and shaded sites within caves are its habitat of choice. *L. magnifica* is extremely poisonous!

T: 22°–26°C, **L:** 80 cm, **AV:** 4, **D:** 4

Oscarella tuberculata, Mediterranean.

Order: HADROMERIDA
Fam.: Spirastrellidae

Spirastrella purpurea LAMARCK, 1814
Purple boring sponge

Hab.: Red Sea, Indo-Pacific.

Sex.: None.

M.: Unknown.

F.: O; plankton.

S.: An aggressive secondary colonizer that bores into and encrusts corals. The surface of the sponge has numerous conical oscula.

T: 20°–26°C, Ø: 40 cm, AV: 4, D: 4

Spirastrella cunctatrix (SCHMIDT, 1868)
Orange ray sponge

Hab.: Atlantic, Mediterranean.

Sex.: None.

M.: Unknown.

F.: O; plankton.

S.: This flat, irregular-shaped encrusting sponge grows on hard substrates, including bivalve and gastropod shells. Its habitats include shaded cave walls, harbor installations, shipwrecks, and areas beneath ledges. The pronounced radial arrangement of the water canals around the oscula gives the sponge its name. Coloration is variable.

T: 10°–18°C, Ø: 20 cm, AV: 4, D: 4

Order: HADROMERIDA
Fam.: Spirastrellidae

Spirastrella inconstans RIDLEY, 1884
Star sponge

Hab.: Indian Ocean, Maldives.

Sex.: None.

M.: Unknown.

F.: O; plankton, detritus.

S.: Oscula are located at the apex of conical, slightly stooped projections. With its broad base, *S. inconstans* anchors itself to dead corals and weathered substrates along the foot of reef slopes in suspension-rich waters. This sponge is so soft, it "flutters" in strong current.

T: 22°–27°C, L: 12 cm, AV: 4, D: 4

Haliclona vetulina DE LAUBENFELS, 1954
Purple star-sponge

Hab.: Red Sea, Indo-Pacific.

Sex.: None.

M.: Unknown.

F.: O; plankton.

S.: A very aggressive habitat competitor in front of all sessile organisms, especially corals. It encrusts surfaces and dead organisms like a skin. The sponge's surface has channels that radiate in a starlike fashion from the oscula and highly visible criss-crossing light spicules. Almost all substrates at all depths are colonized.

T: 20°–26°C, Ø: 40 cm, AV: 4, D: 4

Cliona viridis (SCHMIDT, 1868)
Green boring sponge

Hab.: Mediterranean.

Sex.: None.

M.: Unknown.

F.: O; plankton.

S.: Once *C. viridis* has damaged its substrate, it assumes the lumpy, undulating, encrusting form pictured above. The few large, volcanolike oscula have a light-colored fringe around the opening.

T: 10°–18°C, Ø: 60 cm, AV: 4, D: 4

Cliona celata GRANT, 1826
Yellow boring sponge, sulfur sponge

Hab.: Atlantic, Mediterranean, North Sea, western Baltic Sea. *C. celata* occurs from shallow water to 250 m depth.

Sex.: None.

M.: Unknown.

F.: O; plankton.

S.: The larvae of this sponge penetrate calcareous rocks and other hard substrates, drilling fine, interconnected, reticulate channels. Contact with the surrounding water is maintained via porous papillae and oscula; calcareous particles are expelled through the tiny oscula. Old sponges, having largely destroyed their substrate, grow to the surface, where they appear as raglike to encrusting mats. The excurrent openings are located along the crest of the sponge and are surrounded by a dense warty zone. *C. celata* is sulfur yellow exteriorly and orange to orange-red interiorly.

T: 10°–18°C, L: 20 cm, AV: 4, D: 4

Order: Hadromerida
Fam.: Clionidae

Cliona carteri (RÜTZLER, 1973)
Orange-red boring sponge

Hab.: Atlantic, Mediterranean.

Sex.: None.

M.: Unknown.

F.: O; plankton.

S.: The orange-red boring sponge chemically drills a network of fine channels (0.5–1 mm) into calcareous rocks and bivalve and gastropod shells, only growing along the substrate's surface once the integrity of the substrate has been compromised. During the first stages, the only visible signs of invasion are the irregular holes through which the sponge extends its incurrent and excurrent openings.

T: 10°–18°C, L: a few millimeters, AV: 4, D: 4

Cliona delitrix PANG, 1923
Red boring sponge

Hab.: Caribbean, Florida, Bahamas, Providence Island, San Andrés.

Sex.: None.

Soc.B./Assoc.: It is often found with the colonial anemone *Parazoanthus parasiticus*.

M.: Unknown.

F.: O; plankton.

S.: *C. delitrix* is considered a coral killer. As seen in the above photo, the red boring sponge forms a skinlike layer over corals, in this case *Montastrea cavernosa*. The only prominent feature of the sponge are the large chimneylike oscula. The methods the sponge employs to dissolve the coral's calcium are little understood, especially since acid secretions have not been detected. However, chemically dissolved calcium is released into the surrounding water through the oscula. It is hypothesized that specialized cells of the sponge surround calcium particles and destroy them through assimilation processes.

T: 20°–24°C, Ø: 45 cm, AV: 4, D: 4

Cliona delitrix, Caribbean.

Aplysina ianthelliformis (LENDENFELD, 1888)
Yellow mesh sponge

Hab.: Indo-Pacific, Australia.

Sex.: None.

M.: Unknown.

F.: O; plankton, detritus.

S.: The tough, 2 cm diameter stalk of this characteristically-shaped sponge affixes the animal to a hard substrate. Leathery to rubberlike in texture, the surface is an extensive hexagonal mesh created by fibril bundles. Once in contact with atmospheric oxygen, it quickly loses its yellow color.

T: 22°–26°C, L: 15 cm, AV: 4, D: 4

Order: HADROMERIDA
Fam.: Tethyidae

Aaptos suberitoides DENDY, 1905
Yellow-brown encrusting sponge

Hab.: Indo-Pacific, Malaysia.

Sex.: None.

M.: Unknown.

F.: O; plankton.

S.: The yellow-brown encrusting sponge is an aggressive colonizer that quickly destroys its substrate. Dark brown with independent yellow areas, this sponge has a few scattered, chimneylike, protrusive oscula along its surface. *A. suberitoides* predominantly lives on corals of shallow, protected reef zones.

T: 22°–26°C, Ø: 25 cm, AV: 3, D: 4

Axinella polypoides (SCHMIDT, 1868)
Mediterranean finger sponge

Hab.: Mediterranean.

Sex.: None.

M.: Unknown.

F.: O; plankton, detritus.

S.: This erect sponge is supported by a short axis. Superiorly, it has numerous variable-sized cylindrical branches. Oscula are found at the center of radiating channels that create a starlike pattern. Rocky bottoms and cliffs from shallow water to depths of over 100 m are its preferred habitat.

T: 10°–18°C, L: 70 cm, AV: 4, D: 4

Order: Axinellida
Fam.: Axinellidae

Axinella cannabina (ESPER, 1864)
Tree sponge, antler sponge

Hab.: Mediterranean.

Sex.: None.

Soc.B./Assoc.: *A. cannabina* lives in association with the yellow encrusting sea anemone *Parazoanthus axinellae*.

M.: Unknown.

F.: O; plankton, detritus.

S.: The tree sponge solely inhabits warmer sections of the Mediterranean. Its richly branched arboreal form is largely found on rocky substrates of current-swept coastal regions down to a depth of 60 m. Each branch has irregular lumps and ridges with excurrent openings located at their apex.

T: 10°–18°C, L: 90 cm, AV: 4, D: 4

Axinella damicornis (ESPER, 1794)
Elk horn sponge

Hab.: Mediterranean.

Sex.: None.

Soc.B./Assoc.: This sponge lives in association with the encrusting sea anemone *Parazoanthus axinellae*.

M.: Unknown.

F.: O; plankton.

S.: Reminiscent of elk horns, numerous blunt-tipped, flat, intertwining branches are borne on a short axis. *A. damicornis* colonizes hard substrates as well as moving bottoms from shallow waters down to great depths. The buttonlike structures on the sponge's surface are contracted encrusting sea anemones.

T: 10°–18°C, L: 30 cm, AV: 4, D: 4

Order: Axinellida
Fam.: Axinellidae

Axinella verrucosa SCHMIDT, 1862
Finger sponge

Hab.: Mediterranean. This sponge occurs at depths of 10–100 m.

Sex.: None.

Soc.B./Assoc.: It lives in association with the colonial anemone *Parazoanthus axinellae*.

M.: Unknown.

F.: O; plankton, detritus.

S.: *A. verrucosa* inhabits current-swept hard substrates such as rocks within grottos and hollows. There are diverse grow forms, but all are heavily branched to fascicular. Numerous smooth, fingerlike branches originate from a short stem. The colony's coloration is variable, spanning from yellow to orange.

T: 10°–18°C, L: 100 cm, AV: 4, D: 4

Acanthella acuta SCHMIDT, 1862
Cactus sponge

Hab.: Mediterranean.

Sex.: None.

M.: This species is easily kept with *Caulerpa* algae.

F.: O; plankton, plankton substitutes. Special feedings are not necessary.

S.: *A. acuta* has an extremely craggy, thorny surface and short, uneven branches that terminate in pointed conical papillae. The extension of the available surface determines the size of the sponge. Virtually all hard substrates are colonized, regardless of depth. *A. acuta* is predominately orange.

T: 10°–18°C, Ø: 8 cm, AV: 3, D: 4

Order: Axinellida
Fam.: Axinellidae

Acanthella carteri (DENDY, 1897)
Elephant-ear sponge

Hab.: Red Sea, Indo-Pacific, Australia.

Sex.: None.

M.: Unknown.

F.: O; plankton.

S.: As inferred by its common name, *A. carteri* has an erect, broadly lobed growth form which brings to mind an elephant ear. At favorable sites this sponge can achieve immense proportions. It is found to a depth of 60 m along current-swept steep reef slopes and sediment-free reef platforms. Most specimens are yellow to orange. When exposed to atmospheric oxygen, the sponge's pigments quickly oxidize.

T: 22°–26°C, **L:** 50 cm, **AV:** 4, **D:** 4

Cymbastela coralliophila HOOPER & BERGQUIST, 1992
Plate sponge

Hab.: Great Barrier Reef, Australia.

Sex.: Oviparous. Gametes are released synchronously.

M.: The plate sponge is readily maintained. Caution—growth is very fast.

F.: O; plankton.

S.: *C. coralliophila* is dorsoventrally flattened, assuming a discoid shape. Its slender base anchors it to all suitable substrates along the reef, e.g., corals, coral fragments, rocky slopes, sandy lagoons, and deep mud bottoms, predominately in areas subjected to high sedimentation rates. The surface of the sponge has concentric canals with centrally positioned digitate projections. Fine sediments always collect in the irregular undulations and deep clefts of the sponge's periphery. From there, the sediments are carried away by the current or merely shifted.

T: 20°–24°C, Ø: 35 cm, AV: 4, D: 4

Pseudaxinella zeai (sp. nov.) ALVAREZ & VAN SOEST, 1994
Zea's brown encrusting sponge

Hab.: Caribbean, Colombia.

Sex.: None.

M.: Unknown.

F.: O; plankton, detritus.

S.: This low-growing sponge is encountered in deep water on unoccupied colonizable substrates such as dead corals and coral fragments; there it spreads in all directions via runners. Small incurrent openings densely cover the sponge's smooth brown surface, and the oscula are volcanolike projections with a light fringe around the excurrent opening. This new species was described at the time of the original German edition of Marine Atlas, Vol. 2.

T: 20°–24°C, L: 30 cm, AV: 4, D: 4

Ectyoplasia ferox (DUCHASSAING & MICHELOTTI, 1864)
Brown encrusting octopus sponge

Hab.: Caribbean, Bahamas, Florida, Jamaica, Colombia.

Sex.: None.

M.: Unknown.

F.: O; plankton.

S.: Both the growth pattern and coloration of *E. ferox* are variable, the former extremely so. When *E. ferox* lives among gorgonians, it produces long runners, otherwise it assumes an encrusting growth form with cylindrical protuberances. Its surface is covered by numerous conical projections that have an osculum at their apex. Using minute quantities of acid, this sponge bores into calcareous rocks and corals without causing visible damage. The brown encrusting octopus sponge lives in lagoons and on coral reefs, overgrowing corals, gorgonians, and dead reef sections.

T: 20°–24°C, L: 35 cm, AV: 4, D: 4

Order: Spirophorida
Fam.: Tetillidae

Cinachyra australiensis CARTER, 1869
Australian sphere sponge

Hab.: Indo-Pacific, Australia.

Sex.: None.

M.: Unknown.

F.: O; plankton.

S.: The multitude of large, circular, concave incurrent openings are characteristic for this hemispherical to spherical sponge. It lives at a broad range of depths, preferably along shaded locals, and is always covered with fine sediment.

T: 22°–26°C, Ø: 6 cm, AV: 4, D: 4

Yellow tube sponge, *Ectyoplasia ferox*, from the Caribbean.

Cinachyra australiensis

Cinachyra australiensis

Order: HALICHONDRIDA
Fam.: Halichondriidae

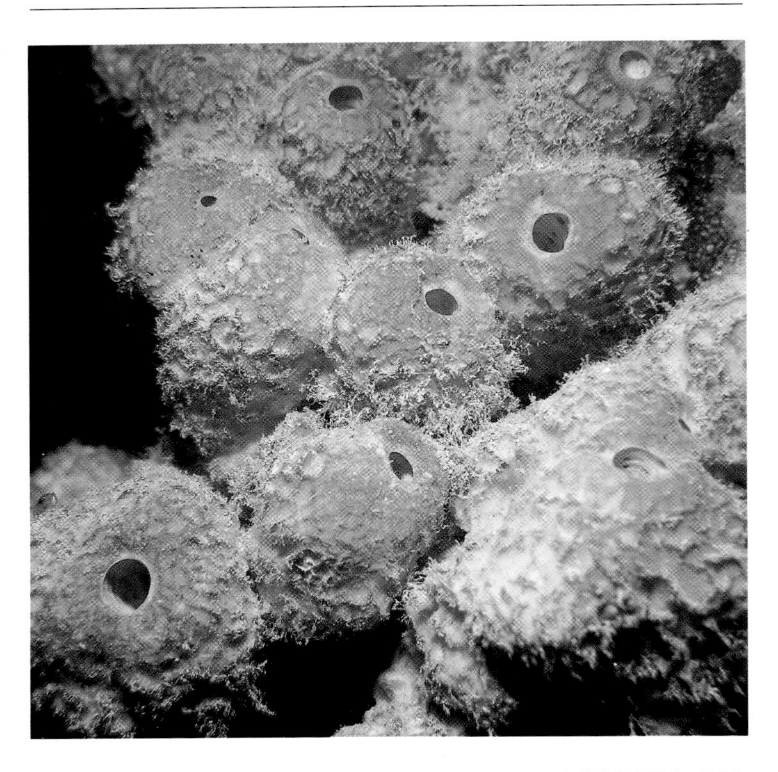

Axinyssa ambrosia (LAUBENFELS, 1934)
Orange-red knot sponge

Hab.: Caribbean, Dutch West Indies, Colombia.

Sex.: None.

M.: Unknown.

F.: O; plankton, detritus.

S.: Several species of this genus live in the tropical west Atlantic. The sponge pictured above was growing on a pier in the harbor of Bonaire. Small, thick, knotlike projections with an excurrent opening at their center cover the encrusting mass, and the surface itself is rough and cleft and partially covered with aufwuchs and fine sediment. Only the areas around the oscula remain uncovered. Orange-red knot sponges have a soft consistency. Most specimens inhabit dimly lit reef sections and man-made structures.

T: 20°–26°C, L: 12 cm, AV: 4, D: 4

Liosina paradoxa THIELE, 1903
Tree sponge

Hab.: Red Sea, Indo-Pacific.

Sex.: None.

M.: Unknown.

F.: O; plankton, detritus.

S.: *L. paradoxa* is erect, and its surface is rough and bumpy. The excurrent openings are located terminally on blunt protuberances which spiral up the length of the sponge. Horizontal hard substrates at the bottom of the reef are the preferred habitat.

T: 22°–26°C, L: 20 cm, AV: 4, D: 4

Hemimycale columella (BOWERBANK, 1862)
Slimy encrusting sponge

Hab.: Atlantic, Mediterranean.

Sex.: None.

M.: Unknown.

F.: O; plankton, detritus.

S.: *H. columella* has a slimy texture, a fleshy consistency, and an encrusting growth pattern. The crater-shaped oscula are in clusters with numerous sievelike incurrent openings grouped around them. It is found from shallow to deep water.

T: 10°–18°C, Ø: 20 cm, AV: 3, D: 4

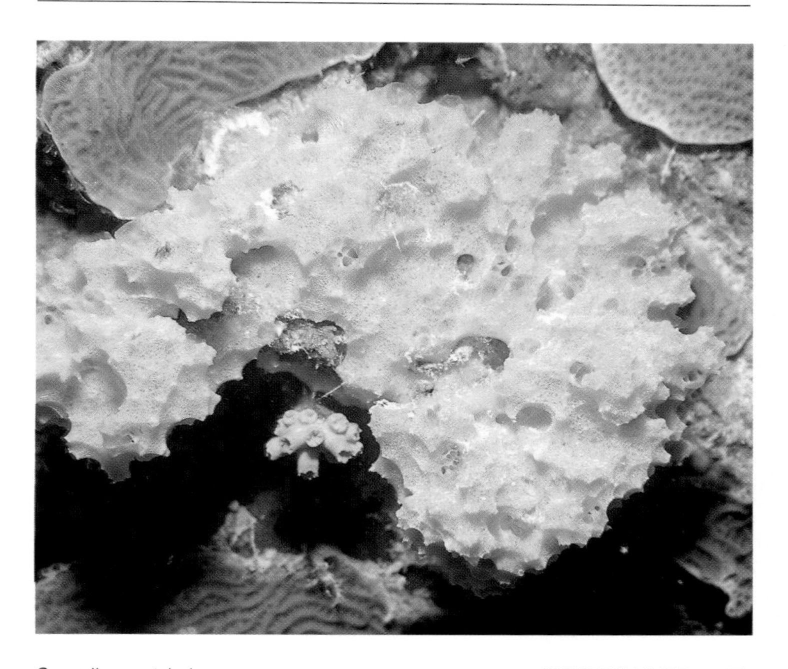

Scopalina ruetzleri (WIEDENMAYER, 1977)
Rützler's clump-sponge, orange clump-sponge

Hab.: Caribbean, Bahamas, Florida, Bonaire, Colombia.

Sex.: None.

M.: Unknown.

F.: O; plankton, detritus.

S.: *S. ruetzleri* encrusts corals and other hard substrates in a rough, irregular, lobate to lumpy mass. Large excurrent openings are absent. It occurs in all reef sections and embracing the roots of *Rhizophora mangle*, a mangrove.

T: 24°–28°C, L: 15 cm, AV: 4, D: 4

Order: Poecilosclerida
Fam.: Esperiopsidae

Neofibularia nolitangere (DUCHASSAING & MICHELOTTI, 1864)
Touch-me-not sponge

Hab.: Caribbean, Bahamas, Florida, Bonaire, Colombia.

Sex.: None.

Soc.B./Assoc.: *N. nolitangere* lives in conjunction with the sponge threadworm *Syllis spongicola*, the brittle star *Ophiothrix suensonii*, the Florida wormsnail *Vermicularia knorrii*, and two gobies, *Gobisoma horsti* and *G. chancei*.

M.: Unknown.

F.: O; plankton, detritus.

S.: This large, massive sponge has an irregular, feltlike surface. It is reddish brown and frequently blanketed in fine sediment. The oscula are generous openings of various shapes, but rarely round. When touched, *N. nolitangere* provokes an allergic reaction manifest by swelling and pain that lasts for weeks. Very sensitive people can suffer respiratory problems, hence the Latin name meaning: "Do not touch!" At practically all depths, flat and steep reef sections are inhabited.

T: 20°–26°C, **Ø**: 60 cm, **AV**: 4, **D**: 4

Monanchora arbuscula (DUCHASSAING & MICHELOTTI, 1864)
Red encrusting sponge

Hab.: Caribbean, Bahamas, Florida, Bonaire.

Sex.: None.

M.: Unknown.

F.: O; plankton, detritus.

S.: *M. arbuscula* is extremely common along almost all reefs and reef sections, where it forms a thin crust on corals, vertical walls, and along the base of large coral colonies. Apparently it does not require much light for growth. Numerous well-defined canals radially diverge from the oscula.

T: 20°–26°C, **Ø**: 10 cm, **AV**: 4, **D**: 4

Neofibularia nolitangere

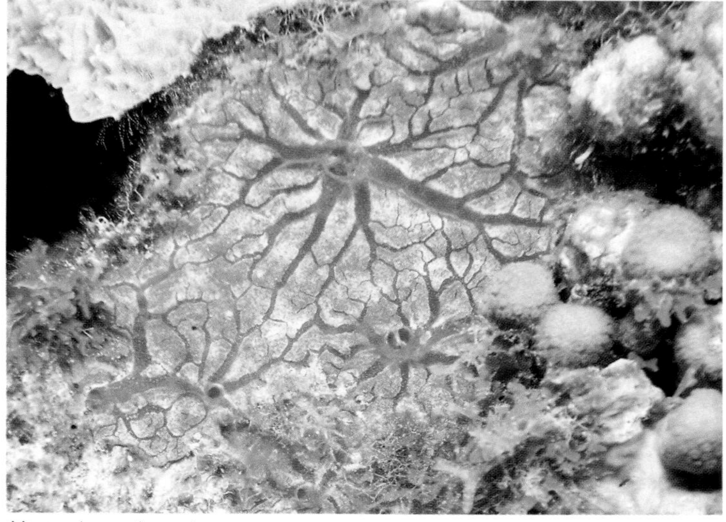

Monanchora arbuscula

Order: Poecilosclerida
Fam.: Myxillidae

Crambe crambe (SCHMIDT, 1864)
Bumpy encrusting sponges

Hab.: Mediterranean, Goglio (Italy).

Sex.: None.

M.: Unknown.

F.: O; plankton, detritus.

S.: Probably the most common sponge in the Mediterranean, this encrusting species inhabits shallow to deep water along dimly lit areas of the rocky littoral, e.g., beneath overhanging ledges and within caves and crevices. Its surface is bumpy due to the irregular arrangement of the excurrent openings. There are no defined canals leading to the oscula. Bumpy encrusting sponges are intense red-orange to red.

T: 10°–22°C, **Ø:** 50 cm, **AV:** 4, **D:** 4

Iotrochota birotulata (HIGGIN, 1877)
Green finger sponge, black bush sponge

Hab.: Caribbean, West Indies, Colombia to Brazil.

Sex.: None.

M.: Unknown.

F.: O; plankton, detritus.

S.: The green finger sponge exhibits an erect growth form consisting of many spiny, intertwined, cylindrical or laterally compressed branches of variable length. The coloration ranges from black and gray to gray-green. It inhabits shallow to deep waters.

T: 20°–26°C, L: 60 cm, AV: 4, D: 4

Iophon laevistylus BERGQUIST, 1967
Tube sponge

Hab.: Indian Ocean, Thailand.

Sex.: None.

M.: Unknown.

F.: O; plankton, detritus.

S.: A multitude of long cylindrical tubes, occasionally branched, originate from a common base. All tubes have a similar diameter and an osculum centered at their tip. Many of the tubes are overgrown exteriorly. *I. laevistylus* is encountered at various depths on hard substrates of reef slopes.

T: 20°–26°C, L: 40 cm, AV: 4, D: 4

Lissodendoryx colombiensis ZEA & VAN SOEST, 1986
Colombian porous sponge

Hab.: Caribbean, Bahamas, Colombia.

Sex.: None.

M.: Unknown.

F.: O; plankton, detritus.

S.: *L. colombiensis* has a rough surface, large pores, and generous-sized, centrally located oscula. It remains free of aufwuchs and sediments as it grows on corals and other hard substrates. The oscula have a few incurrent pores distributed along their fringe. The pictured sponge has been "strangled" among the branches of a gorgonian which has died and been encased in a layer of algae, hydroids, and small sponges.

T: 20°–26°C, L: 40 cm, AV: 4, D: 4

Order: Poecilosclerida
Fam.: Mycalidae

Mycale laevis (CARTER, 1882)
Orange icing sponge

Hab.: Caribbean, Bahamas, Florida.

Sex.: None.

M.: Unknown.

F.: O; plankton, detritus.

S.: The orange icing sponge mostly lives at the base of corals of the genera *Porites* and *Montastrea*. There it protects the coral from sessile and boring invertebrates. Its thin to encrusting grow form proceeds to the living tissue of the coral, only becoming visible at that point. The large crater-shaped excurrent openings have a transparent rim.

T: 20°–26°C, Ø: 20 cm, AV: 4, D: 4

Crella cyathophora CARTER, 1869
Wart sponge

Hab.: Red Sea, Indo-Pacific.

Sex.: None.

M.: Unknown.

F.: O; plankton, detritus.

S.: The encrusting growth pattern of this sponge conforms to the irregularities of its substrate. Incurrent openings are relatively large nubs arranged on the crater walls of the oscula and in the depressions of the sponge's surface. The oscular processes are chimneylike projections. Shaded rock faces under overhanging ledges from shallow water to a depth of 40 m are its habitat of choice. It is milk-gray to brown in color.

T: 20°–24°C, Ø: 15 cm, AV: 4, D: 4

Order: POECILOSCLERIDA
Fam.: Adociidae

Aka coralliphaga (RÜTZLER, 1971)
Yellow boring sponge

Hab.: Caribbean, Bahamas, Curaçao, Honduras, Colombia.

Sex.: None.

M.: Unknown.

F.: O; plankton, detritus.

S.: Though this sponge exhibits several growth forms, a cluster of distinct, vertical, vase-shaped structures projecting above the substrate is the most prevalent. These structures, which are the only visible part of the sponge, are oscula. The larvae settle on coral surfaces and begin their destructive boring process which eventually kills the coral polyps. Secreting acid, they tunnel into the coral head and spread within. Small incurrent openings surround the base of the protruding chimneys. This intense yellow sponge is found at all depths and bores into virtually all coral species.

T: 20°–24°C, L: 3 cm, AV: 4, D: 4

Aka mucosa BERGQUIST, 1965
Black-brown boring sponge

Hab.: Indo-Pacific, Maldives.

Sex.: None.

M.: Unknown.

F.: O; plankton, detritus.

S.: With the exception of the long excurrent chimneys thrusting from the surface of the *Porites* coral, *A. mucosa* is completely hidden from view. While this sponge is not sufficiently aggressive to destroy the coral colony, the coral must make compensatory structural modifications to grow around the sponge. Halos at the point where sponge and coral meet are visible signs of the fight between the two organisms.

T: 22°–26°C, L: 4 cm, AV: 4, D: 4

Order: Poecilosclerida
Fam.: Clathriidae

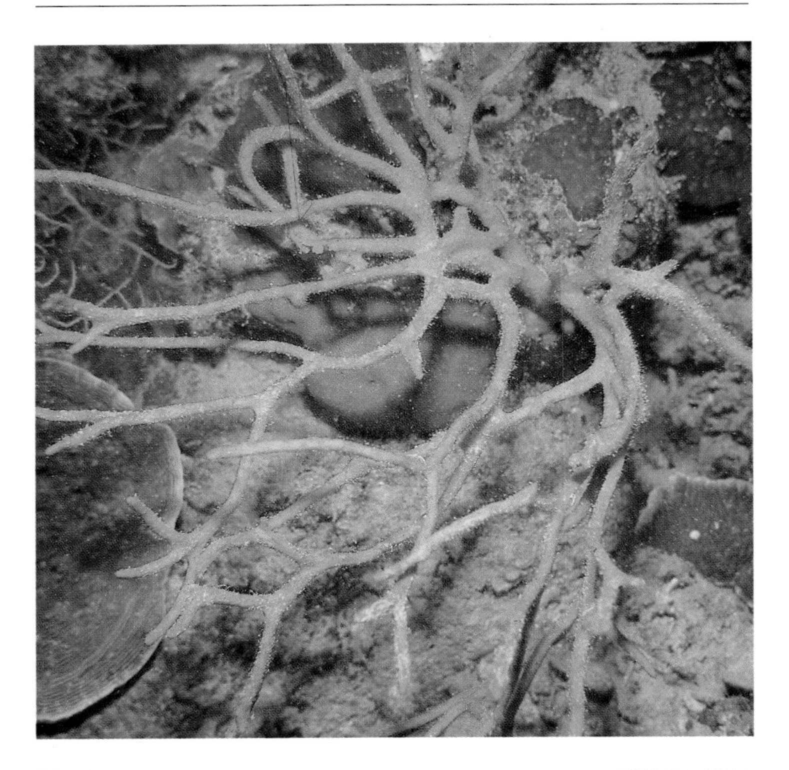

Clathria procera (RIDLEY, 1884)
Orange finger sponge

Hab.: Indo-Pacific, Malaysia.

Sex.: None.

M.: Unknown.

F.: O; plankton, detritus.

S.: *C. procera* inhabits vertical faces of reef slopes, where the long, digitate, terminally forked branches extend far into the open water. The flexible nature of the round branches is an adaptation to the strong currents that buffet *C. procera*.

T: 22°–26°C, **L:** 80 cm, **AV:** 4, **D:** 4

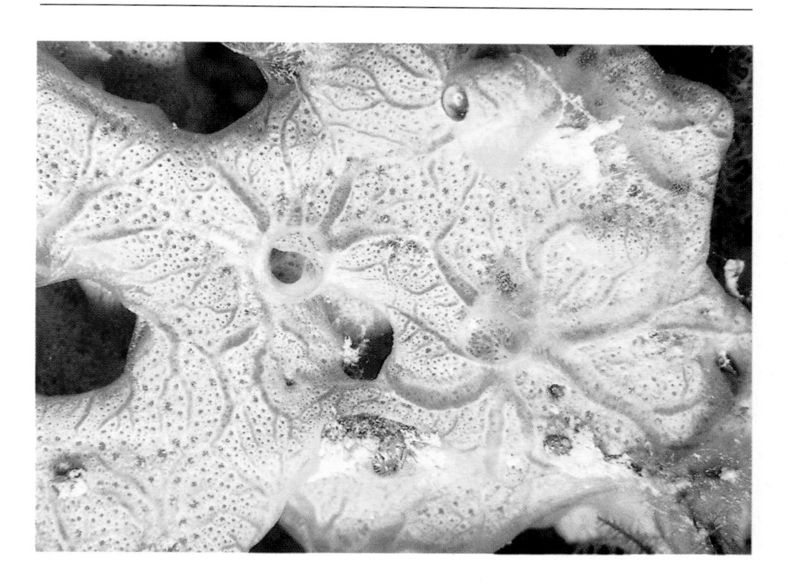

Clathria sp.
Peach encrusting sponge

Hab.: Caribbean, Bahamas, southern Florida, Bonaire.

Sex.: None.

M.: Unknown.

F.: O; plankton, detritus.

S.: This thin, encrusting sponge grows along dead reef sections and walls beneath overhanging ledges. Rootlike interconnected canals radiate from large oscula. Among the canals are many small incurrent openings.

T: 20°–26°C, Ø: 20 cm, AV: 4, D: 4

Order: Poecilosclerida
Fam.: Clathriidae

Phyllospongia dendyi LENDENFELD, 1889
Honey mountain sponge

Hab.: Red Sea, Indo-Pacific, Australia.

Sex.: None.

M.: Unknown.

F.: O; plankton, detritus.

S.: *P. dendyi* exists in a variety of growth forms, although cylindrical to spherical forms predominate. The characteristic surface of the sponge is sculpted into rough ridges and valleys, where sediments and coral sand collect. Its clefts offer ideal hiding places for a number of small animals.

T: 22°–26°C, Ø: 60 cm, AV: 4, D: 4

Rhaphidophlus cervicornis (ESPER, 1794)
Red finger sponge, red antler sponge

Hab.: Red Sea.

Sex.: None.

M.: Unknown.

F.: O; plankton, detritus.

S.: The red finger sponge extends its supple, long, digitate branches into the open water far beyond its anchor point on the fore reef. The flexibility of the branches is an adaptation to strong currents of its habitat. Encountered from shallow water to extremely deep environments, *R. cervicornis* affixes its small base to hard substrates of reef slopes and outcroppings.

T: 22°–26°C, L: 70 cm, AV: 4, D: 4

Amphimedon compressa, Caribbean.

Rhaphidophlus raraechelae (VAN SOEST, 1984)
Orange encrusting sponge

Hab.: Caribbean, Florida, Bonaire.

Sex.: None.

M.: Unknown.

F.: O; plankton, detritus.

S.: *R. raraechelae* forms thin encrustations on dead corals and other colonizable surfaces. It is considered an aggressive colonizer, since it overgrows and summarily kills everything in its vicinity. Canals radially diverge from the few raised oscula. The sponge's surface is always devoid of aufwuchs and sediment.

T: 20°–26°C, Ø: 80 cm, AV: 4, D: 4

Order: POECILOSCLERIDA
Fam.: Anchinoidae

Anchinoe tenacior TOPSENT, 1936
Blue encrusting sponge

Hab.: Mediterranean.

Sex.: None.

M.: Unknown.

F.: O; plankton, detritus.

S.: *A. tenacior* encrusts dead substrates. The surface is covered in round pits, which give the sponge a bullose texture, and converging canals that ultimately terminate at the raised oscula. It is found at virtually all depths on smooth, shaded rock walls and shipwrecks, within caves, and beneath ledge overhangs.

T: 10°–18°C, Ø: 25 cm, AV: 4, D: 4

Fam.: Agelasidae

Agelas oroides SCHMIDT, 1864
Leathery gold sponge

Hab.: Mediterranean.

Sex.: None.

M.: Unknown.

F.: O; plankton, detritus.

S.: Few oscula are found on the leathery, finely textured surface of this thick-bellied sponge. In the literature *A. oroides* is always presented as a "horny sponge," which is incorrect, since it has spongin fibers and calcareous spicules. It settles on rock walls and within crevices and small caves from shallow water to depths of more than 30 m.

T: 10°–18°C, Ø: 15 cm, AV: 4, D: 4

Anchinoe tenacior

Agelas oroides

Order: Poecilosclerida
Fam.: Agelasidae

Agelas conifera (SCHMIDT, 1870)
Brown tube sponge

Hab.: Caribbean, Florida, Bahamas, Colombia.

Sex.: None

Soc.B./Assoc.: *A. conifera* is colonized by the colonial anemones *Parazoanthuis swiftii* and *Parazoanthus puertoricense*.

M.: Unknown.

F.: O; plankton, detritus.

S.: Though *A. conifera* has many growth patterns, a broad base with many partially fused tubes is the most common. Each tube is tall, flexible, and branched with an excurrent opening at the center of the tips. The surface is irregular, rough, and variably colored. It predominantly settles on overhangs and niches of quiescent waters.

T: 20°–24°C, L: 40 cm, AV: 4, D: 4

Agelas conifera

Agelas dispar (DUCHASSAING & MICHELOTTI, 1864)
Orange elephant ear sponge

Hab.: Caribbean, Florida, Bahamas to Brazil.

Sex.: None.

Soc.B./Assoc.: The surface is frequently colonized by hydroids, tunicates, and other small invertebrates.

M.: Unknown.

F.: O; plankton, detritus.

S.: This massive sponge assumes a spherical, discoid, or planar (ear-shaped) form either with or without central oscula, depending on habitat, type of substrate, and water movement. The surface is very rough and firm and honeycombed by many irregularly dimensioned incurrent openings. From shallow to deep water, this species inhabits calm, clear environments. Coloration is variable.

T: 20°–24°C, L: 30 cm, AV: 4, D: 4

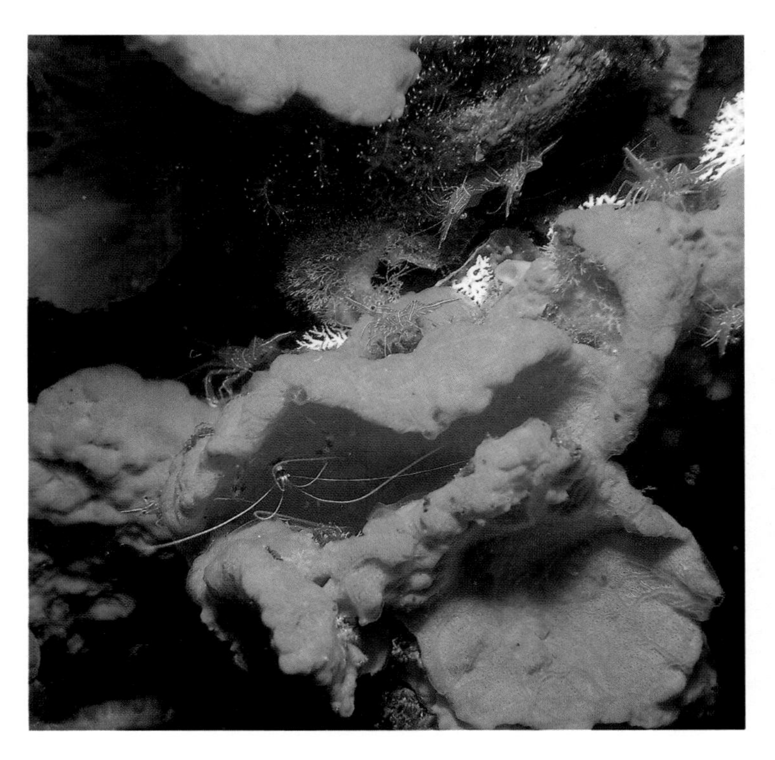

Agelas mauritiana (CARTER, 1862)
Orange leather sponge

Hab.: Indo-Pacific, Malaysia.

Sex.: None.

M.: Unknown.

F.: O; plankton, detritus.

S.: A very aggressive space competitor, *A. mauritiana* forms a thick, leathery crust on shaded rocks and dead corals within crevices and underneath overhangs at virtually all depths. It has exceptionally small oscula and numerous papillae.

T: 22°–26°C, Ø: 30 cm, AV: 4, D: 4

Order: Poecilosclerida
Fam.: Agelasidae

Agelas sp.
Brown tube sponge

Hab.: Caribbean, San Andrés.

Sex.: None.

Soc.B./Assoc.: It associates—cleaner shrimp and small *Gobisoma* gobies—indicate that *Agelas* sp. is used as a cleaner station by many fishes.

M.: Unknown.

F.: O; plankton, detritus.

S.: A cluster of tubes is affixed to the substrate with a broad common base. The outer layer of the tubes is hard, rough, and uneven. The light spots visible on the pictured specimen are scars where nudibranchs have fed. *Agelas* sp. frequently inhabits current-swept reef slopes down to deep water environments.

T: 20°–24°C, **L**: 50 cm, **AV**: 4, **D**: 4

Haliclona fascigera (HENTSCHEL, 1912)
Pink tube sponge

Hab.: Red Sea, Indo-Pacific.

Sex.: None.

M.: Unknown.

F.: O; plankton, detritus.

S.: Erect, variable-length tubes branch from a common base. The tubes' color fades towards the terminal excurrent opening. *H. fascigera* occurs along a broad depth range, colonizing all hard substrates.

T: 22°–26°C, **L:** 80 cm, **AV:** 4, **D:** 4

Order: Haplosclerida
Fam.: Haliclonidae

Haliclona fascigera (HENTSCHEL, 1912)
Blue-gray tube sponge

Hab.: Red Sea, Port Safaga.

Sex.: None.

M.: Unknown.

F.: O; plankton, detritus.

S.: This relatively large, erect, horny sponge lacks hard skeletal elements. Because the skeleton is made entirely of spongin, the body is very flexible. The variable-length tubes have few branches and large oscula. Small bristleworms live on the sponge's surface, taking advantage of its filtering activity. They are not commensals. The sponge lives among corals on reef slopes down to 35 m depth.

T: 22°–26°C, L: 60 cm, AV: 4, D: 4

Haliclona rosea (*mediterranea*?) (BOWERBANK, 1862)
Pink tube sponge

Hab.: Mediterranean, Giglio (Italy).

Sex.: None.

M.: Unknown.

F.: O; plankton, detritus.

S.: Each short, brittle, slender tube has a terminal osculum and a broadened base. Individually or in clusters, this sponge inhabits dimly lit areas, e.g., under outcroppings or within caves, of the rocky littoral zone, from shallow waters down to great depths. It is always free of aufwuchs and sediment.

T: 10°–22°C, L: 10 cm, AV: 4, D: 4

Order: HAPLOSCLERIDA
Fam.: Haliclonidae

Haliclona mediterranea GRIESSINGER, 1971
Mediterranean tube sponge

Hab.: Mediterranean.

Sex.: None.

M.: Unknown.

F.: O; plankton, detritus.

S.: A multitude of branched tubes give rise to numerous tall, stiff, brittle chimneys. Usually the entire colony is anchored to the substrate at just a few points, and crusty plates bridge the distance between individual chimneys. *H. mediterranea* is most commonly found in the shade of overhangs and caves and on deep coral bottoms to depths of more than 50 m. Most specimens are pink hues.

T: 10°–18°C, L: 4–10 cm, AV: 4, D: 4

Haliclona viscosa (TOPSENT, 1928)
Flesh sponge

Hab.: Eastern Atlantic, southern Ireland, Mediterranean.

Sex.: None.

M.: Unknown.

F.: O; plankton, detritus.

S.: *H. viscosa* encrusts rocks and other hard substrates of the sublittoral. Oscula are raised, appearing like short, cylindrical chimneys above the crusty layer. Current-swept coastal sections are its preferred habitat. Most specimens are flesh-colored.

T: 8°–18°C, Ø: 70 cm, AV: 4, D: 4

Order: Haplosclerida
Fam.: Haliclonidae

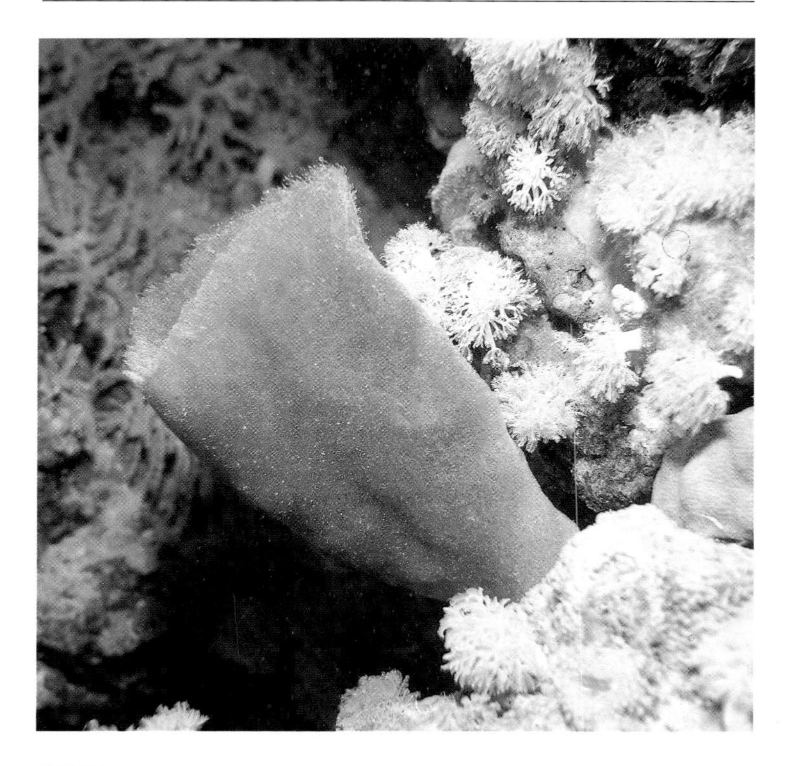

Haliclona sp.
Gray-brown tube sponge

Hab.: Red Sea.

Sex.: None

M.: Unknown.

F.: O; plankton, detritus.

S.: This tube-shaped sponge has a slender base, a soft, very flexible consistency, and a broad, usually ragged-edged osculum. The intricate exterior surface is generally free of sediments. *Haliclona* sp. always presents itself as an individual tube and grows on coral substrates.

T: 22°–26°C, **L**: 60 cm, **AV**: 4, **D**: 4

Niphates digitalis (LAMARCK, 1814)
Pink vase sponge, gray cornucopia sponge

Hab.: Caribbean, southern Florida, Bahamas, Colombia.

Sex.: None.

Soc.B./Assoc.: Usually the sponge's surface is inhabited by the colonial anemone *Parazoanthus parasiticus* and the brittle star *Ophiothrix suensonii*.

M.: Unknown.

F.: O; plankton, detritus.

S.: Most specimens of pink vase sponges are cup-shaped with a large osculum. The growth form pictured above is unusual; it is almost planar and lacks a central osculum. Nearly all hard substrates are suitable anchor points for the slender base of *N. digitalis*. The edges of the sponge are very fragile and virtually transparent, and the surface is very rough.

T: 20°–24°C, L: 15 cm, AV: 4, D: 4

Niphates digitalis

Niphates olemda (LAUBENFELS, 1936)
Tube sponge

Hab.: Indo-Pacific, Malaysia.

Sex.: None.

M.: Unknown.

F.: O; plankton, detritus.

S.: *N. olemda* is made up of soft, pliable, erect tubes that grow in assorted shapes and heights. The tubes have both terminal and lateral oscula, the latter in a flutelike arrangement. It grows on current-swept reefs from shallow water to a depth of 20 m.

T: 22°–26°C, L: 40 cm, AV: 4, D: 4

Order: HAPLOSCLERIDA
Fam.: Niphatidae

Gelloides fibulatus (RIDLEY, 1884)
Thorny horny sponge

Hab.: Indo-Pacific, Malaysia.

Sex.: None.

M.: Unknown.

F.: O; plankton, detritus.

S.: *G. fibulatus* has two growth forms—encrusting and tubular. The latter has short, erect tubes that project from a common base. The surface is rough with irregular spikes which have curved, frayed points. Thorny horny sponges inhabit a broad range of depths along reef slopes and platforms where there is dead coral, but it is rarely encountered on Indo-Pacific reefs.

T: 22°–26°C, L: 35 cm, AV: 4, D: 4

Callyspongia vaginalis (LAMARCK, 1814)
Branching vase sponge

Hab.: Caribbean, Florida, Bermuda, Bahamas.

Sex.: None.

Soc.B./Assoc.: Branching vase sponges are commonly colonized by the colonial anemone *Parazoanthus parasiticus* and the brittle star *Ophiothrix suensonii.*

M.: Unknown.

F.: O; plankton, detritus.

S.: Though its growth form depends on the magnitude of the current it is exposed to, a tubular growth pattern is generally adopted. The tubes are very elastic, vary in length, and either stand singly or in groups. Irregular pits and nubs cover the surface, making the sponge rough. *C. vaginalis* is always found on hard substrates, and reef plateaus and deep reef slopes are preferred sites of colonization.

T: 20°–24°C, **L:** 60 cm, **AV:** 4, **D:** 4

Callyspongia plicifera (LAMARCK, 1814)
Azure cup sponge

Hab.: Caribbean, Florida, Bahamas.

Sex.: None.

Soc.B./Assoc.: The sponge's chimneys are inhabited by the brittle star *Ophiothrix suensonii*.

M.: Unknown.

F.: O; plankton, detritus.

S.: Growth is erect and cylindrical or cup-shaped. Single chimneys are the most common. The surface is very rough and deeply fissured, whereas the edge of the terminal excurrent opening is extremely thin and virtually transparent. It is mainly found at 15–30 m depth on reef slopes, reef ledges, and steep drop-offs. In shallow waters under intense solar radiation, it can fluoresce. Most are predominately pink.

T: 20°–24°C, L: 25 cm, AV: 4, D: 4

Callyspongia plicifera

Order: HAPLOSCLERIDA
Fam.: Callyspongiidae

Callyspongia monilata (RIDLEY, 1884)
Blue finger sponge

Hab.: Red Sea, Indo-Pacific.

Sex.: None.

M.: Unknown.

F.: O; plankton, detritus.

S.: The blue finger sponge has long, thin, very flexible tubular branches. It is a quick secondary colonizer that immediately settles on open areas of the reef. Although it is predominantly found on reef slopes at depths of 30 m and below, *C. monilata* also lives on reef sections exposed to very strong currents.

T: 22°–26°C, L: 60 cm, AV: 4, D: 4

Callyspongia siphonella LEVI, 1973
Siphon sponge

Hab.: Red Sea, Indo-Pacific.

Sex.: None.

M.: Unknown.

F.: O; plankton, detritus.

S.: With its small base, *C. siphonella* anchors to rocks or corals. Several parallel tubes branch from the base, giving rise to a cluster. The skeleton is entirely made up of spongin. There are no spicules or other skeletal elements, and the tubes are very flexible. *C. siphonella* occurs to a depth of 20 m.

T: 22°–26°C, L: 30 cm, AV: 4, D: 4

Callyspongia schulzi (KIESCHNIK, 1990)
Schulz's vase sponge

Hab.: Indo-Pacific, Malaysia.

Sex.: None.

M.: Unknown.

F.: O; plankton, detritus.

S.: The tubes, which are short, flexible, and vase-shaped, are supported on a broad base and blanketed in blunt and pointed conical nubs. Dead corals are colonized at virtually all depths.

T: 22°–26°C, **L:** 10 cm, **AV:** 4, **D:** 4

Callyspongia sp.
Brown vase sponge

Hab.: Red Sea, Port Safaga.

Sex.: None.

M.: Unknown.

F.: O, plankton, detritus.

S.: Tubes exist singly or in groups. Their surface is very rough with uneven clefts and many round or irregular, pointed tubercles. Neither algae nor invertebrates are found living on or within this sponge. It grows on dead corals and among live corals on reef slopes and platforms down to a depth of 30 m. At favorable sites some specimens can grow to immense proportions.

T: 22°–26°C, **L:** 30 cm, **AV:** 4, **D:** 4

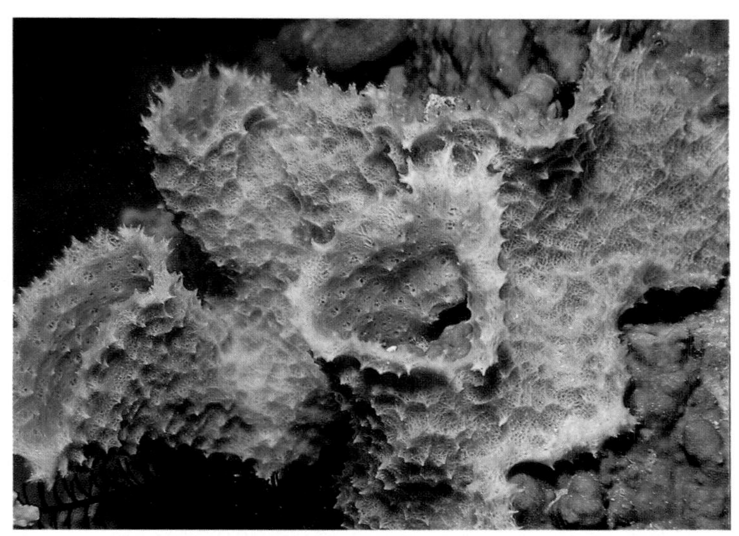

Callyspongia schulzi

Callyspongia sp.

Callyspongia sp.
Blue encrusting sponge

Hab.: Indian Ocean, Maldives.

Sex.: None.

M.: Unknown.

F.: O; plankton, detritus.

S.: Its many round excurrent openings are chimneylike and just a few millimeters higher than the sponge's surface. In view of its encrusting growth form, this species probably colonizes dead coral substrates along current-swept sites.

T: 22°–26°C, Ø: 20 cm, AV: 4, D: 4

Petrosia ficiformis POIRET, 1789
Bread sponge

Hab.: Mediterranean to the Cape Verde Islands.

Sex.: None.

Soc.B./Assoc.: This species is grazed upon by the leopard nudibranch, *Peltodoris atromaculata*.

M.: Unknown.

F.: O; plankton, detritus.

S.: A variety of growth forms may be assumed by this species. It is a tough and brittle sponge with a high proportion of siliceous spicules. Normally *P. ficiformis* is brown-purple superiorly and greenish inferiorly. Bright areas on the sponge's upper side are grazing paths of the leopard nudibranch. The few excurrent openings are reticulate. Most specimens colonize stones, rocks, and caves of shallow waters.

T: 10°–18°C, Ø: 25 cm, AV: 4, D: 4

Order: Haplosclerida
Fam.: Petrosiidae

Xestospongia muta (SCHMIDT, 1870)
Giant barrel sponge, basket sponge, tub sponge

Hab.: Caribbean, Florida, Bermuda, Bahamas to Brazil.

Sex.: None.

M.: Unknown.

F.: O; plankton, detritus.

S.: This is the largest sponge of the Caribbean. Full-grown specimens can grow to a height of almost 2 m—a fully equipped diver can virtually disappear into an osculum. The surface is richly cleft, wrinkled, and hard. Gigantic specimens live on reef sections exposed to strong current, while small individuals can be found on hard substrates of protected reefs. Coloration is variable. Specimens found at depths beyond 120 m may be light-colored. Growth is exceedingly slow.

T: 20°–24°C, L: 40 cm, AV: 4, D: 4

Xestospongia rosariensis (ZEA & RÜTZLER, 1983)
Orange tube sponge

Hab.: Caribbean, Rosarios Islands, Puerto Rico.

Sex.: None.

M.: Unknown.

F.: O; plankton, detritus.

S.: Anchored to hard substrates with its broad base, *X. rosariensis* grows cylindrical and branched. The tubes are rough yet soft, laterally compressed, and covered with bumps, groves, and irregular dark furrows. The terminal oscula have a light rim. Reef slopes down to great depths are colonized.

T: 20°–24°C, L: 60 cm, AV: 4, D: 4

Xestospongia muta

Xestospongia rosariensis

Xestospongia testudinaria (LAMARCK, 1814)
Large vase sponge

Hab.: Indo-Pacific, Malaysia.

Sex.: None.

M.: Unknown.

F.: O; plankton, detritus.

S.: *X. testudinaria* has a huge body and a broad base. Its surface is hard and rough with irregular ridges and folds along its length, and the terminal osculum is large with a smooth interior. Hard, almost horizontal surfaces of lower reef slopes are colonized.

T: 22°–26°C, L: 80 cm, AV: 4, D: 4

Cacospongia scalaris SCHMIDT, 1868
Leather sponge

Hab.: Mediterranean.

Sex.: None.

M.: Unknown.

F.: O; plankton, detritus.

S.: The leather sponge is massive, occasionally lobate, with an irregular, strongly cleft surface. Round excurrent openings are located along the crests of the humps. Due to the arrangement of spongin fibers, it is tough and leathery. *C. scalaris* is generally encountered in algae stands, along shaded areas, and on coral substrates at great depths.

T: 10°–18°C, Ø: 50 cm, AV: 4, D: 4

Order: DICTYOCERATIDA
Fam.: Spongiidae

Carteriospongia foliascens (PALLAS, 1766)
Foil sponge

Hab.: Indo-Pacific, Maldives.

Sex.: None.

M.: Unknown.

F.: O; plankton, detritus.

S.: An unusual sponge. Off a small base, flat, foliaceous discs emerge singly or in groups in a spiral pattern. There is no central osculum. The surface is slightly grooved with a multitude of small papillae. Over time, the funnel-shaped depressions fill with sediment and become colonizing sites for other organisms. *C. foliascens* attaches to coral gravel in deep water environments.

T: 22°–26°C, Ø: 40 cm, AV: 4, D: 4

Spongia agaricia (PALLAS, 1816)
Elephant-ear sponge

Hab.: Mediterranean.

Sex.: None.

M.: Not known. *S. agaricia* is unsuitable for aquaria because it grows too large.

F.: O; plankton, detritus.

S.: Although elephant-ear sponges exhibit a number of different grows patterns, plate-, fan-, and funnel-shaped forms predominate. Excurrent openings are abundantly distributed within the funnel or unilaterally on the fan. Because of the tight net of spongin fibers, the sponge is very pliable. Taxonomists consider this species merely a variety of the bath sponge *Spongia officinalis*—essentially a growth form adapted to deeper water. Exceptional specimens can attain a diameter of 100 cm. This species is also collected, cleaned, and used as bath and industrial sponges.

T: 5°–18°C, Ø: 60 cm, AV: 4, D: 4

Order: DICTYOCERATIDA
Fam.: Spongiidae

Spongia officinalis LINNAEUS, 1759
Bath sponge

Hab.: Cosmopolitan in temperate latitudes. Gulf of Mexico, Mediterranean, Japan, Philippines. It occurs from shallow waters to depth of more than 250 m.

Sex.: None.

M.: Unknown.

F.: O; plankton, detritus.

S.: This massive horny sponge exhibits different growth forms and colors. Its skeleton consists solely of spongin; there are absolutely no calcareous spicules. The surface is dark with a few slightly raised oscula, while the interior is reddish. It is chiefly found on rocky substrates, commonly at sites with dim light and a moderate current. *S. officinalis* is collected in the eastern Mediterranean, cleaned, and sold as bath and industrial sponges.

T: 10°–18°C, Ø: 25 cm, AV: 4, D: 3–4

Ircinia fasciculata PALLAS, 1766
Encrusting leather sponge

Hab.: Mediterranean.

Sex.: None.

M.: Unknown.

F.: O; plankton, detritus.

S.: *I. fasciculata* assumes various shapes—flat and encrusting, round, oval, and lobate. Its surface is either smooth or riddled with chimneylike oscula. Unlike *Spongia* species, *I. fasciculata* has a very tough skeleton of spongin fibers. Along a broad range of depths, encrusting leather sponges colonize rock substrates of cave walls and under ledges.

T: 10°–18°C, Ø: 25 cm, AV: 4, D: 4

Coral reef slope in the Red Sea.

Ircinia campana (LAMARCK, 1816)
Stinking vase sponge

Hab.: Caribbean, Florida, Bahamas, Colombia to Brazil.

Sex.: None.

M.: Unknown.

F.: O; plankton, detritus.

S.: This soft, flexible, easily compressed sponge has a vaselike shape and large oscula. The sponge's surface is rough and covered in conical projections. Suspension-rich lagoons and areas around mangroves are its preferred habitats; there it anchors to hard substrates with its slender base.

T: 20°–24°C, L: 30 cm, AV: 4, D: 4

Ircinia campana (LAMARCK, 1816)
Bell sponge

Hab.: Caribbean, Bahamas, Florida, Colombia to Brazil.

Sex.: None.

M.: Unknown.

F.: O; plankton, detritus.

S.: *I. campana* is vase-shaped and has a large osculum. Due to the presence of numerous conical papillae, the surface is rough. The sponge itself is soft, very flexible, and easily compressed. With its narrow base it adheres to hard substrates in suspension-rich lagoons and areas around mangroves.

T: 22°–24°C, Ø: 30 cm, AV: 4, D: 4

Ircinia strobilina (LAMARCK, 1816)
Pillow stinking sponge, black ball sponge

Hab.: Caribbean, Bahamas, Florida, Bonaire, Colombia to Brazil.

Sex.: None.

M.: Unknown.

F.: O; plankton, detritus.

S.: This species forms very large colonies. Large, dark excurrent openings are arranged in aggregations on the superior pole of this massive, almost spherical sponge. The surface is covered by irregular flat areas interspaced with white-tipped conical papillae. Black ball sponges are tough animals that have the consistency of a pencil eraser. They live on or among corals at all depths that can be reached by divers. Whitespotted filefish frequently nibble on these sponges.

T: 20°–26°C, Ø: 25 cm, AV: 4, D: 4

Sarcotragus muscarum (SCHMIDT, 1868)
Black leather sponge

Hab.: Mediterranean.

Sex.: None.

M.: Unknown.

F.: O; plankton, detritus.

S.: *S. muscarum* has a massive growth form, an irregular surface, and small excurrent openings. This leathery sponge is encountered on rocky substrates and among algae stands from shallow water, where it can become very voluminous, to great depths.

T: 10°–18°C, Ø: 15 cm, AV: 4, D: 4

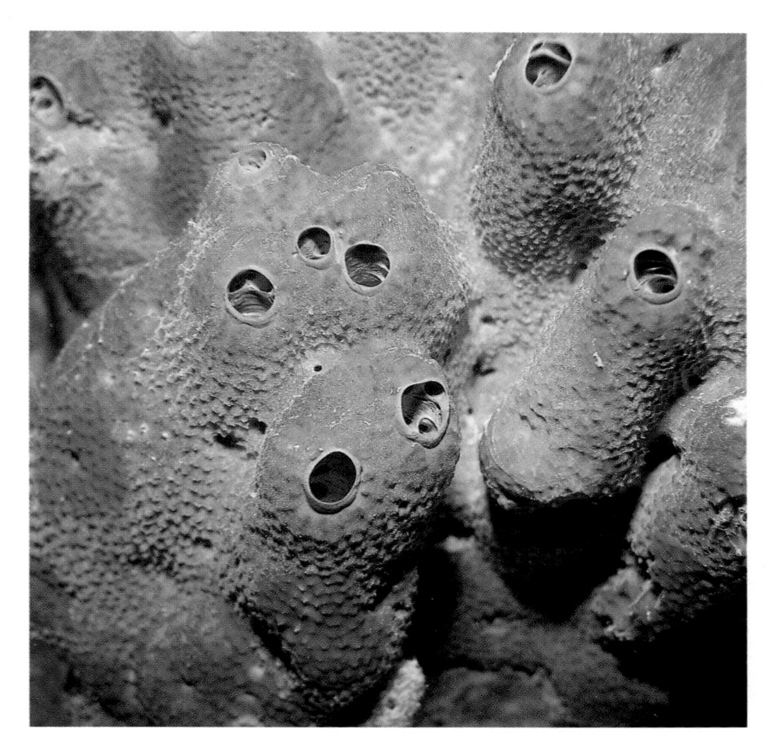

Smenospongia aurea (HYATT, 1875)
Brown rubber sponge

Hab.: Caribbean, Bahamas, Cuba, Haiti, Bonaire.

Sex.: None.

M.: Unknown.

F.: O; plankton, detritus.

S.: These amorphous mats have small, conical papillae covering their surface and a fine, almost transparent membrane fringing the elevated excurrent openings. The consistency of *S. aurea* is soft and rubberlike; out of water the body collapses. Any hard surface is a suitable substrate—even piers and mangroves.

T: 20°–26°C, Ø: 60 cm, AV: 4, D: 4

Order: Verongiida
Fam.: Aplysinidae

Aplysina aerophoba SCHMIDT, 1862
Golden sponge

Hab.: Atlantic, Mediterranean, Bay of Biscay.

Sex.: None.

M.: Unknown.

F.: O; plankton, detritus.

S.: Golden sponges are bright yellow to brown-yellow in color, fleshy, tough, and slick. The cylindrical chimneys are erect and about 6 cm in diameter with a lumpy surface and a flat top where the excurrent canal terminates. From shallow water to a depth of 15 m, *A. aerophoba* inhabits rocky substrates and seagrass lawns of quiescent bays, where it commonly grows in dense groups.

T: 10°–18°C, **L:** 15 cm, **AV:** 3, **D:** 4

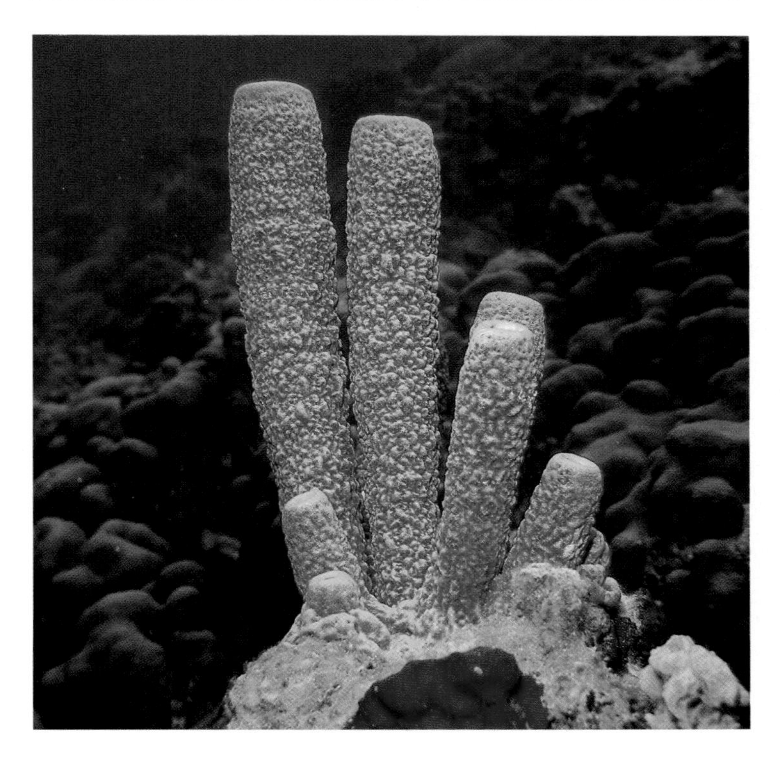

Aplysina archeri (HIGGIN, 1875)
Stovepipe sponge

Hab.: Caribbean, Bahamas, Florida, Bonaire.

Sex.: None.

M.: Unknown.

F.: O; plankton, detritus.

S.: This species produces single tubes or clusters of tubes that can grow to a height of more than 2 m. Its surface is rough and rubbery. Though normally purple-red, *A. archeri* can be found in a variety of hues. Located in deep, calm waters, the sponge can be found attached to horizontal substrates or suspended from vertical walls into the open water.

T: 20°–26°C, L: 50 cm, AV: 4, D: 4

Order: VERONGIIDA
Fam.: Aplysinidae

Aplysina cavernicola (VACELET, 1959)
Yellow cave sponge

Hab.: Mediterranean.

Sex.: None.

M.: Unknown.

F.: O; plankton, detritus.

S.: Yellow cave sponges have oval, round-tipped cylindrical chimneys that branch from a common base. In contrast to *Aplysina aerophoba*, this species does not have zooxanthellae or cyanobacteria in its cells, which means it can live beneath ledges and in grottos, caves, and other shaded locals of the rocky littoral.

T: 10°–18°C, L: 12 cm, AV: 4, D: 4

Aplysina cauliformis (CARTER, 1882)
Raw pore rope sponge

Hab.: Caribbean, Bahamas, southern Florida.

Sex.: None.

M.: Unknown.

F.: O; plankton, detritus.

S.: *A. cauliformis* is found on reef slopes and plateaus. Its slender base is either erect and little branched or extremely curved and intricately branched. The diameter and color of the lengthy tubes is variable. The oscular rims are pale.

T: 20°–26°C, L: <100 cm, AV: 4, D: 4

Order: VERONGIIDA
Fam.: Aplysinidae

Aplysina fistularis (PALLAS, 1766)
Yellow tube sponge

Hab.: Caribbean, Bahamas, Florida.

Sex.: None.

M.: Unknown.

F.: O; plankton, detritus.

S.: This species presents itself in a variety of shapes, depending on depth and current. Yellow to orange tubes project from a common base. In turbulent shallow waters the tubes are short and stout, but in deep waters, especially on steep reef slopes, the tubes grow quite tall. The sponge's surface is ornate. *A. fistularis* secretes a purple pigment that stains human skin for several days.

T: 20°–26°C, L: 60 cm, AV: 4, D: 4

Aplysina lacunosa (PALLAS, 1766)
Giant tube sponge, convoluted barrel sponge

Hab.: Caribbean, Bahamas, Colombia.

Sex.: None.

M.: Unknown.

F.: O; plankton, detritus.

S.: The cylindrical extensions of *A. lacunosa* have a maximum diameter of 10 cm, a pitted and irregular surface of deep crevices and grooves, and a contrasting yellow excurrent opening that is located centrally at the tube's apex. The tubes stand alone or in groups and can grow to great heights. From shallow water to a depth of 30 m, giant tube sponges grow upon stones or corals, often in suspension-rich water.

T: 20°–24°C, L: 50 cm, AV: 4, D: 4

Order: VERONGIIDA
Fam.: Aplysinidae

Verongula rigida (ESPER, 1794)
Pitted sponge

Hab.: Caribbean, Bahamas, Florida, Curaçao, Bonaire.

Sex.: None.

M.: Unknown.

F.: O; plankton, detritus.

S.: Projecting from an encrusting sponge mass are irregular conical chimneys with a central excurrent opening. The surface has a honeycomb design that is often covered by sediment. The body is soft, elastic, and rubbery. Pitted sponges colonize dead corals, stones, and steep reef slopes from shallow water to great depths.

T: 20°–26°C, Ø: 20 cm, AV: 4, D: 4

Pseudoceratina crassa (HYATT, 1875)
Branching tube sponge

Hab.: Caribbean, Florida, Bahamas to Brazil.

Sex.: None.

M.: Unknown.

F.: O; plankton, detritus.

S.: Growth and color are extremely variable. Rubberlike tubes—sometimes long, sometimes short—project from a common base; they have a rough, tuberculate surface. Terminal excurrent openings usually have a light-colored border. *P. crassa* occurs among corals and on dead substrates along reef slopes and rock walls from shallow water to great depths.

T: 20°–26°C, L: 50 cm, AV: 4, D: 4

Pseudoceratina crassa

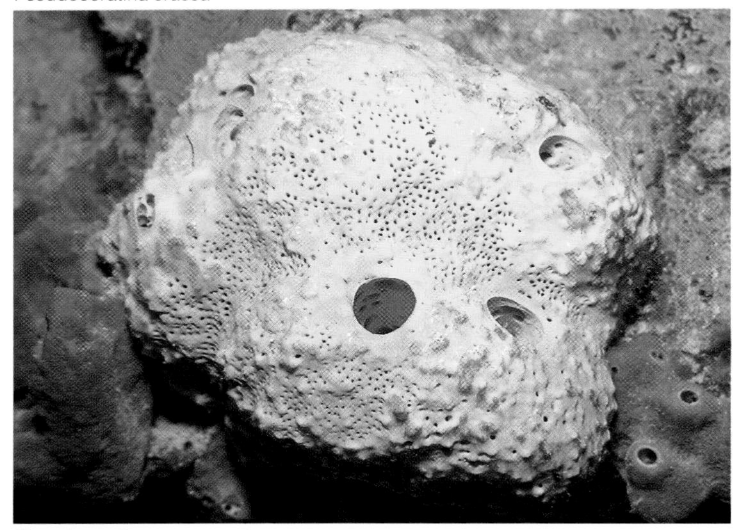

Pseudoceratina crassa

Pseudoceratina durissima (CARTER, 1886)
Tough tube sponge

Hab.: Red Sea, Indo-Pacific.

Sex.: None.

M.: Unknown.

F.: O; plankton, detritus.

S.: Although generally lumpy and rough, *P. durissima*'s surface features are highly variable, and its tubes are polymorphic. It is olive green with light green spots. On corals and other hard substrates from shallow water to moderate depths are where this species is generally encountered.

T: 20°–24°C, **L:** 20 cm, **AV:** 4, **D:** 4

Order: VERONGIIDA
Fam.: Druinellidae

Pseudoceratina sp.
Yellow-brown leather sponge

Hab.: Indo-Pacific, Malaysia.

Sex.: None.

M.: Unknown.

F.: O; plankton, detritus.

S.: *Pseudoceratina* sp. is a highly aggressive space competitor that forms lumpy crusts over live corals. Its knurled runners have a terminal osculum. Upper reef slopes, even those found in sediment-rich waters, are its preferred habitat.

T: 22°–26°C, Ø: 20 cm, AV: 4, D: 4

Haliscara caerulea VACELET & DONADAY, 1967
Encrusting star sponge

Hab.: Caribbean, Florida, Bahamas, Bequia.

Sex.: None.

M.: Unknown.

F.: O; plankton, detritus.

S.: The encrusting star sponge forms a crustlike layer over dead corals and other hard substrates in shaded locals. The star pattern—the result of canals radiating from the excurrent openings—is irregularly distributed over the sponge's surface.

T: 20°–26°C, Ø: 25 cm, AV: 4, D: 4

Class Hydrozoa

Hydrozoans, members of the phylum Cnidaria, characteristically have cnidocytes, a polypoid or medusoid form (or in some cases alternating generations of the two) and three body layers.
Of the two forms that hydrozoans may assume, the polyp is the simplest. Its tubelike body adheres to the substrate with its pedicel or hydrocaulus end, while the pole bearing the tentacles and mouth, the hydranth end, is typically directed towards the open water. The one opening acts as both mouth and anus and is encircled by extensions of the body wall, the tentacles.

Histologically, the body wall of hydrozoans has three distinct layers, the epidermis, gastrodermis, and mesoglea.
The epidermis serves a protective role in the organism. Cnidocytes are one type of epidermal cell. Nematocysts, possibly the most complex secretory product know to the animal kingdom, are a product of the cnidocytes.
Cnidocytes, or stinging cells, are double-walled capsules. The end exposed to the environment has a modified cilium, a cnidocil, that acts as a trigger. Upon tactile stimulation of the cnidocil (e.g., passing prey), the capsule is fired. At this point, the capsule inverts and the coiled, hollow thread from within is expelled and spines along its surface erected.
Depending on the type of nematocyst, there are two basic modes by which they function in prey procurement. The first—and most dangerous—is by immobilizing the prey by piercing the integument and injecting a paralyzing toxin. This type can also be dangerous, sometimes even lethal, to humans. The second type of nematocyst is rather innocuous by comparison, as it does not possess any known toxins. It merely sticks to prey, forming a sort of tow rope to pull the prey towards the oral opening of the hydrozoan. These are called desmonemes.
After discharge, nematocysts are regenerated within two days by nearby interstitial cells.
Amazingly enough, some nudibranchs that feed on hydrozoans can ingest nematocysts without digesting them. These same

nematocysts are then employed towards the defense of the nudibranch. After ingestion, the nematocysts are transported via the intestine to the cerata, which are dorsal appendages on the nudibranch. There they are fired when the sea slug feels threatened. Since they "steal" nematocysts from other animals to use for their own defense, these nudibranchs are called kleptocnidae.
The inner gastrovascular cavity of the hydrozoan is lined by the gastrodermis. Its main function is digestion. Using enzymes secreted by the enzymatic-gland cells, ingested foodstuffs are largely digested extracellularly. Indigestible remains are expelled through the mouth.
The third layer of the body wall, the mesoglea, lies between the epidermis and the gastrodermis. It principally provides support for the organism.

Medusae are the free-swimming complements to the sessile polyps. Despite a whole range of special morphological characteristics, they can still be described as short, broadened polyps with the mesoglea transformed into a gelatinous swimming body. The body is often called a bell or umbrella because of its shape. The dorsal surface of the bell is called the exumbrella, while the ventral side is termed the subumbrella. Located along the edge of the umbrella and covered with batteries of cnidocytes are the tentacles. Not only do the tentacles contain the cnidocytes for defense, they are also the site of two types of sensory organs—the ocelli and the statocysts. The former detect light, whereas the latter aid in orientation.
Typically, medusae are solitary animals that live in open waters, although there are a few species that adopted a sessile or semisessile lifestyle. Due to their high water content (90%), medusae are usually the same density as their aqueous environment, a condition that enables them to drift effortlessly. Propulsion is rhythmic and pulsating and often limited to the vertical plane. Forward locomotion, that is, movement along the horizontal plane, is generally passive (with the currents).
Many animals of this class only have a polyp phase, but very few have a medusoid generation without a polyp generation. The ma-

jority have what is termed an alternation of generations, passing through both an asexual and a sexual phase in their life cycle that corresponds to the polyp and medusa respectively.
Polyps reproduce asexually by budding. Often the daughter polyps remain attached to the mother polyp, giving rise to colonies. In this case, all the individual polyps are interconnected through tube-shaped stolons called hydrorhizae. Erect stolons can form stalks which support new polyp growth.
Colonial hydrozoans often have a division of labor through specialized polyps, i.e., to better perform certain functions, a group of polyps have been morphologically transformed (polymorphism). Only defense polyps (dactylozooids) and feeding polyps (gastrozooids) maintain the "original" polyp morphology. Polyps that have adapted to serve certain functions, such as gonozooids (reproductive polyps), have lost their mouth and tentacles.
Many polyps give rise to medusae by budding, in which case the medusae continuously separate. It is through these pelagic medusae that the species becomes distributed throughout the seas. The medusae are dioecious and reproduce sexually through eggs and sperm. The fertilized egg develops into a planula larva. After an initial free-swimming phase, it adheres to a substrate and develops into a polyp, thereby completing the alternation of the generations cycle.

The epidermis of most hydroids secretes an elastic sheath of chitin, called a perisarc. The extent to which the polyp is covered by this sheath determines part of its taxonomic classification. Athecate hydroids are those whose perisarc does not cover the hydranth. If the perisarc surrounds the hydranth, forming a hydrotheca, then the hydroids are considered to be thecate.
The medusae of athecate hydroids are called anthomedusae. They are bell-shaped with ocelli (light sensory organs) on the fringe of their umbrella and gonads on the manubrium. In contrast, medusae of thecate polyps are flat.

Fire corals (Milleporidae) and lace corals (Stylasteridae) are a special kind of athecate hydroid.

Fire corals are very similar to stony corals in appearance, habitat, and the fact that they are hermatypic; that is, they are reef-building corals capable of synthesizing a hard skeleton from calcium absorbed from their aqueous environment. The long, mouthless defense polyps of fire corals are equipped with batteries of penetrating nematocysts. These nematocysts are even able to puncture human skin, resulting in an excruciating sting. Several defense polyps surround a stout feeding polyp. All are capable of retracting into cavities of the calcareous skeleton. Ampullae are small medusae that develop in special indentions of the skeleton; they lack tentacles and a mouth. These transparent, inconspicuous medusae serve in a reproductive capacity.

Lace corals also secrete hard calcareous skeletons, but they are mostly small, delicately wrought, and richly branched in comparison to fire corals. Like fire corals, lace corals have several defense polyps surrounding one feeding polyp. They are dioecious. Since their life cycle does not include a medusoid phase, gametes are produced in special chambers, which are visible as small knots in some species.
Lace corals are typically encountered in shaded sites under overhangs or along the fringe of submarine caves.

Hydrozoans in the Aquarium

Many small hydroid polyps frequently enter the marine aquarium inadvertently as aufwuchs on invertebrates and plants. For example, the hydroid *Hydractinia echinata* is commonly introduced with the snail *Buccium undatum*. With the exception of some Mediterranean *Eudendrium* spp., hydroids have an abbreviated life span in aquaria.

Occasionally fire corals are imported, but because of the minute size of the polyps and their virulent stings, fire corals hold little appeal to aquarists. Furthermore, long-term maintenance is extremely difficult, as the zooxanthellae within their tissues demand intense illumination. An unfortunate side effect of the required intense illumination is the increased algae growth, as algae quickly overgrow these hydrozoans. Dead, these hydrozoans are valued objects *d'decor*. *Millepora dichotoma,* the reticulated fire coral, is a particular favorite.

Hydrozoans in their natural biotope predominantly feed on plankton. It therefore follows that plankton substitutes and *Artemia* nauplii would constitute an appropriate diet in the aquarium. Nevertheless, their care should be left to specialists.
Of all the members of the cnidaria, jellyfishes are the most infrequently kept. Aquaristically, the scyphozoans are of negligible importance, largely because their maintenance is practically impossible.

ORDER SIPHONOPHORA

Members of the order Siphonophora are characteristically large, pelagic organisms that either passively drift in the current or actively swim. Made up of modified polypoid and medusoid persons, they display the highest degree of polymorphism in the phylum Cnidaria.

Within the colony, individual polyps and medusae have adapted morphologically to fulfill diverse functions. Medusae are present as natant bodies, whereas different polyps assume the responsibility of defense, food acquisition, nutrition, and reproduction.
Only feeding polyps (gastrozooids) are capable of procuring nutrients. Unlike other Cnidaria, siphonophores do not have oral tentacles on their gastrozooids. Instead, these organisms have one long, contractile tentacle originating from the base of the gastrozooid; it serves the same purpose. The terminal ends of the short lateral branches emerging from the tentacles are endowed with batteries of large, highly toxic nematocysts.
Other than the lack of a mouth, foliaceous defense polyps (dactylozooids) are similar to feeding polyps.
Gonozooids, or reproductive polyps, support fascicular gonophores; reproductive persons, like medusae, have gonads and a manubrium (stomach stalk).
Because the phylogenetic origin of the colony is a medusa and the swimming bell is considered a modified medusa, it understandably has four radial canals and a circular canal; however, swimming bells do not have tentacles, sensory organs, or a mouth.

Siphonophora primarily feed on planktonic crustacea, worms, many types of larvae, snails, and small fishes.
The air chamber (float, pneumatophore) contains a gland that secretes gas into the float, enabling the organism to move vertically in the water column. The pneumatophore of some Siphonophora floats on the water surface and is therefore visible to animals viewing it from above.
It is highly probable that the Siphonophora developed from athecate hydroids.

Class Scyphozoa

Scyphozoans are frequently referred to as jellyfish. There are both polypoid and medusoid forms, whereby the former are only a few millimeters in size and therefore rather inconspicuous in relation to the latter. Both types have a characteristic morphology and are related through metagenesis.

The polyp, called a scyphistoma in this class, is always solitary, never colonial. It is comprised of a body with an oral disc surrounded by a crown of tentacles. The centrally located gastrovascular cavity is subdivided into four gastral pouches by four radial septa. Scyphistoma live in the littoral and sublittoral zones of the sea and feed on plankton. They reproduce asexually. New, independent polyps are formed by budding of the body wall.

The asexual reproduction of polyps into medusae occurs by characteristic transverse budding, termed strobilation, which is unique to this class. The body of the polyp constricts underneath the oral disc towards the base, and the tentacles degenerate and develop into 8 broad lobes with terminal sensory organs, leaving the polyp looking like a stack of dishes. Afterwards, the young medusa (ephyra) separates and develops into an adult medusa (scyphomedusa).

Unlike the diminutive scyphistoma, the scyphomedusa achieves a significant size. Some species have an umbrella 2 m in diameter. While a scyphomedusa can basically be compared morphologically to a hydromedusa, the former has ameboid cells in the mesoglea, gonads in the gastrodermis, and lacks a true velum.

The umbrella of scyphomedusae can be flat, discoid, or dome-shaped. Between the epidermis and the gastrodermis lies a gelatinous mesoglea with cellular inclusions that give the medusa a cartilaginous consistency. Though their texture and form may lead one to believe otherwise, jellyfishes are largely composed of water (94%).

The fringe of the umbrella has eight indentions and resulting lobes called lappets. These contain the sensorial grooves, statocysts (organs of equilibrium), ocelli (simple photoreceptors), and a variable number of tentacles.

Suspended from the subumbrella (ventral side of the umbrella) is the manubrium, the connection between the rectangular mouth and the stomach. The mouth is cross-shaped, branched, frayed, or ruffled. The fringe of the mouth is often elongated into oral arms or long appendages. In the order Rhizostomae, the oral arms are fused together, forming a weir basket of sorts.

Scyphomedusae are the sexual generation of the class Scyphozoa. Their four gonads either hang into the gastrodermis or alongside the oral arms. The latter is the most common arrangement in large species.

The gametes of most dioecious species are released through the mouth, and the eggs are fertilized in the open water. The resulting zygote becomes a planktonic planula larva. After approximately 10 days, the planula larva adheres to a substrate and gives rise to the scyphistoma. Through an asexual process of transverse fission (strobilation), immature medusae (ephyrae) are produced. Many species engage in a primitive type of brood care in which the eggs develop in the ovary or between the oral arms.

Ranging from harmless to extraordinarily virulent, the nematocysts of scyphomedusae are situated on the tentacles, oral arms, and even the epidermal layer of the umbrella. Prey is captured over this large surface area and transported to the edge of the umbrella with cilia. From there the oral arms carry the prey to the mouth.

After the food is placed in the mouth, it passes to the central stomach. Leading off the central stomach are four gastric pouches, each of which has glandular gastric filaments—the equivalent of the mesenteral filaments of corals. Enzymes secreted from these glands digest ingested foods extracellularly.

Scyphomedusae predominantly prey on jellyfishes, small fishes, and many planktonic organisms using their cnidocyte-covered tentacles and oral arms. Only members of the order Rhizostomae are planktivores.

Scyphozoans are perennial swimmers in the sea, transversing great distances with the aid of marine currents. With rhythmic swimming motions regulated by ganglia located next to the sensory organs on the margin of the bell, they can move relatively quickly. Many species live on the water surface as well as in deeper strata, depending on time of day, temperature, and meteorological conditions. An entire group has adapted to life in the deep sea.

ORDERS RHIZOSTOMAE AND SEMAESTOMAE

Although Rhizostomae and Semaestomae are closely related orders, two significant differences separate the two. First, Rhizostomae do not have tentacles along the margin of its umbrella. Second, the oral arms of Rhizostomae are fused, creating a porous, convoluted tube. This morphological feature corresponds to the filtering lifestyle adopted by these medusae. Microplankton is filtered from the water as it passes through pores in the oral arms, enters the stomach, and is assimilated. In some genera the oral apparatus is enlarged by additional extensions. Sometimes there are clublike appendages densely covered with cnidocytes and adhesive cells. Most of the approximately 80 known species are pelagic. Frequently juvenile fishes live among their oral arms. It is unknown whether this constitutes commensalism, since the nematocysts are too weak to safeguard the young fishes. The genus *Cassiopeia* leads a semisessile life. While it is capable of moving about, the majority of its life is passed resting on the bottom, tentacles oriented into the open water and the exumbrella upon the substrate ("upside down").
Reproduction and development has not been determined for all genera. However, a scyphistoma generally seems to be present. The species *Rhopilema esculenta* is eaten in Japan and China.

Semaestomae medusae are likely to be the most typical and familiar representatives of this class. The approximately 50 known species are widely distributed in temperate seas. Most are pelagic, inhabiting the upper water strata. They can appear in extensive schools that cover great expanses of the water surface. Some of their members are astoundingly large, reaching a diameter in excess of 2 meters. Almost all Semaestomae medusae are transparent to a varying degree with a colorful design. Their umbrella is flat to bowl-shaped and lacks a velum. Characteristically, medusae of the order Semaestomae have 4 oral arms on their manubrium. The oral arms are elongated, flexible, flaglike, and covered with frilly dermal fringes. Many species practice brood care.

Taxonomy

Phylum: CNIDARIA — Cnidarians

- Class: HYDROZOA — Hydrozoans
  - Order: ATHECATA (= Anthomedusae) — Athecate hydroids
    - Suborder: Capitata
      - Family: Tubulariidae
        - Genera: *Ralpharia*
          Tubularia
      - Family: Halocordylidae
        - Genus: *Halocordyle*
      - Family: Milleporidae — Fire corals
        - Genus: *Millepora*
    - Suborder: Filifera
      - Family: Eudendriidae
        - Genus: *Eudendrium*
      - Family: Stylasteridae — Lace corals
        - Genera: *Distichopora*
          Stylaster
  - Order: THECATA (= Leptomedusae) — Thecate hydroids
    - Family: Campanulariidae
      - Genus: *Obelia*
    - Family: Aequoreidae
      - Genus: *Aequorea*
    - Family: Haleciidae
      - Genus: *Halecium*
    - Family: Sertulariidae
      - Genera: *Sertularia*
        Sertularella
    - Family: Plumulariidae
      - Genera: *Dentitheca*
        Halopteris
        Kirchenpaueria
    - Family: Aglaopheniidae
      - Genera: *Aglaophenia*
        Gymnangium
        Lytocarpus
        Macrorhynchia
        Nemertesia

Family: Solanderiidae
Genus: *Solanderia*

Order: Siphonophora — Siphonophores
Suborder: Physophorida
Family: Physaliidae
Genus: *Physalia*
Family: Forskaliidae
Genus: *Agalma*

Class: Scyphozoa — Jellyfishes

Order: Rhizostomeae
Family: Rhizostomatidae
Genus: *Rhizostoma*
Family: Cepheidae
Genera: *Cephea*
Cotylorhiza
Physophora
Family: Mastigiidae
Genus: *Phyllorhiza*
Family: Cassiopeidae
Genus: *Cassiopeia*
Family: Stomolophidae
Genus: *Stomolophus*
Family: Thysanostomatidae
Genus: *Thysanostoma*

Order: Semaestomae
Family: Pelagiidae
Genera: *Chrysaora*
Pelagia
Family: Cyaneidae
Genus: *Drymonema*
Family: Ulmaridae
Genus: *Aurelia*

Millepora alcicornis, Caribbean.

Ralpharia magnifica WATSON, 1980
Solitary magnificent hydroid

Hab.: Indo-Pacific, Australia.

Sex.: None.

M.: Unknown.

F.: C; zooplankton.

S.: This solitary hydroid is the largest species found in Australian waters. It has long, erect, terminal hydrants, each of which has up to 150 long, white, delicate tentacles, and elongated to oval gonophores among its numerous tentacles. *R. magnifica* is found in shaded sites on a variety of substrates. Frequently it lives in small colonies of up to 50 individuals.

T: 22°–26°C, L: 5 cm, AV: 4, D: 3–4

Tubularia indivisa LINNAEUS, 1761
Undivided hydroid

Hab.: Atlantic, Norway.

Sex.: None.

M.: Unknown.

F.: C; zooplankton.

S.: *T. indivisa* forms extensive colonies on hard substrates such as stones, rocks, wood, buoys, and unattached items down to great depths. The long, erect stalk (hydrocaulus) thrusts into the open water from an encrusting mass and supports the hydranth and its two whorls of pinnate tentacles. Between the two whorls are the gonophores. Individual colonies arise through budding stolons. Fertilized eggs pass through an actinula larval stage before metamorphosing into a polyp.

T: 10°–18°C, L: 6 cm, AV: 4, D: 3–4

Halocordyle disticha (GOLDFUSS, 1820)
Christmas tree hydroid

Hab.: Caribbean, Florida, Bahamas, Cuba, Colombia, Venezuela. Cosmopolitan.

Sex.: None.

M.: Unknown.

F.: C; zooplankton.

S.: Pinnate, alternate, thorny branches are borne on the central stalk. The white polyps are "naked" and appear along light dots at the edges of the branches. The hydrants are only 3–4 mm long. Gonophores are long, oval, and medusoid; they release tiny, free-swimming medusae. This hydroid colony grows in clumps on dead corals, sponges, and gorgonians under overhangs, along steep walls, and on shipwrecks that are exposed to strong water movement. It is found at various depths. *Cratena pilata*, a nudibranch, grazes on *H. disticha*.

T: 20°–26°C, L: 8 cm, AV: 4, D: 3–4

Millepora dichotoma LINNAEUS, 1758
Reticulated fire coral, branched fire coral

Hab.: Red Sea, Indo-Pacific.

Sex.: None.

Soc.B./Assoc.: *M. dichitoma* lives in association with the tubeworm *Spirobranchus giganteus*.

M.: Occasionally small colonies are imported from the Red Sea or the Indo-Pacific. An aquarium devoid of filamentous algae is demanded, as is quality illumination for their zooxanthellae. A current must transverse the colony. Skeletons of this fire coral are utilized as objects *d'decor*.

Light: Sunlight zone.

F.: C; zooplankton.

S.: The skeletons are very brittle. *M. dichitoma* is a secondary colonizer that quickly settles on reef edges and free sections of reef tops and platforms. Their defense polyps are capable of producing an intense sting which causes a painful burning sensation upon contact. Do not touch!

T: 22°–26°C, **L:** 60 cm, **AV:** 3 (protected genus), **D:** 3

Millepora alcicornis, Caribbean.

Millepora alcicornis, Caribbean.

Millepora alcicornis LINNAEUS, 1758
Branching fire coral

Hab.: Caribbean, Florida, Bermuda, Bahamas, Puerto Rico, Colombia.

Sex.: None.

M.: Some years ago live colonies were still being imported, but now only skeletons are available to aquarists. They are sold and used as decorative objects for the aquarium.

F.: C; zooplankton.

S.: Two growth forms may be assumed—encrusting and erect and highly branched. Branches are long and variable in cross-section with light tips. The branching fire coral grows along reef edges and crests, commonly encrusting dead gorgonians and other substrates. The tiny, transparent medusae that are produced and released from the polyps lack tentacles as well as a mouth. Painful stings are administered upon contact.

T: 20°–24°C, L: 20 cm, AV: 4 (protected genus), D: 4

Millepora complanata LAMARCK, 1816
Plate fire coral, blade fire coral

Hab.: Caribbean, Florida, Bahamas, Bermuda, Curaçao, Colombia.

Sex.: None.

Soc.B./Assoc.: *M. complanata* lives in association with *Spirobranchus* tubeworms.

M.: Unknown.

F.: C; zooplankton.

S.: Colonies of *M. complanata* are made up of broad, flat, undulant, erect plates that have an almost smooth surface and dentate to sinuate edges. The plates project from a common base and are partially fused and intertwined. This fire coral colonizes shallow reef sections exposed to strong currents. Its colonies are not very extensive and never form continuous stands. Neither a mouth nor tentacles are present on the tiny, transparent medusae that are produced and released by the polyps. *M. complanata* is capable of dealing a painful sting!

T: 20°–24°C, **L:** 25 cm, **AV:** 4 (protected genus), **D:** 3–4

Millepora platyphylla HEMPRICH & EHRENBERG, 1834
Plate fire coral

Hab.: Red Sea, Indo-Pacific.

Sex.: None.

Soc.B./Assoc.: *M. platyphylla* lives in association with tubeworms of the genus *Spirobranchus*.

M.: See *Millepora dichotoma*.

F.: C; zooplankton.

S.: Its broad, stout, vertical plates have a rough, knotty, undulant surface. In shallow water, this hydrozoan is hermatypic (reef-forming). It may be dominant to surface-covering on exposed locals along upper reef slopes. The polyps are capable of rendering a painful sting! Tiny, transparent medusae that possess neither tentacles nor a mouth are produced.

T: 22°–26°C, Ø: 60 cm, AV: 3 (protected genus), D: 3

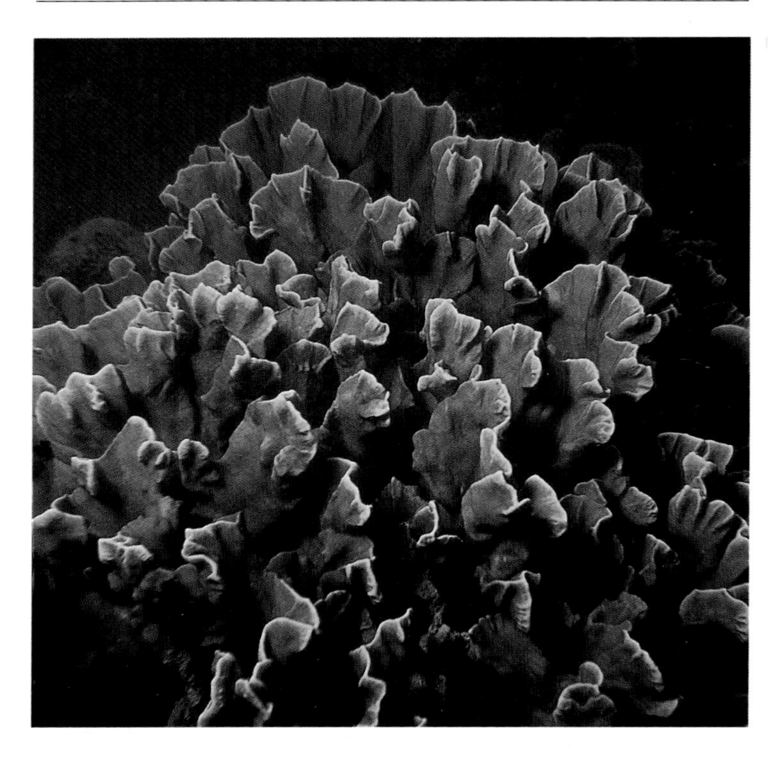

Millepora squarrosa LAMARCK, 1816
Encrusting fire coral, box fire coral

Hab.: Caribbean, Florida, Bahamas, Bermuda, Colombia.

Sex.: None.

Soc.B./Assoc.: *M. squarrosa* lives in association with tubeworms of the genus *Spirobranchus*.

M.: Unknown. Occasionally the calcareous skeletons are used as decorative objects for aquaria.

F.: C; zooplankton.

S.: Box fire corals form extensive encrustations on rocks and dead corals within seagrass lawns. The short, erect colonies have numerous, light-tipped folds. It is a hermatypic coral that grows in shallow waters along reef edges and flats. Neither a mouth nor tentacles are present on the tiny transparent medusae that are released from the polyps. Painful stings are administered upon contact.

T: 20°–24°C, **Ø**: 100 cm, **AV**: 3 (protected genus), **D**: 3–4

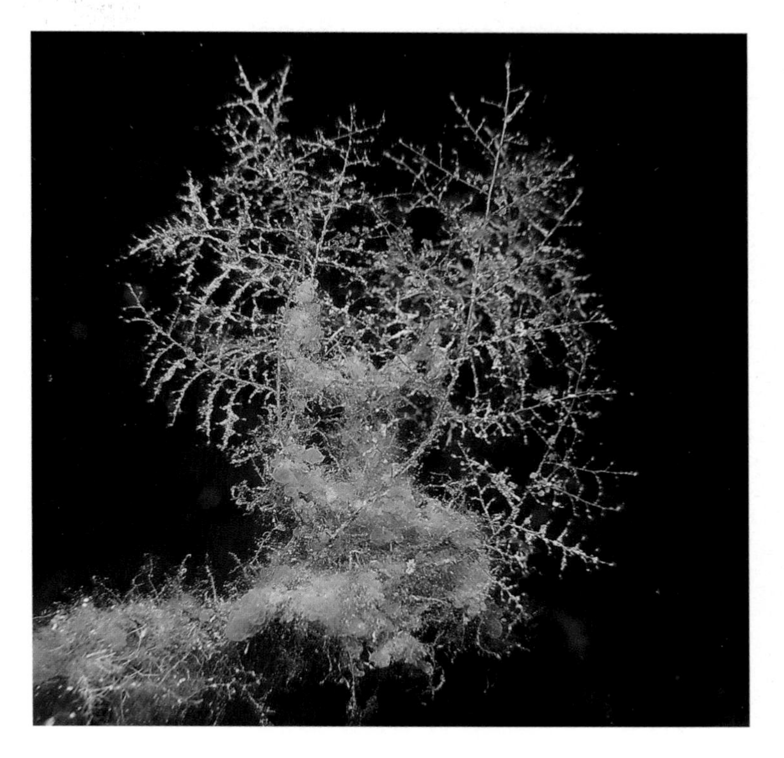

Eudendrium rameum PALLAS, 1766
Stick hydroid

Hab.: Mediterranean.

Sex.: None.

M.: Unknown.

F.: C; zooplankton.

S.: The creeping hydrorhiza of these erect colonies is capable of affixing to any substrate. The stout central stalk has many delicate intertwined branches. Each polyp is light, bottle-shaped, and naked (lacks hydrotheca) with a crown of fine, threadlike tentacles. Current-swept shaded sites, i.e., caves and under overhanging ledges, are where it is most frequently encountered.

T: 10°–20°C, L: 15 cm, AV: 4, D: 3–4

Distichopora nitida VERRILL, 1864
Indo-Pacific hydrocoral

Hab.: Indian Ocean, southwest Pacific.

Sex.: Dioecious.

M.: Unknown.

F.: C; zooplankton.

S.: These stout, fan-shaped, and richly-branched colonies grow such that their broad side faces the current. *D. nitida* is typically found at cave entrances and under ledges from shallow water down to great depths. They lack zooxanthellae. Skeletons of these animals are sold as souvenirs.

T: 22°–26°C, L: 12 cm, AV: 4, D: 4

Stylaster sp.
Lace coral

Hab.: Indo-Pacific, Malaysia.

Sex.: Dioecious (separate sexes).

M.: Unknown.

F.: C; zooplankton.

S.: Very bright and richly branched, this species grows on the branches of dead gorgonians situated in the shade of shallow water ledges. The colony lacks zooxanthellae.

T: 22°–26°C, L: 3 cm, AV: 4, D: 4

Order: Athecata
Fam.: Stylasteridae

Stylaster elegans VERRILL, 1864
Elegant lace coral

Hab.: Indo-Pacific, Australia.

Sex.: Colonies have separate sexes (dioecious).

M.: Unknown.

F.: C; zooplankton.

S.: This is a finely structured and exceptionally delicate and brittle lace coral. The central stalk has fine lateral, light-tipped branches. Though the coloration of the colony varies, it is mostly pink. With its broad side always directed into the prevailing current. *S. elegans* is encountered beneath overhanging ledges and at the entrance of caves and within the cave proper. It does not have zooxanthellae living within its tissues.

T: 22°–26°C, **L:** 8 cm, **AV:** 4, **D:** 4

Stylaster roseus (PALLAS, 1766)
Rose lace coral

Hab.: Caribbean.

Sex.: Colonies are dioecious.

M.: Unknown.

F.: C; zooplankton.

S.: *S. roseus* is a relatively small, richly branched lace coral. Its branches, each elliptical in cross-section, are arranged in one plane. It is frequently purple at the base, fading to white at the new tips. The colony is very brittle and does not have zooxanthellae. From shallow water down to depths in excess of 3,000 m, *S. roseus* is generally found at cave entrances, under overhanging ledges, and other shaded niches.

T: 20°–24°C, **L:** 5 cm, **AV:** 4, **D:** 4

Stylaster elegans

Stylaster roseus

Obelia geniculata (LINNAEUS, 1758)
Bell polyp

Hab.: Atlantic, Mediterranean. Cosmopolitan.

Sex.: None.

M.: Colonies are small and unattractive and therefore only hold the interest of true specialists.

F.: C; zooplankton.

S.: *O. geniculata* colonizes stones, algae, harbor installations, floating debris, and other hard substrates. Its small bell- to wineglass-shaped polyps, which grow alternately up the zigzagging branches, are interconnected through a stolon network. The life cycle includes a medusoid as well as a polypoid form. The nudibranch *Miesea evelinae* preys on this hydrozoan.

T: 10°–18°C, L: 3 cm, AV: 3, D: 3–4

Aequorea aequorea (FORSSKÅL, 1775)
Many-ribbed hydromedusa

Hab.: Mediterranean, Atlantic. From western Africa to Scotland.

Sex.: Medusae have separate sexes and exhibit sexual dichromatism. Males are bluish, whereas females are pink.

M.: This species is suitable for aquarium maintenance.

F.: Plankton.

S.: Cap-shaped and slightly constricted inferiorly, this medusa has 60–400 long tentacles arranged along the margin of its umbrella. Many simple radial channels lead from the edge of the gastric cavity to the umbrella margin close to the tentacles. A large number of statocysts are distributed among the radial canals. The stomach is very broad, funnel-shaped, and lipped. With a diameter of 18–20 cm, this is one of the largest leptomedusae known. *A. aequorea* is bioluminescent.

T: 10°–26°C, Ø: 15 cm, AV: 4, D: 4

Order: THECATA
Fam.: Haleciidae

Halecium halecinum (LINNAEUS, 1758)
Pinnate polyp

Hab.: Atlantic, Mediterranean, Norway.

Sex.: None.

M.: Unknown.

F.: C; zooplankton.

S.: *H. halecinum* is a pinnate colony that has alternate polyps along its sympodial stalks and gonangia at the tips of its branches. It attaches to hard substrates with its creeping stolons (hydrorhizae). As for all members of Thecata, the polyps are surrounded by hydrothecae. This hydroid colony is found from shallow water to great depths.

T: 10°–18°C, L: 12 cm, AV: 4, D: 3–4

Halecium sp.
Halecium hydroid

Hab.: Mediterranean, Giglio (Italy).

Sex.: None.

M.: Unknown.

F.: C; zooplankton.

S.: Growth is arborescent or fascicular, consisting of an erect, hornlike central stalk and many delicate branches. Polyps emerge alternately off the branches. Colonies grow at a depth of 40 m on rocky substrates in areas with good water circulation.

T: 10°–18°C, **L:** 10 cm, **AV:** 4, **D:** 4

Sertularia sp.
White weed

Hab.: Caribbean, Colombia.

Sex.: None.

M.: Unknown.

F.: C; zooplankton.

S.: White weed colonizes dead section of whip gorgonians at great depths in areas exposed to a strong current. The relatively small polyps are arranged alternately along its long filamentous branches.

T: 20°–24°C, **L:** 8 cm, **AV:** 4, **D:** 4

Sertularella speciosa CONGDON, 1907
Branching hydroid

Hab.: Caribbean, Florida, Bahamas, Cuba, Colombia.

Sex.: None.

M.: Unknown.

F.: C; zooplankton.

S.: The scant branches extend alternately and in one plane from the central stalk; the colony is pinnate. The polyps are large and sting strongly. Hard substrates in fast currents are their habitat of choice. The branching hydroid can be found at a wide range of depths on shaded sites, e.g., under ledges and on shipwrecks, harbor installations, and dead gorgonians. It often grows in turbid waters of harbors.

T: 20°–24°C, L: 6 cm, AV: 4, D: 3–4

Sertularella diaphama, Caribbean.

Sertularella diaphama (BILLARD, 1925)
Feather hydroid

Hab.: Caribbean, Florida, Bahamas, Colombia.

Sex.: None.

M.: Unknown.

F.: C; zooplankton.

S.: Colonies are pinnate to fascicular with alternating branches in one plane. At variable depths, but most often between 30–40 m, feather hydroids are found growing on corals and other hard substrates. *S. diaphama* is capable of administering a strong sting!

T: 20°–24°C, L: 25 cm, AV: 4, D: 4

Dentitheca dendritica STECHOW, 1920
Feather bush hydroid

Hab.: Caribbean, Florida, Bahamas, Colombia.

Sex.: None.

Soc.B./Assoc.: An appropriate tankmate for the colonial anemone *Parazoanthus tunicans*.

M.: Unknown.

F.: C; zooplankton.

S.: The colonies are fan-shaped or bushlike with a central stem and many primary and secondary branches. To maximize the amount of zooplankton that comes in contact with their densely grouped polyps, the broad plane of the pinnate branches faces the current. Sections of reef slopes and reef bottoms exposed to strong water currents are their habitat of choice. Polyps are capable of administering a potent sting!

T: 20°–24 °C, L: 60 cm, AV: 4, D: 4

Kirchenpaueria sp.
Feather hydroid

Hab.: Atlantic, Norway.

Sex.: None.

M.: Unknown

F.: C; zooplankton.

S.: Colonies are erect, pinnate, and usually occur in small groups. The alternate lateral branches are densely covered with polyps. Shaded sites, e.g., caves and beneath ledges, exposed to strong currents are preferred. *Kirchenpaueria* sp. is found at a broad range of depths.

T: 10°–18°C, L: 4 cm, AV: 4, D: 4

Aglaophenia allmani NUTTING, 1900
Allman's hydroid

Hab.: Caribbean, Florida, Bahamas, Colombia to Brazil.

Sex.: None.

M.: Unknown.

F.: C; plankton.

S.: Colonies are arborescent with a stout central stalk supporting lateral branches that are virtually the same thickness as the central stalk. The branches emerge irregularly from the central stalk, yet they are oriented in one plane. The light polyps are small and toxic. When touched, they cause a strong burning sensation. Allman's hydroids occupy almost all Caribbean reef biotopes.

T: 20°–24°C, L: 10–15 cm, AV: 4, D: 4

Aglaophenia cupressina LAMOUROUX, 1812
Feather hydroid

Hab.: Indo-Pacific, Malaysia, Australia.

Sex.: None.

M.: Unknown.

F.: C; zooplankton.

S.: The colonies are erect, pinnate to fernlike, and have numerous branches. Each "feather" or "frond" has a central stalk with slender, lateral branches which are broadened like the blade of a leaf. The polyps are arranged in rows along these branches. Due to the abundance of nematocysts, *A. cupressina* produces an extremely painful sting upon contact. The overall color of the colony depends on the color of the zooxanthellae living within its tissues. The literature does not differentiate between yellow-brown and light-colored species.

T: 22°–26°C, L: 60 cm, AV: 4, D: 3–4

Order: THECATA
Fam.: Aglaopheniidae

Aglaophenia kirchenpaueri (HELLER, 1868)
Feather hydroid

Hab.: Atlantic, Mediterranean.

Sex.: None.

M.: Unknown.

F.: C; zooplankton.

S.: Rootlike stolons, or hydrorhizae, affix these erect, pinnate colonies to various substrates. On a central stalk, delicate alternating polyp-bearing branches are borne. This feather hydroid prefers shaded areas, particularly under overhangs. Individual colonies or clusters are found within a broad range of depths.

T: 10°–18°C, L: 4 cm, AV: 4, D: 4

Aglaophenia sp.
Feather hydroid

Hab.: Indo-Pacific, Malaysia.

Sex.: None.

M.: Unknown.

F.: C; zooplankton.

S.: An erect, pinnate species that anchors itself to hard substrates with rootlike stolons. Each central stalk supports a multitude of delicate, polyp-bearing branches. Growth is predominately in bunches or clusters. Caution! *Aglaophenia* sp. is toxic and capable of producing a strong sting.

T: 22°–26°C, L: 8 cm, AV: 4, D: 4

Order: THECATA
Fam.: Aglaopheniidae

Gymnangium eximium (ALLMAN, 1874)
Feather bunch hydroid

Hab.: Red Sea, Indo-Pacific, Polynesia.

Sex.: None.

M.: Unknown.

F.: C; zooplankton.

S.: A multitude of thin pinnate branches are produced by this strong, erect, horny-stemmed colony. All are oriented transverse to the current. *G. eximium* is most often found growing among corals along current-swept reef slopes. The colonies are toxic, producing a potent sting when touched!

T: 22°–26°C, L: 25 cm, AV: 4, D: 4

Gymnangium montagui (BILLARD, 1924)
Feather hydroid

Hab.: Atlantic, Brittany.

Sex.: None.

M.: Unknown.

F.: C; zooplankton.

S.: This is one of the largest hydroid colonies of the north Atlantic. Young colonies settle in the immediate vicinity of the parent animals, producing clumps consisting of many individual hydroids. Sometimes these groups cover large extensions. Each colony is erect and pinnate with many lateral branches originating from one central stalk. Its habitat is current-swept areas along the rocky littoral.

T: 10°–18°C, L: 15 cm, AV: 4, D: 3–4

Lytocarpus philippinus (KIRCHENPAUER, 1872)
Philippine stinging moss

Hab.: Canary Islands to Brazil, Red Sea, Indo-Pacific.

Sex.: None.

M.: Unknown.

F.: C; zooplankton.

S.: Colonies are erect and arborescent. The stalwart central stalk bears numerous delicate branches, each of which resembles a feather. At the tips of the lateral branches are many light polyps with very poisonous nematocysts. Because these nematocysts are able to penetrate human skin, they cause excruciating stings. *L. philippinus* predominantly lives in shallow waters along reefs.

T: 22°–26°C, L: 60 cm, AV: 4, D: 4

Macrorhynchia philippina (KIRCHENPAUER, 1872)
Philippine stinging hydroid

Hab.: Red Sea, Indo-Pacific, Malaysia, Fiji.

Sex.: None.

M.: Unknown.

F.: C; zooplankton.

S.: Colonies are fascicular with numerous erect, pinnate branches, all of which emerge from a central stalk and are aligned in one plane. The polyps line the many alternating secondary branches. Although the majority grow on gorgonians, dead substrates are also colonized.

T: 22°–26°C, L: 8 cm, AV: 4, D: 4

Gymnangium longicauda, Caribbean.

Gymnangium longicauda (NUTTING, 1900)
Feather hydroid

Hab.: Caribbean, Florida, Bahamas, Bonaire. *G. longicausda* lives at depths of 3–60 m.

Sex.: None.

M.: Unknown.

F.: C; plankton.

S.: Colonies are pinnate and grow on hard substrates. The pale, alternating branches are in one plane. They have small, light polyps that sting strongly, causing severe burns to human skin.

T: 24°–29°C, L: 12 cm, AV: 4, D: 4

Macrorhynchia sp., Caribbean.

Macrorhynchia sp.
Nettling hydroid

Hab.: Caribbean, Florida, Bonaire. *Macrorhynchia* sp. lives at depths of 3–60 m.

Sex.: None.

M.: Unknown.

F.: C; plankton.

S.: The nettling hydroid is toxic, capable of administering severe burns to human skin. The low-growing colonies are just slightly taller than their substrate. All primary branches have thin, densely placed secondary branches which bear polyps. While not particular about the substrate, this species prefers current-swept reef sections in the shade of steep reef slopes.

T: 24°–29°C, L: 15 cm, AV: 4, D: 4

Order: THECATA
Fam.: Plumulariidae

Halopteris carinata, Caribbean.

Halopteris carinata ALLMAN, 1877
Thread hydroid

Hab.: Caribbean, Bahamas, Florida, Bequia. The thread hydroid occurs at depths of 5–40 m.

Sex.: None.

M.: Unknown.

F.: C; plankton.

S.: The thread hydroid lives in groups. Every erect, threadlike colony originates from a thin fibriform mass. The short, polyp-bearing lateral branches are attached alternately to the central stalk. *H. carinata* colonizes all suitable substrates, especially sponges, algae, coral rubble, and shipwrecks. It causes a slight sting when touched.

T: 24°–29°C, L: 13 cm, AV: 4, D: 4

Fam.: Solanderiidae

Solanderia gracilis, Caribbean.

Solanderia gracilis (DUCHASSAING & MICHELOTTI, 1860)
Seafan hydroid

Hab.: Caribbean, Bahamas, Florida, Bequia. At depths of 5–40 m.

Sex.: Colonies have separate sexes (dioecious).

M.: Unknown.

F.: C; plankton.

S.: *S. gracilis* displays a greater similarity to fan corals than it does to colonial hydroids, which explains why it was originally described as such. The red to red-purple main stalk is stout and supports strong branches which are further branched towards the edge of the colony. All branches are arranged in one plane and bear rows of light, thin, short polyps. This hydroid grows on vertical walls in clear, swift waters.

T: 24°–29°C, L: 35 cm, AV: 4, D: 4

Nemertesia antennina (LINNAEUS, 1758)
Antenna hydroid

Hab.: Atlantic, Mediterranean.

Sex.: None.

M.: Unknown.

F.: C; zooplankton.

S.: Clusters of erect, pinnate stalks are affixed to the substrate with delicate, intertwining, fibriform hydrorhizae. Each strand is reminiscent of an antenna with rosettes. The whorled branches are short and curved, and the polyps are encased in vase-shaped hydrothecae. Colonies settle on hard and soft substrates to depths beyond 100 m.

T: 10°–18°C, L: 12 cm, AV: 4, D: 4

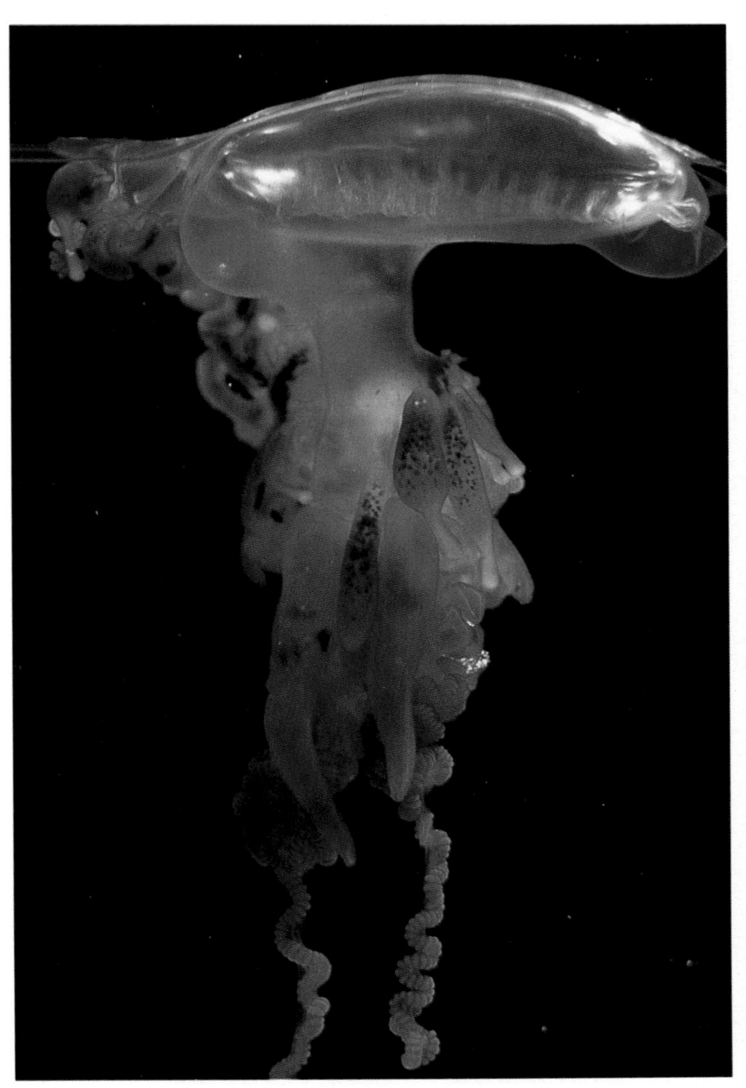

Physalia physalis. Text is on the following page.

Order: Siphonophora
Fam.: Physaliidae

Physalia physalis (photo on previous page) (LINNAEUS, 1759)
Portuguese man-of-war

Hab.: Cosmopolitan.

Sex.: *P. physalis* has separate sexes (dioecious).

Soc.B./Assoc.: The small man-of-war fish, *Nomeus gronovii*, may be found among the tentacles.

M.: Unknown.

F.: C; small fishes.

S.: The Portuguese man-of-war is a very complex hydrozoan colony made up of a nitrogen-, argon-, and oxygen-filled blue-violet pneumatophore* (float), which may grow to be as long as 30 cm, and numerous delicate tentacles. Extending up to 30 m below the floating pneumatophore, the tentacles are highly contractile and full of nematocyst batteries that enable the animal to capture and ingest small fishes and other organisms. Contact with the Portuguese man-of-war can result in excruciating, possibly fatal, stings. Occasionally *P. phsalis* is found in schools. The loggerhead turtle, *Caretta caretta*, preys upon these hydrozoans.

T: 10°–26°C, **L:** 30 cm (pneumatophore), **AV:** 4

* Gas bladder of some siphonophores.

Channels between the Exuma Cays, Bahamas.

Fam.: Forskaliidae

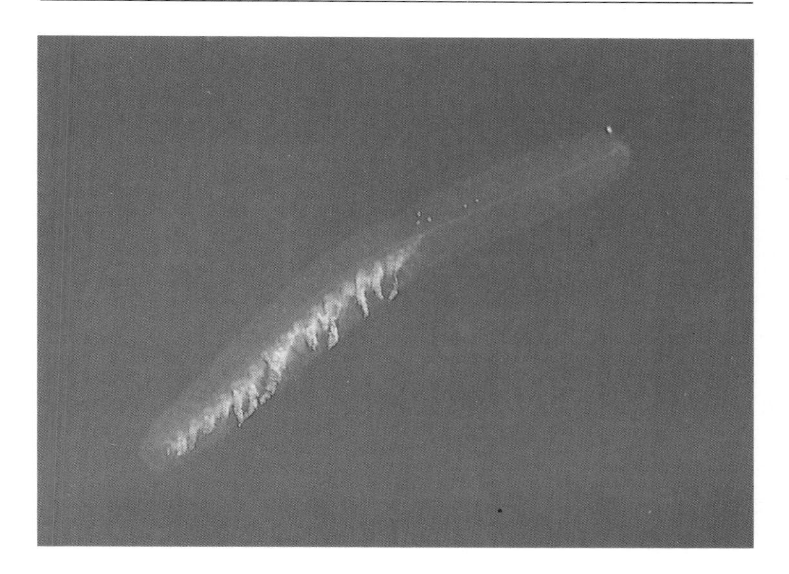

Agalma sp.
Jellyfish

Hab.: Western Atlantic, Cuba.

Sex.: *Agalma* sp. has separate sexes (dioecious).

M.: Unknown.

F.: Plankton.

S.: Thanks to its swimming bells, this siphonophore actively swims through its environment, predominantly horizontal. The colony resembles a garland with a small pneumatophore and a large section of numerous swimming bells beneath; the latter are arranged so densely that they can be likened to the scales of a pinecone. The animal colony is located beneath the medusa and consists of several hundred individuals. *Agalma* sp. captures fishes with its long tentacles. Toxic! Contact results in painful stings.

T: 20°–26°C, Ø: 15 cm, AV: 4, D: 4

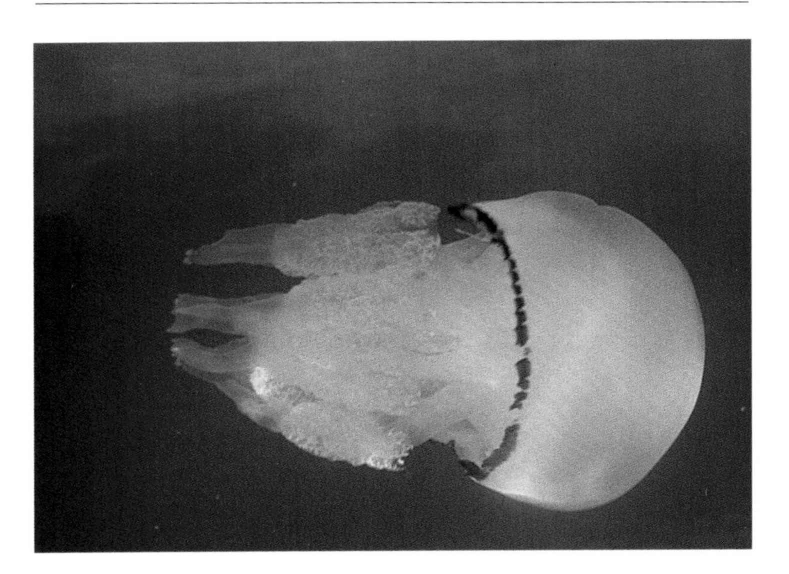

Rhizostoma pulmo AGASSIZ, 1862
Stiff arms jellyfish

Hab.: Atlantic, Mediterranean, North Sea, western Baltic Sea.

Sex.: Dioecious.

M.: Unknown.

F.: Plankton.

S.: *R. pulmo* is an extraordinarily large jellyfish with a steep umbrella. Although the margin of the umbrella lacks tentacles, it has 80–90 bright blue to purple lobes. The 8 oral arms are long with smooth, unbranched tips. Cauliflowerlike dermal lobes can be found in the center of the proximally fused oral arms. Encountered both in the high seas and along coastlines, it inhabits the upper water column throughout the year. Stiff arms jellyfish occasionally congregate and form large schools. Various juvenile fishes accompany them. Mild stings may result upon contact. Turtles of the genus *Dermochelys* are known predators.

T: 10°–18°C, Ø: 60 cm, AV: 4, D: 4

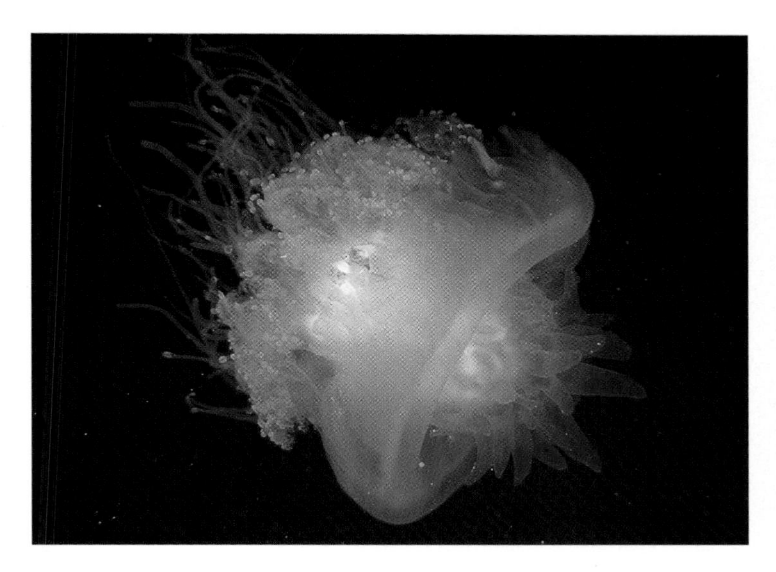

Cephea cephea (FORSSKÅL, 1775)

Hab.: Tropical Indo-Pacific, Red Sea.

Sex.: Dioecious.

M.: Unknown.

F.: Plankton, copepods, bristleworms, and arrow worms.

S.: This species is relatively flat with regular growths along the superior side of the umbrella. Each of the growths has approximately 30 conical warts or tubercles of variable length thereon. The margin of the umbrella is cleft into 80–90 lappets. While the 8 thick oral arms are fused proximally, the distal ends are richly branched and have long, thin oral filaments.

T: 22°-26°C, Ø: 14 cm, AV: 4, D: 4

Cotylorhiza tuberculata (MACRI, 1778)
Fried-egg jellyfish

Hab.: Mediterranean.

Sex.: *C. tuberculata* has separate sexes (dioecious). Eggs are fertilized internally, and the larvae are incubated in "pockets" before they are released into the open water, where they develop into sessile polyps.

M.: Not suitable for aquaria.

F.: C; microplankton.

S.: Flattened along its margin, this golden yellow jellyfish has a dome at the center of its umbrella, giving it the appearance of a large fried egg. There are many, variable-length tentacles attached to eight lappets. The 8 oral arms are brittle, short, and fused proximally. Numerous blue- or purple-tipped appendages are located between each of the oral arms. *C. tuberculata* is frequently found in aggregations in the open waters of the high seas and along coastlines; it migrates vertically. Various juvenile fishes are associated with *C. tuberculata*.

T: 10°–18°C, Ø: 35 cm, AV: 4, D: 4

Physophora hydrostatica (FORSSKÅL, 1775)
Physophore

Hab.: Cosmopolitan in warm waters; Atlantic to Indo-Pacific.

Sex.: Dioecious.

M.: Unknown.

F.: C; plankton.

S.: There are two rows of five pairs of swimming bells arranged on a stem. Though one end of the stem has a gas bladder, it is insufficiently buoyant to support the animal in the water column; active swimming movements are required to maintain the animal colony's position. Feeding, defense, and capture polyps as well as reproductive medusae and a crown of reddish palps are found on the opposite end of the stem. The palps hang vertically in the water, slowly sinking when the swimming bells relax. They are equipped with batteries of nematocysts.

T: 16°–26°C, **L:** 10 cm, **AV:** 4, **D:** 4

Phyllorhiza punctata LENDENFELD, 1884
Spotted jellyfish

Hab.: Indo-Pacific, Australia to Thailand.

Sex.: *P. punctata* has separate sexes (dioecious).

M.: Unknown.

F.: Plankton.

S.: *P. punctata* has a pronounced hemispherical umbrella with numerous round, light dots. The exumbrella is thick and finely granular. Oral arms are broad and cauliflowerlike with long windowlike secondary mouths. Distally, the oral arms have long, blunt-tipped filaments. It swims close to coasts and enters estuaries, sometimes appearing in great densities.

T: 22°–26°C, Ø: 50 cm, AV: 4, D: 4

Cassiopeia andromedra ESCHSCHOLZ, 1829
Suction cup jellyfish

Hab.: Red Sea, Suez Canal, Indo-Pacific, Micronesia, Hawaii, Australia.

Sex.: *C. andromedra* is hermaphroditic.

M.: Only small animals occasionally enter the trade. This species is relatively suitable for aquaria and is very tolerant of water pollution.

F.: The presence of zooxanthellae in the oral tentacles minimizes its need for food, making maintenance easier. In nature it feeds on detritus, small crustacea, unicellular organisms, and diatoms.

Light: Sunlight zone.

S.: Older animals settle on the substrate of lagoons, in seagrass lawns, and on sand and mud substrates of mangrove bays. Using the exumbrella to apply suction, *C. andromedra* fastens itself to the substrate. The oral arms have a multitude of branches that extend into the open water. When the thin edge of the umbrella is contracted, fresh water, plankton, and detritus are fanned towards the oral arms. This species is capable of swimming short distances. In some mangrove bays and estuaries, thousands may be encountered side by side along the bottom.

T: 20°–29°C, Ø: 25 cm, TL: from 150 cm, WM: m, WR: b, AV: 3–4, D: 3–4

Order: Rhizostomeae
Fam.: Cassiopeidae

Cassiopeia xamachana BIGELOW, 1892
Upside-down jellyfish

Hab.: Caribbean, Bahamas, Florida, Colombia.

Sex.: *C. xamachana* has separate sexes.

M.: Occasionally silver dollar-sized animals are available in pet stores. They are rapid growing and might even reproduce in the aquarium.

F.: O; plankton, algae, very small crustacea, and small worms.

S.: *C. xamachana*'s uncommon habits are demonstrative of a semisessile lifestyle. Only juvenile medusae are free-swimming. In contrast, adult animals frequently lie on muddy bottoms in shallow waters of lagoons and mangrove bays in great densities, exumbrella oriented toward the substrate and the richly branched oral arms extended into the open water. To prevent being swept away by the current, they search for a depression in the substrate and anchor themselves therein with suction applied by the exumbrella. The mesoglea of the umbrella contains a rich population of zooxanthellae. This jellyfish stings strongly.

T: 22°–24°C, Ø: 40 cm, AV: 4, D: 3–4

Stomolophus meleagris AGASSIZ, 1862
Cannonball jelly

Hab.: Caribbean, Florida, Colombia.

Sex.: The sexes are separate (dioecious).

M.: Unknown.

F.: Plankton.

S.: These jellyfish have short, dentate, milky-colored oral tentacles and a hemispherical bell. The latter exists in a variety of colors. Along the edge of the bell and extending towards the apex are white stripes of variable length and chocolate brown spots. At certain times of the year huge numbers of these organisms occur near coastlines, plaguing commercial fisheries. Cannonball jellies are good swimmers and voracious planktivores.

T: 22°–24°C, Ø: 8 cm, AV: 4, D: 4

Order: Rhizostomeae
Fam.: Thysanostomatidae

Thysanostoma loriferum (EHRENBERG, 1835)

Hab.: Red Sea, Indo-Pacific.

Sex.: Dioecious.

M.: Unknown.

F.: Plankton.

S.: The exumbrella of this jellyfish is hemispherical with a finely grained surface which is rough to the touch. The pink to reddish border of the umbrella is in distinct contrast to the practically transparent dome. Though the oral arms are thick and ruffled close to the dome, they are long, tubelike structures with swollen tips distally. Prey is captured using the distal section of the oral arms. The intricate system of intracirculatory canals and the ring canal into which they lead become apparent when the animal is closely scrutinized. Normally *T. loriferum* lives in open water near the coast, but in rough seas it may retreat to shallow lagoons.

T: 22°–29°C, Ø: 20 cm, AV: 4, D: 4

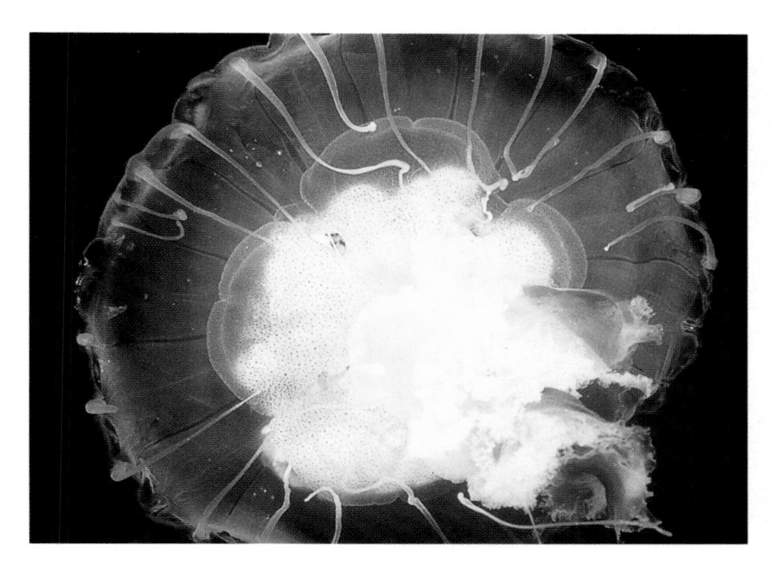

Chrysaora hysoscella LINNAEUS, 1758
Compass jellyfish

Hab.: Atlantic, Mediterranean, North Sea.

Sex.: Dioecious.

M.: Unknown; it is unsuitable for aquaria.

F.: Plankton.

S.: The umbrella, which can attain a diameter of 40 cm, is discoid, sinuate, and reddish in color. The sixteen peculiar V-shaped brown radial stripes along the umbrella's margin are reminiscent of the graduations on a compass. *C. hysoscella* has 32 small marginal lappets, 24 short tentacles, and 4 long, reddish oral arms. The compass jellyfish lives in the open water either singly or in great numbers near the surface.

T: 12°–22°C, Ø: 40 cm, AV: 4, D: 4

Order: Semaestomae
Fam.: Pelagiidae

Pelagia noctiluca (FORSSKÅL, 1775)
Pink jellyfish, warty jellyfish

Hab.: Mediterranean, Red Sea, Atlantic.

Sex.: *P. noctiluca* has separate sexes (dioecious). Without passing through a polyp phase, medusae develop from fertilized eggs, planula larvae, and ephyra larvae respectively.

M.: Not suitable for aquaria.

S.: This relatively small species has four short oral arms and eight transparent, nematocyst-bearing tentacles which can grow to be more than 10 m long. Mushroom-shaped, the umbrella has red, pink, or purple dots and nematocyst laden warts. Upon tactile stimulation, the surface of the exumbrella luminesces, making *P. noctiluca* especially visible at night. Upon contact with humans, the extraordinarily toxic tentacles produce painful blisters and fever.

T: 10°–18°C, Ø: 10 cm, AV: 4, D: 4

Pelagia noctiluca

Pelagia noctiluca

Drymonema dalmatinum HAECKEL, 1880
Stinging cauliflower

Hab.: Mediterranean, Adriatic Sea, Strait of Gibraltar, western Africa.

Sex.: Dioecious.

M.: Unknown.

F.: Plankton.

S.: These huge pelagic organisms can grow to be more than 1 m in diameter. Oral arms are heavily evaginated, the gonads hang sacklike from the subumbrella, and the exumbrella has up to 140 radial canals. The multitude of long tentacles originates from the central zone of the subumbrella, and they are extended in all directions as the animal hunts. The animals themselves meander through the water to cover even more territory in their search for food.

T: 8°–24°C, **Ø:** 50 cm, **AV:** 4, **D:** 4

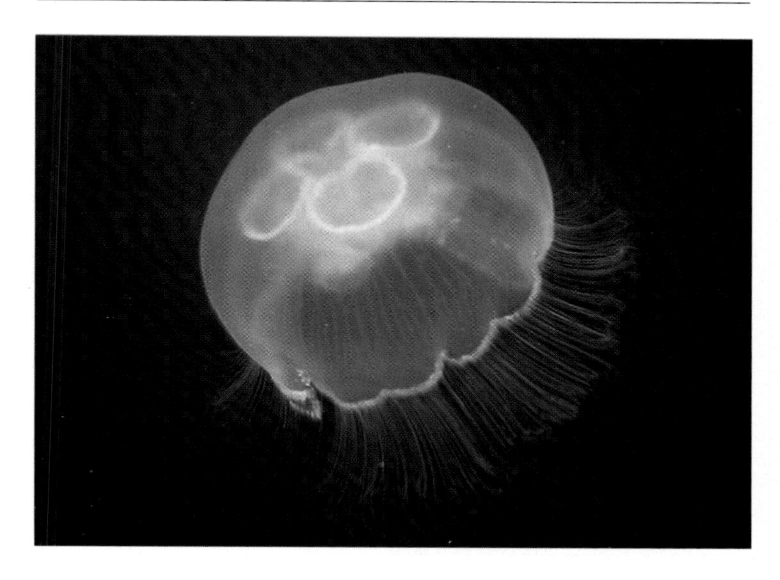

Aurelia aurita (LINNAEUS, 1758)
Moon jelly

Hab.: Cosmopolitan. *A. aurita* is found worldwide.

Sex.: *A. aurita* has separate sexes.

M.: Raising moon jellyfish in the aquarium is possible when polyps are placed into the rearing tank together with their substrate. Each polyp should be specifically fed Resulting ephyra larvae can be transferred into community aquaria after one month. Breeding itself is very laborious and difficult, especially since large specimens cannot be kept in aquaria.

F.: Plankton.

S.: The moon jelly is the most familiar jellyfish of this family. Its umbrella is plate- to bowl-shaped, up to 40 cm in diameter, with many short, delicate tentacles along its periphery. It is sometimes reddish or blue-purple. The transparency of the umbrella allows the pale red to purple horseshoe- to ear-shaped reproductive organs to be easily seen. The oral arms are broad and folded. *A. aurita* is frequently seen schooling close to the water surface. Despite being mostly water (98%), this organism is commercially fished and exported to Japan where it is consumed. The average life expectancy is 3–4 months. It is mildly toxic.

T: 0°–29°C, Ø: 40 cm, AV: 4, D: 3–4

The brown giant jellyfish, *Chrysaora melanogaster,* of California.

The purple jellyfish, *Pelagia panopyra*, of California.

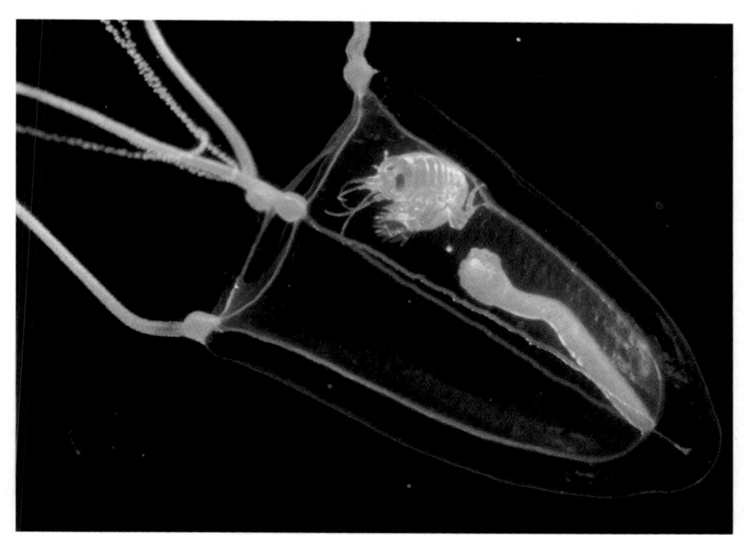

Parasitic isopods on *Sarsia* sp. (Arctic).

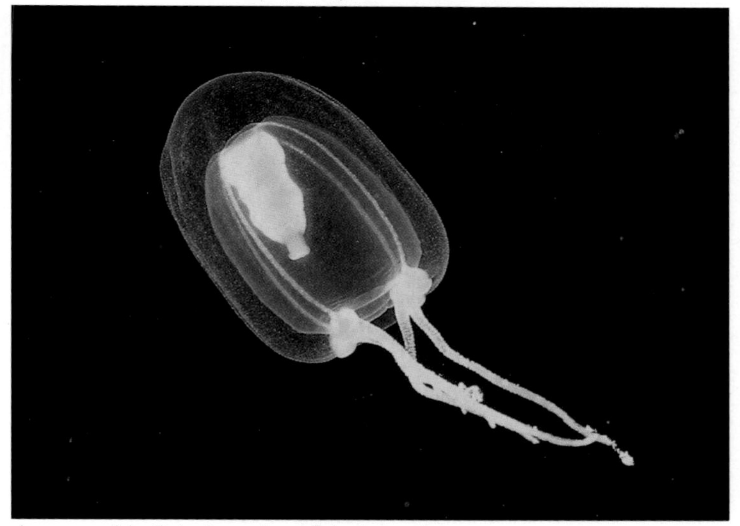

A very small Arctic hydromedusa, *Enhydra medusa*.

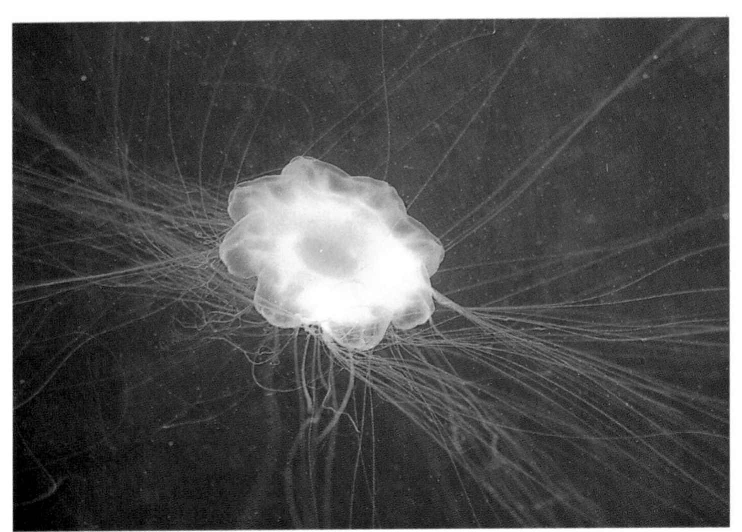

The lion mane jellyfish, *Cyanea* sp., can reach 3 m in diameter.

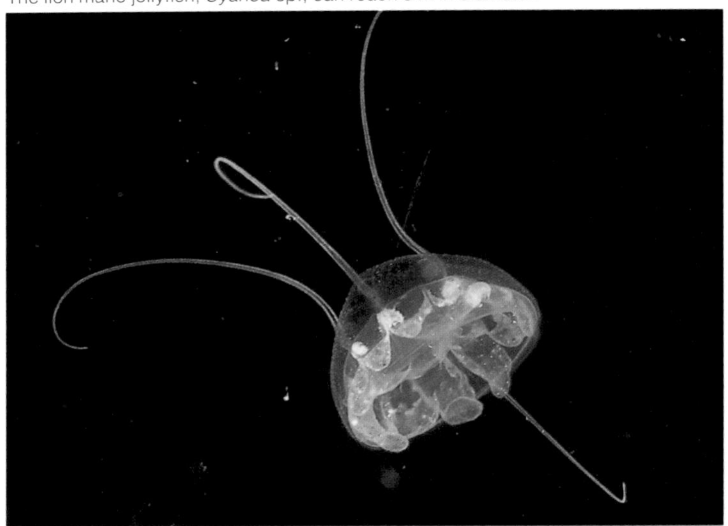

The small Arctic jellyfish *Aegina* sp.

Dendronephyta sp.

ORDER ALCYONACEA

Organisms of the order Alcyonacea are colonial sessile animals that have a leatherlike to fleshy texture. Distributed worldwide, these organisms assume a variety of growth forms—arborescent to fingerlike, branched, or lobed. They principally occur in tropical waters from the intertidal zone to depths of more than 200 m. Few have adapted to life in the deep sea.

Despite the broad spectrum of shapes, all soft corals have the same morphology. Covering the entire surface of the colony are relatively small polyps with 8 pinnate tentacles and 8 septae whose role is food acquisition (autozooids). The 8 septae subdivide the polyp's gastric area into 8 equal-sized compartments, all of which are interconnected by occasional channels. The channels distribute nutrients to the entire colony. Prey—chiefly small planktonic organisms—are stunned and captured by the tentacles, then carried to the mouth of the polyp.

In addition to the autozooids, there are polyps with degenerate tentacles, a single-chambered gastric cavity, and a cilia-lined siphonoglyph. They are called siphonozooids, and their role is to pump water into the body of the colony. The purpose is twofold—oxygen is carried in for the colony and the water creates hydrostatic pressure that keeps the colony erect and taunt. Through modified polyps, soft corals achieve a simple division of labor.

Unlike stony corals, soft corals do not produce a hard external skeleton. Instead, the majority of species have irregular isolated calcarious bodies (spicules or ossicles) loosely embedded in the coenenchyme. Large spicules are easily seen in fully expanded animals that have a virtually transparent coenenchyme.

Most soft corals are considered nocturnal animals. Siphonozooids and autozooids are generally dormant during the day, and members of this order are flaccid, inconspicuous lumps on the substrate until the siphonozooids begin actively pumping water into the colony, making the colony an erect, notable entity. But since numerous colonies are fully expanded and actively preying upon planktonic organisms during the day, it seems that autozooids react to stimuli other than photoperiod.

Soft corals abound in a rainbow of colors, though hues of yellow, orange, red, olive, rust, and purple predominate.
Fossil records show that there was an abundance of species and forms of soft corals in the very early periods of our earth's history. Of the 800 species alive today, most are found in dim locals such as caverns or beneath rocky outcroppings, utilizing a niche inappropriate for hermatypic corals which all require sunlight to grow.

Order: ALCYONACEA

SOFT CORALS IN THE AQUARIUM

by Joachim Großkopf,
Nuremberg

Soft corals are regularly kept in marine aquaria with considerable success despite the fact that instructions for care are greatly hindered because of the extreme difficulty of identifying individual species within the families Alcyoniidae, Nephtheidae and Xeniidae—a virtual impossibility for laypersons. Maintenance requirements are so extreme for these species that collective care instructions contain just broad generalities, particularly for those species which do not have symbiotic zooxanthellae.

Illumination

All species that have zooxanthellae in their tissues can be maintained and propagated with common sources of illumination such as metal halides (HQI, CW 10.000, NDL, D), actinic tubes, and/or daylight florescent tubes. Over the long term, metal halide lights do not garner better results than fluorescent daylight tubes.
Imported soft corals that live in symbiosis with zooxanthellae are generally collected from depths of less than 10 m. Those that dwell in deeper waters, usually members of the genera *Cladiella* and *Alcyonium*, are rare exceptions to the rule.
Colonies from deeper water require a long time to adapt to aquarium illumination. Months may pass before they become accustomed to their new environment and begin to grow. There are several reasons, but the principle problem is the time it takes the symbiotic zooxanthellae to adapt to the discrepant spectral composition of the illumination present in their captive habitat versus that found in their native environment. The proportion of UV radiation in the illumination may also play a role. Excess UV radiation is a problem that confronts shallow water colonies as well. If for no apparent reason colonies remain closed, an overabundance of UV radiation may be responsible.
Metal halide bulbs are particularly bad in this regard, and animal colonies lacking sufficient UV protection may suffer burns, some so severe, as to prove lethal.

When maintained under insufficient sunlight, such as is the case in many collection stations, soft corals spontaneously loose a measure of their UV protection. This is particularly true of extremely pale, whitish colonies. Special care for these specimens may be warranted, especially when they are first introduced into the aquarium.
Animals that fluoresce strongly are the least sensitive to UV radiation. Under actinic lights, their bioluminescent properties can readily be appreciated.
As stated, the animal's immunity to UV rays fluctuates, depending on the intensity of the lighting. Fluorescent daylight tubes are advantageous in that they rarely damage colonies.
Until the natural UV protection has been reestablished, it is often necessary to place a glass—not acrylic—plate above the colony, particularly when metal halides are used. This action may have to be implemented for several weeks. If the new additions are extremely pale to whitish in color, the glass may need to be 1 cm thick to filter the radiation adequately and thus yield the needed degree of protection.
Illumination—though important—loses value when compared to water quality, since strong illumination is unable to totally compensate for poor water quality. While some of the more robust soft corals can withstand inferior water quality, even polluted water, if illuminated with metal halides, their full beauty only becomes apparent under optimal conditions.
Most soft corals which do not have symbiotic zooxanthellae in their tissues must be placed in very shaded zones of the aquarium, as UV radiation is particularly detrimental to these animals over extended periods of time.

Water Values

The tolerated parameters for water quality vary from species to species. Imported soft corals fair best within the following ranges: density 1.021–1.025 (25° C); temperature 21°–29° C; pH 7.8–8.5; nitrate 0–20 mg/l; phosphate 0–1.0 mg/l; calcium 200–450 mg/l.

Some soft corals, such as *Sinularia* and *Sarcophyton*, will continue to grow under high levels of phosphate (5–8 mg/l) and nitrate (>100 mg/l); however, the polyps do not open completely.

Current

Many imported Alcyoniidae and Nephtheidae grow best when exposed to a good current as long as it is not direct and punctual. If the current is inadequate, polyps will remain closed and the organism's ability to clean itself of algae, sediments, and other foreign substances—a common attribute of this order—is curtailed. Xeniidae can generally survive in weak currents. Tall genera such as *Cespitularia* and *Heteroxenia* are especially intolerant of a direct, strong current.

Feeding

Species that possess symbiotic zooxanthellae do not need to be fed. Some *Xenia* spp. are incapable of capturing and digesting solid foods. Any food given on their behalf will foul the water, possibly to such an extent that tolerated ranges are exceeded.
Colonies without zooxanthellae, i.e., those that can live in dimly lit aquaria, must be fed live zooplankton every day. In consequence, the water often has an abundance of phosphate and nitrogenous wastes, stressing the tank's inhabitants and making long-term care virtually impossible. Successful care awaits the advent of filtration systems that reliably remove phosphates and nitrates from the water.

Aquarium Maintenance

Many soft corals grow to appreciable sizes. Correspondingly, roomy aquaria with a volume in excess of 300 l facilitate care. Only encrusting and branched forms, e.g., *Sinularia*, are suitable for smaller aquaria. Nephtheidae often achieve large, voluminous dimensions, particularly the east African soft coral *Litophyton arboreum*; it grows to a considerable size in captive environments. Large soft corals can easily be divided and maintained for years in appropriate-size aquaria. Unlike *Sinularia* species, mushroom soft corals are severely damaged, sometimes irreversibly, by cutting or pruning.

Xeniidae rarely grow larger than 20 cm, but under favorable conditions they quickly spread by runners or division. Some species of *Heteroxenia* will regularly release larvae, unfortunately the mother colony usually dies soon thereafter. Because *Xenia* and *Anthelia* multiply quickly, they can fill small aquaria past their biological carrying capacity within a short time.

Interspecific compatibility among soft corals and between soft corals and other types of cnidarias varies greatly from case to case. Usually congeners are tolerant of each other and do not sting even when in direct contact. However, surprises and exceptions are not uncommon.

Many soft corals are commonly attacked by parasites. The magnitude and frequency of the infestations seems to be seasonal. Parasites that eat into the colonial tissue are particularly unpleasant. Therefore, carefully inspect newly acquired soft corals at night for unwanted pests. Symptoms of parasitism include small holes and wounds, discolored areas, ablated (cut off) polyps, shortened tentacles, and partially closed polyps. Slimy or brownish zones on the colony are usually symptomatic of microscopic parasites and opportunistic diseases. These sections of the colony must be excised from the healthy tissue. Nudibranchs, crabs, bristleworms and small fishes are common ectoparasites, whereas sea spiders (Pycnogonida) are rare presences.

Night is an opportune time to check soft corals for ectoparasites, since many parasites sequester themselves at the base of the colony during the day. Unfortunately, some bristleworms that parasitize soft corals can hardly be controlled. It is sensible to remove the infested colony from the aquarium to prevent the spread of these unwanted animals.

ORDER STOLONIFERA

Until recently, organ pipe corals were considered part of the order Alcyonacea (soft corals). Today they have been placed in their own order. The fact that their polyps are separate entities and not embedded in a common soft body distinguishes them from the soft corals. Their odd structure—a series of bright red calcareous tubes connected by tiers of horizontal plates—further divides this order from other octocorals. The organ-pipe-like tubes are formed as numerous needles (spicules) produced in the mesoglea fuse. Only the uppermost level of the colony is covered by the polyp's endodermis and ectodermis. An extensive system of canals runs through the connecting plates and links the polyps.
Each of the highly elongated polyps has 8 pinnate, moss-green tentacles which remain extended from their tube all day and night.

Organ Pipe Corals in the Aquarium

Organ pipe corals are under the protection of the Washington Species Protection Act; hence a special certificate (CITES) is needed to import and maintain these corals in an aquarium.
While their husbandry was previously considered very difficult and dried skeletons were the most common method to include these organisms in the aquarium, continued advances in aquarium technology in the areas of filtration and illumination has improved the success rate. Nevertheless, care of organ pipe corals remains an affair best left to specialists.
Organic solutes absorbed over the entire surface of the polyps constitute the diet of organ pipe corals. Zooplankton is not consumed. Since their calcareous skeleton continues to grow in captive environments, regular dose of "kalkwasser" (see Vol. 1) must be administered. Care is best provided in a reef tank, with special attention given to spacing in relation to other anthozoans. They are extremely intolerant and quickly perish in the presence of filamentous algae.

Taxonomy

Phylum: Cnidaria	Cnidarians
Class: Anthozoa	Anthozoans
Subclass: Octocorallia	Octocorals
Order: Stolonifera	Organ pipe corals
Family: Tubiporidae	
Genus: *Tubipora*	
Family: Clavulariidae	
Genus: *Clavularia*	
Order: Telestacea	Telestaceans
Family: Telestidae	
Genus: *Telesto*	
Family: Coelogorgiidae	
Genus: *Coelogorgia*	
Order: Alcyonacea	Soft corals
Family: Alcyoniidae	
Genera: *Alcyonium*	
Cladiella	
Lobophytum	
Metalcyonium	
Parerythropodium	
Sarcophyton	
Sinularia	
Family: Nidaliidae	
Genus: *Siphonogorgia*	
Family: Nephtheidae	
Genera: *Dendronephthya*	
Lemnalia	
Capnella	
Scleronephthya	
Family: Xeniidae	Flower soft corals
Genera: *Anthelia*	
Heteroxenia	
Xenia	

Tubipora musica LINNAEUS, 1758
Organ pipe coral

Hab.: Red Sea, Indo-Pacific.

Sex.: None.

M.: Just a few years ago, the organ pipe coral was considered impossible to maintain in an aquarium. Today a few specialists have successfully cared for this coral over extended periods of time thanks to improved marine salt mixes and advances in aquarium technology. Nevertheless, because this is a protected species, maintenance should be waived. Even the red skeletons which act as an attractive focal point in marine aquaria are uncommon now.

F.: C; plankton.

S.: Growth is predominantly massive and hemispherical. The colony is made up of numerous parallel calcareous tubes which are united by tiers of transverse plates. The structure is reminiscent of the pipes of an organ. As it continuously grows in height, it does not become buried in the sand. The tubes house the polyps, each of which bears eight pinnate tentacles. Not surprisingly, jewelry is fashioned from the attractive red to maroon colored skeleton. Even attentive divers have been known to overlook live colonies. Understandable, given the fact that live colonies are a rather unattractive, unnoteworthy moss green.

T: 22°–26°C, Ø: 25 cm, TL: from 80 cm, WM: m, AV: 2–4, D: 3–4

Tubipora musica, aquarium photograph.

Clavularia viridis, aquarium photograph.

Clavularia viridis QUOY & GAIMARD, 1833
Green tube coral

Hab.: Indo-Pacific.

Sex.: None.

Soc.B./Assoc.: It is easily maintained in reef aquaria with small damsels and shrimp.

M.: Adding calcium through kalkwasser is advisable. Neither filamentous algae nor high nitrate concentrations are endured.

Light: Intense light.

B./Rep.: *C. viridis* multiplies by lateral shoots.

F.: Most nutrients are provided by its symbionts, zooxanthellae. Liquid food supplements can be offered.

S.: The primary polyp produces an almost horizontal stoloniferous mat along the substrate. It later produces secondary polyps. Neighboring polyps fuse in the mesoglea, forming a mat from which the tubular polyps emerge. Calcium spicules are frequently embedded in the gelatinous mesogleal layer.

T: 20°–25°C, **Ø:** 20 cm, **TL:** from 120 cm, **WM:** s, **WR:** t, **AV:** 2, **D:** 3

Order: TELESTACEA
Fam.: Telestidae

Telesto multiflora LAACKMANN, 1909
Bouquet telesto

Hab.: Indo-Pacific, Australia.

Sex.: None.

M.: Unknown.

F.: C; plankton.

S.: The bouquet telesto anchors its fascicular growth form to hard substrates using rootlike runners. Shaded sites, i.e., under ledges, are preferred. Pale elongated polyps sprout from a dark encrusting layer; they are generally open during day and night. *T. multiflora* is found both in clear and suspension-rich water from shallow to deep depths. As attested by its common name, the long dense branches resemble a bouquet.

T: 20°–26°C, L: 30 cm, AV: 4, D: 4

Telesto riisei (DUCHASSAING & MICHELOTTI, 1860)
Common telesto

Hab.: Cosmopolitan in tropical and subtropical seas.

Sex.: None.

M.: Successfully transporting and caring for *T. riisei* will prove challenging. Place the colony in the bottom half of the tank in a dark, current-exposed area far removed from stinging anthozoans. Algae, when present, commonly blanket the colony.

F.: C. It lacks zooxanthellae; therefore, daily meals of substitute plankton and *Artemia* nauplii are required.

Light: Dim light.

S.: From shallow waters to depths of more than 50 m, this fascicular colony is generally found in dim to very dark habitats, i.e., beneath outcroppings and harbor installations, in caves, and on buoys and ship bottoms. It is sometimes encountered in highly polluted water. Each pale polyp has eight long, pinnate polyps. Branches have longitudinal canals and originate from a rootlike base. Coloration is highly variable.

T: 20°–26°C, L: 30 cm, TL: from 100 cm, WM: m–s, WR: b, AV: 3, D: 3–4

Coelogorgia palmosa EDWARDS & HAIME, 1758

Hab.: Indo-Pacific.

Sex.: None.

M.: Although *C. palmosa* is occasionally imported from the Indian Ocean, most specimens come from eastern Africa. Place the colony along a current-exposed area, preferably in a reef aquarium containing various gorgonians and soft corals from the genus *Dendronephthya.*

Light: Dim light.

F.: C; substitute plankton, *Artemia* nauplii, plankton.

S.: *C. palmosa* is almost bushlike and similar to stalwart hydroids or fascicular horny corals. Branches are numerous, strenghtened somewhat by calcareous inclusions, and whiplike. Polyps are arranged along their length. Colonies grow attached to hard substrates of shaded, current-exposed sites, i.e., under ledges and at cave entrances; extensive populations may become established in small caves.

T: 20°–26°C, L: 30 cm, TL: from 100 cm, WM: m–s, WR: b, AV: 3, D: 4

Alcyonium digitatum LINNAEUS, 1758
Dead man's fingers

Hab.: North Sea, Atlantic to the Bay of Biscay.

Sex.: None.

M.: *A. digitatum* is long-lived when fed sufficiently. It can be maintained for years, but growth is not always satisfactory.

Light: Moderate light.

B./Rep.: Pinch or cut lateral branches to propagate.

F.: C; substitute plankton, plankton, *Artemia* nauplii.

S.: Colonies that have five or more branches resemble a hand. Dead man's fingers attach to stones, rocks, bivalve shells, and other hard substrates, but never at depths of less than 20 m. Water uptake and release, accomplished by the siphonozooids, is influenced by the tides. Coloration varies, though most are white, yellow, pink, or purple. The 1 cm polyps are pale and transparent.

T: 8°–14°C, L: 10 cm, TL: from 80 cm, WM: m, WR: b, AV: 2–3, D: 2–3

Order: Alcyonacea
Fam.: Alcyoniidae

Alcyonium glomeratum (HASSAL, 1843)
Red finger coral

Hab.: Atlantic.

Sex.: None.

M.: Unknown.

F.: C; plankton.

S.: Red finger corals are branched, fingerlike colonies that anchor to hard substrates down to great depths. Vertical rock walls or shipwrecks situated in current-swept regions are preferred. Coloration varies, but orange and red-orange hues predominate. In the winter the white polyps retract, and the colonies are covered by a thin purple-red mucous film. The taxonomic classification of this species is not yet definite.

T: 8°–12°C, L: 30 cm, AV: 4, D: 3–4

Alcyonium palmatum PALLAS, 1766
Large sea hand

Hab.: Mediterranean.

Sex.: None.

M.: *A. palmatum* is very sensitive to pressure. When it is introduced without its substrate, it quickly perishes. If this species is fed an adequate diet, it is a long-lived, moderate-growing aquarium inhabitant.

Light: Moderate light zone.

B./Rep.: The large sea hand has been successfully reproduced in captivity.

F.: C; substitute plankton, plankton, *Artemia* nauplii.

S.: Colonies grow on rock as well as sand and mud bottoms, stones, and shell fragments; they rarely anchor in the bottom itself. When colonies fill with water each day, they are five times larger than they are when deflated or "resting." Most specimens are pink to dark red.

T: 10°–18°C, L: 30 cm, TL: from 100 cm, WM: s–m, WR: m, AV: 2, D: 3

Alcyonium sp.
Indo-Pacific sea hand

Hab.: Tropical Indo-Pacific.

Sex.: None.

M.: With correct care and dim illumination, colonies are easily maintained.

Light: Dim light.

F.: C; plankton, substitute plankton. When colonies are housed in well-lit aquaria, they should be fed at night also.

S.: *Alcyonium* sp. is similar to European species. The Indo-Pacific sea hand can be found on shallow, shaded reef sites. This low-growing colony has conical, occasionally bifurcated, branches that emerge from a common base. Polyps are elongated, transparent, and predominantly nocturnal. During the day it is an amorphous lump.

T: 22°–28°C, L: 15 cm, TL: 150 cm, WM: m–s, WR: b, AV: 3–4, D: 3

Cladiella sp., aquarium photograph.

Cladiella sp.
Tree leather coral

Hab.: Tropical Indo-Pacific.

Sex.: None.

M.: Members of this genus make appropriate aquarium subjects. Provide a moderate current for an hour at a time. Other soft corals can be housed with *Cladiella* sp., yet heterospecifics should not come in contact. Favorable lighting promotes growth. Refrain from overstocking fishes, crabs, or shrimp.

Light: Sunlight zone; use metal halides.

B./Rep.: Affix a cut piece of a colony to the substrate. After a few weeks it adheres and starts a new colony.

F.: C. Zooxanthellae are present; therefore, one meal of substitute plankton each week suffices.

S.: *Cladiella* sp. lives on current-exposed reefs. Mucus, not skin, is shed as the animal contracts. Tiny skeletal spicules are encased in the body.

T: 20°–26°C, L: 25 cm, WM: m–s, WR: t–m, AV: 3–4, D: 2–3

Lobophytum sp.
Soft coral

Hab.: Red Sea, Indo-Pacific.

Sex.: None.

Soc.B./Assoc.: The colony has zooxanthellae.

M.: Unknown. Aquarium maintenance is contraindicated because of size.

Light: Sunlight zone.

F.: C; plankton. The entire surface of the body absorbs dissolved organic compounds.

S.: When exposed to air—e.g., during low tide, as shown in the above photo—the colony becomes flaccid and convoluted, closely following the contours of the substrate. As the incoming tide submerses the colony, it absorbs water, thereby unfolding and standing erect once again. During low tide, the colony is exposed to intense solar radiation and high salinities and undergoes a certain degree of desiccation. The water that remains in the body helps the colony endure these extreme situations.

T: 20°–45°C, **Ø:** over 100 cm, **WM:** m, **WR:** t–m, **AV:** 2, **D:** 2

Lobophytum sp., aquarium photograph.

Lobophytum sp.
Folded leather coral

Hab.: Sri Lanka.

Sex.: None.

M.: This soft coral is somewhat less demanding than others. It is frequently gnawed on by snails and free-living bristleworms. A moderate, indirect current is best.

Light: Sunlight zone. Use metal halide illumination that penetrates to the aquarium's substrate.

F.: C. Since it lives in symbiosis with zooxanthellae, daily feedings are unnecessary. About 2–3 times a week, substitute zooplankton can be given. Feed in moderation.

S.: Colonies are stooped and convoluted. *Lobophytum* sp. occurs along the upper reef slope and the reef platform, occasionally emersed. Solar radiation, high temperatures, and high salinities are tolerated for short periods of time.

T: 20°–40°C, Ø: 15 cm, TL: from 100 cm, WM: m–s, WR: t–m, AV: 2–3, D: 3

Order: ALCYONACEA
Fam.: Alcyoniidae

Metalcyonium sp.
Indo-Pacific soft coral

Hab.: Tropical Indo-Pacific.

Sex.: None.

M.: Unknown.

F.: C; plankton.

S.: This predominantly nocturnal species has conical runners and a small, stooped form. Polyps are elongated, pale, and transparent; the tentacles have pinnate pinnules. Shaded, hard substrates along current-swept reef sections from shallow to moderate depths are its preferred habitat.

T: 20°–26°C, L: 8 cm, WM: w–m, WR: t–m, AV: 3, D: 3–4

Parerythropodium coralloides (PALLAS, 1766)
False coral

Hab.: Mediterranean, Atlantic, Bay of Biscay.

Sex.: None.

M.: Maintenance is possible, though there were no reports found in the literature.

F.: Plankton.

S.: *Eunicella* and *Paramuricea*, two genera of gorgonians, are the primary substrate for this species—though it is occasionally encountered on tunicates and algae. The growth pattern is encrusting, and the resulting cardinal red structure is frequently mistaken for *Corallium rubrum* by amateurs. It has not been determined just how this species conquers its substrate despite the stinging polyps, but its success can probably be attributed to a more aggressive nature. *P. coralloides* always colonizes from the base of its gorgonian to the tips of the branches. Vertical distribution extends from shallow water down to great depths.

T: 7°–24°C, Ø: 25 cm, AV: 4, D: 4

Order: ALCYONACEA
Fam.: Alcyoniidae

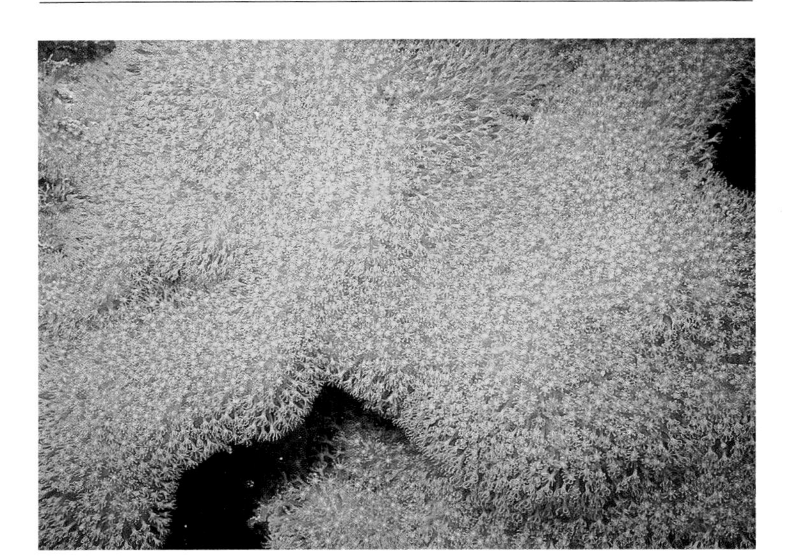

Parerythropodium fulvum (FORSSKÅL, 1775)
Sulfur coral, yellow encrusting leather coral

Hab.: Indo-Pacific.

Sex.: None.

Soc.B./Assoc.: Zooxanthellae are endosymbionts of *P. fulvum*. Parasitic copepods, *Monomolgus unihastatus*, live on the colony.

M.: An appropriate tankmate for other anthozoans, but at least 10 cm of distance between it and stony corals and disc and colonial anemones must be maintained at all times. Filamentous algae will hinder maintenance.

Light: Sunlight zone.

F.: *P. fulvum* chiefly subsists on dissolved organic substances. Large polyps also feed on small substitute plankton.

S.: Sulfur corals form expansive mats on dead coralline rocks in shallow water. These mats are flat—just a few millimeters thick—covering up to several square meters of substrate. All underlying life is smothered. The numerous, very short, densely arranged golden polyps are open during the day as well as at night. *P. fulvum* is not found in turbulent waters; it inhabits lagoons and the leeward side of coralline islands.

T: 22°–28°C, **Ø:** over 100 cm, **TL:** from 100 cm, **WM:** w–m, **WR:** 0, **AV:** 4, **D:** 3–4

Sarcophyton glaucum (QUOY & GAIMARD, 1833)
Mushroom leather coral, cup leather coral

Hab.: Red Sea, Indo-Pacific.

Sex.: None.

Soc.B./Assoc.: An assortment of copepods, such as *Perosyna indonesica, Paradoricola spinulatus, Anisomolgus ensifer*, and *Alcynomolgus sarcophyticus*, is found on this coral.

M.: The mushroom leather coral has been a common addition to reef tanks for quite some time. It is a hardy, easily maintained species.

Light: Sunlight zone.

B./Rep.: By cutting and tying the entire "mushroom cap" and 3 cm of "stalk" to a substrate, *S. glaucum* can be artificially propagated in the aquarium. After 3–4 weeks the organism has attached itself, and a new mushroom cap develops on the remaining stalk. Generally this species can be housed with fishes, shrimp, and hermit crabs.

F.: C; plankton, plankton substitutes.

S.: As implied by their common name, these soft corals look like mushrooms. A lobed, extremely durable, leathery "cap" grows on a thick main axis. Its surface has long, totally retractable polyps. Hard substrates of reef tops, fore reefs, lagoons, and reef slopes are where most mushroom leather corals are found. Coloration varies according to biotope. On reef slopes they frequently grow in groups.

T: 20°–26°C, L: 20 cm, TL: from 100 cm, WM: m, WR: m, AV: 2, D: 3

Order: ALCYONACEA
Fam.: Alcyoniidae

Sarcophyton trocheliophorum MARENZELLER, 1886
Elephant ear coral

Hab.: Red Sea, Indo-Pacific.

Sex.: None.

Soc.B./Assoc.: *S. trocheliophorum* has symbiotic zooxanthellae. It is susceptible to stinging tube anemones, anemones, disc anemones, and bladder corals. Large creeping rhizoids and the blades of algae are also sources of peril for this species: the former constricts the coral, and the latter casts shade, precluding extension of the polyps. The copepod *Paramolgus spathophorus* is an epizoite.

M.: The elephant ear coral is highly suited for captive care. It is impervious to fishes, shrimp, hermit crabs, and even algae, since it regularly "sheds."

Light: Sunlight zone; use metal halide illumination.

B./Rep.: By slicing off a section along the edge of a lobe, this species can be artificially propagated. Fasten the piece to a substrate. There it will grow and establish a new colony; however, the new growth will atypical. The cut on the mother colony heals within 14 days.

F.: Dissolved organics are absorbed from the water. Live zooplankton, infusoria, frozen plankton, and *Artemia nauplii* can be fed.

S.: *S. trocheliophorum* is fleshy, thick, and leathery. While a hard skeleton is lacking, there are spicules embedded in the tissue to support the colony. Edges are undulating lobes, and the surface is covered with long, tube-shaped polyps. Of the two pictured colonies, the one on the left has its polyps extended, while the other has closed polyps. The elephant ear coral occurs from extreme shallow water to shallow reef tops, intertidal pools, reef slopes, and reef plateaus. It is extremely adaptable in regard to high temperatures, fluctuating salinities, and intense solar radiation.

T: 28°C, **Ø:** 60 cm, **TL:** from 100 cm, **WM:** s–m, **WR:** b, **AV:** 2–3, **D:** 2–3

Sarcophyton trocheliophorum

Sinularia dura (PRATT, 1903)
Lobed leather coral

Hab.: Red Sea, Indo-Pacific.

Sex.: None.

Soc.B./Assoc.: *Paramolgus ostentus* and *P. eniwetokensis*, two species of copepods, live in association with *S. dura.*

M.: Lobed leather corals fare well in captive environments. Conditions outlined for *Sarcophyton* spp. will, for the most part, suit this species as well. A certain distance between *S. dura* and disc anemones must be maintained.

Light: Fluorescent tubes are adequate sources of illumination only if actinic tubes are used concurrently.

B./Rep.: By cutting sections from the outer lobes and fastening them to a substrate, this species can be easily propagated. After about 14 days the pieces adhere. The growth pattern is encrusting.

F.: Fine plankton substitutes.

S.: Lobed leather corals form small, low-growing colonies on hard substrates, for example, dead corals and vertical walls under overhangs, frequently at such densities that they overlap. The lobed projections have delicate radial striations and many invaginations along their fringe. *S. dura* "sheds" by releasing mucus.

T: 20°–26°C, **Ø:** 8 cm, **TL:** 80 cm, **WM:** w–m, **AV:** 3, **D:** 3–4

Sinularia polydactyla (EHRENBERG, 1834)
Many-fingered leather coral

Hab.: Red Sea, Indo-Pacific, western Pacific.

Sex.: None.

Soc.B./Assoc.: *S. polydactyla* has zooxanthellae in its tissues. *Paradoricola squaminger* and *Meringomolgus fascetus*, two copepods, live among the digital projections.

M.: This species is long-lived and grows well in the aquarium. If metal halide illumination is used, meticulous acclimation is required.

Light: Sunlight zone. Either metal halide or fluorescent lights are appropriate; however, the latter alternative must be supplemented with an actinic tube.

B./Rep.: Unknown.

F.: C; plankton, plankton substitutes, *Mysis*, tablet foods.

S.: *S. polydactyla*'s broad, low-growing, dense growth typically covers dead corals and other dead substrates in shallow water. The stem of the colony is stout and lacks polyps. Occasionally branched and of different lengths and diameters, the fingerlike projections are densely covered with small, pale polyps. Coloration is variable and largely contingent on the color of the symbiotic zooxanthellae.

T: 22°–29°C, Ø: over 30 cm, TL: from 100 cm, WM: m, WR: t, AV: 2, D: 3–4

Sinularia sp., aquarium photograph.

Sinularia sp.
Leather coral

Hab.: Sri Lanka.

Sex.: None.

M.: *Sinularia* sp. is easily kept in aquaria, though growth is slow. Placement in regard to current and illumination has to be determined on an individual basis. A strong current is needed. Metal halide illumination requires acclimation.

Light: Sunlight zone; metal halide illumination.

B./Rep.: Branched sections can be divided and fastened to clean substrates. Due to the species' well-developed ability to regenerate, the cuts quickly heal.

F.: The nutrients supplied by the zooxanthellae make daily feedings unnecessary. Any substitute plankton sold in pet stores is an appropriate supplement.

S.: Lobed and fingerlike projections are simultaneously produced. *Sinularia* sp. periodically sheds.

T: 22°–29°C, Ø: 12 cm, TL: 100 cm, WM: s, WR: t–m, AV: 2–3, D: 2–3

Siphonogorgia sp.
Soft sea fan

Hab.: Red Sea, Indo-Pacific.

Sex.: None.

M.: Unknown.

B./Rep.: Unknown.

F.: C; plankton.

S.: The erect, richly branched form so resembles a sea fan, inexperienced persons are often confused. Branches originate from the strong main stem which is anchored to the substrate with a broad base. The main stem and the primary branches are smooth with very few, if any, polyps. Only the numerous terminal branches have polyps in various colors. This soft coral inhabits deep water, predominantly at the foot of reef slopes and small plateaus. Areas exposed to swift currents are preferred. Most colonies are rust red.

T: 22°–29°C, **L:** 80 cm, **TL:** 120 cm, **WM:** w–m, **AV:** 4, **D:** 4

Order: ALCYONACEA
Fam.: Nephtheidae

Dendronephthya klunzingeri (STUDER, 1887)
Klunzinger's soft coral

Hab.: Red Sea, Indian Ocean.

Sex.: None.

M.: Unfortunately, specimens sold in pet stores are usually not attached to a substrate, and anchoring them in an aquarium will prove challenging. Furthermore, at night as they fill with water, these soft corals tend to fall over. To avoid bruising, they must be propped upright. Place the corals along elevated sites rather than on the bottom substrate, as *D. klunzingeri* easily doubles over. Clean water is a necessity. Due to their problematic care—even after they have attached—maintenance should be reserved for specialists.

Light: Dim light.

F.: Plankton substitutes. Limit meals to times when the polyps are open!

S.: *D. klunzingeri* does not have zooxanthellae. It anchors itself to hard substrates via rootlike runners. During the day the colony mostly lives in a collapsed state, while at night it expands by taking in water. Polyps are arranged in clusters and are incapable of retracting. The planktonic organisms which constitute its diet are captured by the polyps. This species is able to grow to more than 1 m in size, and it presents itself in a range of hues. Colored calcium needles give the colony form. Dim, protected areas of deep water environments, e.g., niches and beneath overhanging ledges, are where most specimens are encountered.

T: 20°–29°C, L: 80 cm, TL: from 100 cm, WM: m, WR: m, AV: 2, D: 3–4

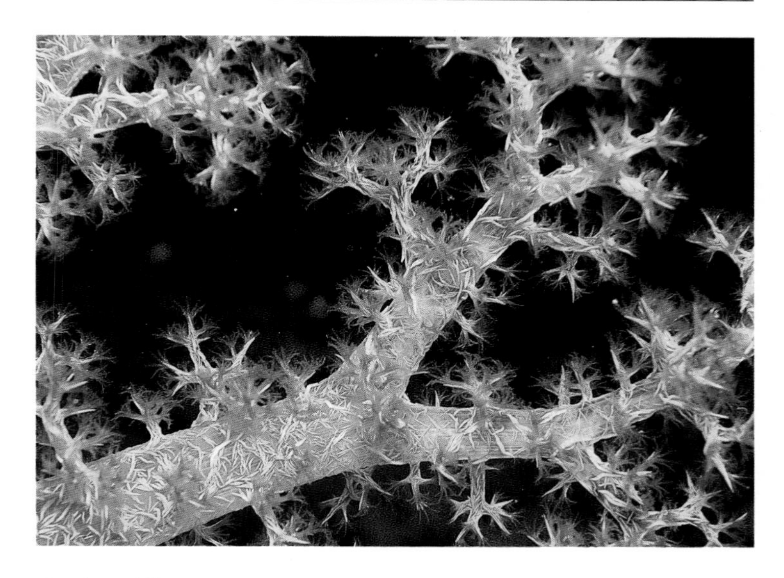

Dendronephthya sp.
Cauliflower coral

Hab.: Red Sea, Indo-Pacific.

Sex.: None.

M.: The cauliflower coral is best placed in an ample cave or under a ledge of a reef aquarium. Housing this soft coral with fishes, shrimp, or crabs diminishes its chances of survival.

Light: Dimly lit zones.

B./Rep.: Unknown.

F.: Expanded colonies must be fed zooplankton twice a day.

S.: Growth is arboreal or fascicular. This is one of the most colorful colonial animals of tropical coral reefs. The colony's shape can be significantly altered by the intake or expulsion of water. Metabolites are removed as the water is pumped from the colony. As *Dendronephthya* sp. prefers hard substrates in shaded sites such as caves and beneath ledges, it does not compete with light-dependent stony corals. The primary stem has few if any polyps and relatively short, thick lateral branches exclusively along its upper section. The lateral branches bifurcate further, and clusters of polyps are supported by the tips of the branches. Numerous loosely embedded spicules can be seen in the stem.

T: 20°–29°C, **L:** 60–80 cm, **TL:** from 100 cm, **WM:** m, **WR:** b, **AV:** 3, **D:** 3–4

Lemnalia africana (MAY, 1898)
Kenya tree

Hab.: Indian Ocean, eastern African coast.

Sex.: None.

Soc.B./Assoc.: Soft corals of the genera *Sarcophyton, Sinularia, Lobophytum,* and *Cladielle* make good tankmates, but best growth is achieved when *L. africana* is kept in a species tank with small crustacea and benthic fishes.

M.: As long as chemical and biological parameters are correct, *L. africana* can be maintained in an aquarium, but care requires a lucky touch. Disc anemones of the genus *Actinodiscus* are contraindicated. The coral population should be kept to a minimum, and a significant distance between this species and corals should be respected.

Light: Moderate light. During acclimation, dim metal halide illumination.

F.: C; plankton. Zooxanthellae make supplemental feedings largely unnecessary. However, mussel milk or hydrated TetraTips can be offered 2–3 times per week.

S.: *L. africana* resides in the sublittoral zone. At rest it lies collapsed on the substrate, only extending to combat oxygen deficiency or capture plankton. Water streams into the colony through the siphonozooids, erecting the colony until it stands like a small tree with a long "trunk" and many branches.

T: 20°–29°C, L: 50 cm, TL: from 150 cm, WM: s, WR: b, AV: 3, D: 3–4

Capnella sp.
Broccoli soft coral, tree soft coral

Hab.: Red Sea, Indo-Pacific.

Sex.: None.

Soc.B./Assoc.: This soft coral lives in symbiosis with zooxanthellae.

M.: If one can handle the problems that arise due to exuberant growth and the organism's gigantic dimensions, successful husbandry is possible. Since many organisms fall prey to stings of *L. arboreum*, disc anemones and other stinging cnidarians make inappropriate tankmates.

Light: Moderate light zone. Careful acclimation is needed when metal halides are used.

B./Rep.: Besides budding, propagation is also possible by fixing cuttings to the substrate or weighing a branch down with a stone.

F.: Most nutrients are absorbed in their dissolved state from the water. Supplemental foods can be offered in the form of substitute plankton, *Artemia* nauplii, and "natural microscopic plankton."

S.: Broccoli soft corals are erect treelike corals that have heavily forked lateral branches and a short, smooth base devoid of polyps. It is the delicate tertiary branches that are covered with nonretractable polyps. At a depth of 6 m, this species is often found covering expansive areas on the reef slope, where it colonizes hard substrates such as stones and corals. Coloration varies according to habitat.

T: 20°–29°C, L: 80 cm, TL: from 100 cm, WM: s–m, WR: b, AV: 3, D: 3–4

Order: ALCYONACEA
Fam.: Nephtheidae

Scleronephthya sp.
Small-tree soft coral

Hab.: Red Sea, Indo-Pacific.

Sex.: None.

M.: Unknown.

F.: C; plankton.

S.: This filigree, richly branched soft coral prefers to live on the ceiling of dark caves and on shaded rock walls under overhanging ledges. Several primary branches and their multitude of delicate secondary branches hang into the open water. The primary branches lack polyps; the nonretractable polyps are solely found on the terminal branches.

T: 20°–29°C, L: 30 cm, TL: 100 cm, WM: s–m, WR: b, AV: 4

Anthelia glauca (LAMARCK, 1816)
Pinnate bouquet soft coral

Hab.: Red Sea, Indo-Pacific.

Sex.: None.

Soc.B./Assoc.: *Doridicola antheliae*, a copepod, lives on this coral.

M.: *A. clauca* can easily be kept in aquaria. It will quickly overgrow the tank's decorative infrastructure, calcareous red algae, and even *Caulerpa*. Do not place this coral too close to other anthozoans.

Light: Moderate light zone.

B./Rep.: With appropriate care, propagation is rapid; less aggressive, sessile tankmates are quickly overgrown.

F.: C; plankton substitutes.

S.: The extremely elongated polyps—up to 5 cm in length—branch from a flat, inconspicuous base. Each of the polyps has 8 delicate, pinnate tentacles which open and close rhythmically. The colony is moderate-sized, only covering a few square decimeters (10 cm^2) of substrate. The delicate growth form is indicative of their deep quiescent habitat.

T: 20°–29°C, Ø: 20 cm, TL: from 100 cm, WM: w–m, WR: b, AV: 3–4, D: 3–4

Heteroxenia fuscescens (EHRENBERG, 1834)
Palaver coral

Hab.: Red Sea, Indo-Pacific.

Sex.: None.

M.: Unknown.

Light: Sunlight zone.

B./Rep.: Unknown.

F.: C; plankton.

S.: *H. fuscescens* forms a gray to gray-brown carpetlike covering over its substrate. The colony consists of short, unbranched stems which support relatively long retractable polyps. Eight pinnate tentacles synchronously open and close up to 40 times per minute, 24 hr a day. These constant pumping motions not only act to capture prey, but also serve a respiratory function. The general impression is that all polyps are communicating with each other, or "palavering."

T: 20°–29°C, **Ø:** 60 cm, **TL:** 150 cm, **WM:** w–m, **WR:** b, **AV:** 4, **D:** 4

Xenia umbellata LAMARCK, 1816
Bouquet encrusting coral

Hab.: Red Sea, Indo-Pacific.

Sex.: None.

Soc.B./Assoc.: The tissues of this species contain symbiotic zooxanthellae. Various parasitic nudibranchs and *Paradoridicola glabripes*, a copepod, are found on *X. umbellata*.

M.: Once acclimated, the bouquet encrusting coral does well in captive environments. Caution—some fishes, e.g., surgeon and angel fishes, are unsuitable tankmates.

Light: Moderate light zone.

B./Rep.: Sexual reproduction involving free-swimming larvae has been observed in the aquarium. However, the mother colony died soon after releasing the larvae. Under intense illumination propagation is also realized through runners.

F.: Most of this coral's nutritional requirements are satisfied by the metabolic products of its symbiotic zooxanthellae and dissolved organics that are absorbed through the skin. Substitute plankton is an appropriate supplemental food.

S.: *X. umbellata* looks like a cluster of mushrooms. The 8 pinnate tentacles are borne on 5 cm polyps which, in turn, are supported by an unbranched stem. Polyps move rhythmically as they absorb oxygen from the surrounding water. Depending on its location on the reef and its symbiotic algae, the color of the colony varies.

T: 22°–29°C, Ø: 12 cm, TL: from 100 cm, WM: w–m, WR: m–b, AV: 3–4, D: 3–4

Xenia sp., aquarium photograph.

Xenia sp.
Encrusting coral

Hab.: Red Sea, tropical Indo-Pacific.

Sex.: None.

M.: Encrusting corals grow very well in aquaria. An indirect moderate current is best. Filamentous algae must be avoided. Excessive nitrate concentrations lead to a decrease in the pumping activity of the polyps.

Light: Sunlight zone. Place in a brightly lit area. Metal halide and actinic lights are required.

F.: The great part of its nutritional requirement is supplied by its symbiotic partners, the zooxanthellae. It is not an active planktivore.

S.: The length of the polyps of *Xenia* sp. is awe inspiring. Understandably, the polyps cannot be completely retracted. The pumping motions of the polyps are for respiration, not to capture plankton.

T: 22°–29°C, Ø: 8 cm, TL: from 100 cm, WM: m, WR: t–m, AV: 3–4, D: 3–4

ORDER GORGONARIA

Gorgonians are cosmopolitan, but the highest diversity of species and forms is found in tropical and subtropical climates. Of the more than 1,200 known species, the majority inhabits coastal waters at depths down to 300 m. Few species colonize the deep sea.

Gorgonians are colonial animals that have a tough, yet very flexible central axial rod, which is covered by a living rind. It is this rind that gives the order its German common name, *Rindenkorallen* (= rind corals). The common denomination for the order in English-speaking countries is gorgonians or horny corals. Both of these names are derived from the "horny" axial skeleton which is composed of gorgonin, a fibrous proteinaceous substance that contains iodine. Calcareous inclusions temper gorgonians, hardening and stiffening the skeleton. The less calcium deposited in the central axial rod, the more supple and smooth the animal colony. The quantity of calcium inclusions and the subsequent characteristics it renders to colonies enables these organisms to survive, even flourish, in inhospitable environs such as areas exposed to surf or strong coastal currents.

Some gorgonian corals have negligible amounts of gorgonin, while others may totally lack this substance. In these species, numerous calcium needles (spicules) impart an inflexible, stiff shape. Spicules are systematically arranged in the axial rod as well as in the living rind layer (coenenchyme). Species identification largely depends on the morphology of these spicules.

Numerous polyps are embedded in the coenenchyme. From species to species, polyp morphology is quite similar. Each has up to eight pinnate tentacles and, in comparison to the coenenchyme, a contrasting color. Most are capable of completely retracting their polyps into the coenenchyme. Eight septae (mesenteries) subdivide the gastric cavity into eight equal compartments, and the gastrovascular cavities are interconnected by intricately branched canals.

Gorgonians affix to hard substrates using a broad holdfast. In habitats with a paucity of hard substrates, e.g., at great depths, the

organisms are capable of anchoring themselves to the soft bottom with rootlike runners.
A startling range of growth patterns are exhibited by gorgonians—unbranched, pinnate, fascicular, arborescent, fanlike, and brush-shaped. Branches, when present, further distinguish species by being long or short, flexible or stiff, stout or delicate.
Typically, gorgonians have a stem that attaches to the substrate and delicate lateral branches that radiate outward. A pinnate growth pattern is the result of two branches emerging opposite to one another. When the branches interconnect into a mesh, a fanlike colony results. Such growth forms are generally oriented broadside to the prevailing current, exposing the organisms to the maximum volume of water.
Gorgonians are choice colonizing substrates for many invertebrates, particularly sponges, bryozoans, bivalves, snails, colonial tubeworms, and fire corals. The site and orientation of the host guarantees their epizoans constant high oxygen concentrations and ample nutrients. Especially at night when many animals climb to exposed sites of the coral reef to seine plankton, gorgonians are regularly scaled by feather and brittle stars.
Few animals prey upon gorgonians. Some snails number among this minority, yet their rate of predation rarely exceeds the coral's ability to regenerate.
Development and multiplication occurs through planula larvae which settle on appropriate substrates once their planktonic phase is concluded. After metamorphosis, a small new gorgonian appears.
Most colonies are colorful shades of yellow, orange, and red due to carotenoid inclusions in the coenenchyme.
The proportion of spicules in the axial rod determines classification to suborder. If the axial rod is made up of pure gorgonin with few calcium inclusions, then the animals are placed into the suborder Holaxonia. However, if the axial rod contains numerous, strongly fused calcareous inclusions, then the animals belong to the suborder Scleraxonia.

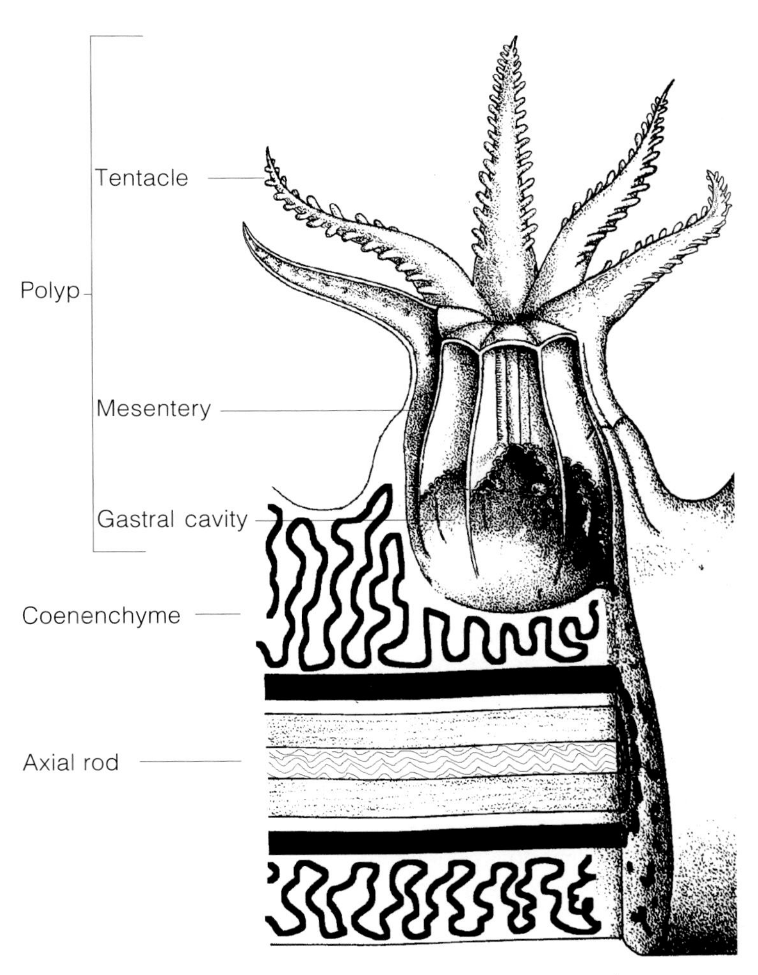

Simplified anatomy of the branch of a gorgonian coral.
Modified from BARNES, 1987.

Order: Gorgonaria

Gorgonians in the Aquarium by Joachim Großkopf, Nuremberg

Husbandry of Mediterranean gorgonians can be traced to the advent of the marine aquarium hobby. Though these species are no longer sold, tropical species were introduced in 1970. Gorgonians from tropical seas generally survive and grow for extended periods of time in captive environments when their very particular requirements are met. Since about 1980, species from the Caribbean have also become available. Initially true exotics, these animals can now be found every day in aquarium stores.

Gorgonian skeletons, fortunately, are no longer sold in pet stores. Since pure gorgonin axes largely dissolve in salt water, using them as decorative infrastructure is rather inapt.

In regard to requirements and care, gorgonians must be divided into two large groups: species **with** symbiotic zooxanthellae and species **without** zooxanthellae.

Coloration is generally a good indication of which category gorgonians fall into. Most species that have zooxanthellae are beige or brown. However, some Caribbean species of the genera *Pterogorgia, Pseudopterogorgia*, and *Plexaura* deviate substantially from this norm, as only their polyps are beige or brown while their coenenchyme is purple-blue. Illumination, current, and position are the principle factors that must be taken into account when caring for these animals. Keep the current brisk, but not too powerful; colonies exposed to strong water movement will not open their polyps. Caribbean species with particularly large polyps (*Plexaurella*) will even fare well in weak currents.

Self-cleaning gorgonians periodically require a strong current. These animals will contract for a few days while they secrete a slimy skin. At first, this skin is firmly adhered to the colony; only later as the colony swells with water does the skin loosen and detach. If the current is insufficient, the skin does not separate and algae overgrow the colony. At that point, if not sooner, the aquarist must simultaneously remove the skin mechanically and increase the current.

Anchoring gorgonians firmly to the substrate is vital to their well-being. This is best accomplished by placing the colony on a fist-size stone riddled with holes. Do not be alarmed if the colony's holdfast is damaged, as these wounds typically heal quickly. Healthy colonies which have been completely severed from their holdfast are capable of growing a new one. To foment this process, place the animal in a small, deep cavity with a diameter only slightly exceeding that of the axial rod. The coral must be introduced to the level of the living tissue. If this last process is not respected, it may take several months for the base to regenerate. Until the colony becomes attached, it may be necessary to secure it to the substrate with a piece of nylon thread. Insure that the colony is held immobile until such time as it has regrown its base and affixed itself to the substrate.

Since gorgonians do not have a particularly strong sting, they can be readily associated with other cnidarians in an aquarium. However, placing them in direct contact with corals and molluscs is not recommended. If, for example, a gorgonian falls on a stony coral, the potent stings of the stony coral will prove lethal to the sections of the gorgonian in contact with the stony coral.

Feeding gorgonians that have zooxanthellae is unnecessary. A few species, such as *Briareum asbestinum,* capture small solid foods. Liquid suspensions like mussel milk are unsuitable and even considered detrimental. Those species which are capable of metabolizing foodstuffs in addition to metabolites from their zooxanthellae grow significantly faster when they receive regular, directed meals. All animal-based plankton substitutes commiserate to polyp size are suitable for these species, e.g., crumbled flakes, *Artemia* nauplii, fish eggs, *Cyclops,* and bosmids. Always keep in mind that water pollution and nitrate and phosphate concentrations should never increase. One meal a week normally suffices for a colony. With optimal lighting, most species tolerate up to 100 mg/l of nitrate and 5.0 mg/l of phosphate, though ideal levels fall far below these values. The water should have a specific gravity between 1.020 and 1.025 and a temperature of 20°–29°C. During the summer, these values may be exceeded for brief periods, but temperatures above 31°C may prove lethal.

Caribbean gorgonians chiefly inhabit shallow waters of coral reefs. Hence, metal halide illumination is best suited to their needs. While fluorescent illumination is adequate, growth may be slower. The shape of the reflector is a very important consideration when using metal halide illumination, as round reflectors create a spotlight effect. Only gorgonians with fully opened polyps should be placed under such intense illumination. If the polyps close and refuse to open despite sufficient current and a peaceful environment (i.e., absence of molesting fishes or anthozoans), UV radiation from the metal halide bulbs is the probable culprit. Either the UV exposure must be reduced by installing a glass cover pane, or the colonies must be relocated. Despite all these warnings and precautions, problems of this kind are infrequent. The photoperiod when using metal halides should not exceed 8–12 hours. Use fluorescent tubes the remainder of the day.

Gorgonians without zooxanthellae are less commonly sold in pet stores. These gorgonians are characteristically very colorful. All their nutritional needs are met by the plankton they capture, which explains why the majority are nocturnal. Unfortunately, the daily meals these species require load the aquarium water, making a powerful protein skimmer, an anaerobic denitrification filter, and a rapid turnover rate a necessity. In some cases the aquarist will also be obligated to use an ozonizer and/or a carbon filter. Though all substitute plankton of animal origin is suitable foodstuff, live plankton such as *Brachionus* should be fed when possible. Gorgonians without symbiotic zooxanthellae are less tolerant of suboptimal water conditions than those that have zooxanthellae, i.e., the water temperature should not exceed 28°C and fluctuations in salinity should be contained between 1.021 and 1.025. Nitrate and phosphate concentrations must be kept as low as possible. Correspondingly, the calcium level is extremely critical; keep it at about 400 mg/l.
When fastening the colony to the substrate, use the procedure described for species with zooxanthellae. As these corals have little regenerative power, injuries must be kept to a minimum when placing the colony. Never bury the colony or parts thereof in the sand; these sections will die within days. Because this group of gorgonians requires a strong current before their polyps will be-

gin to open, the point of attachment must be sufficiently stalwart to withstand the force exerted by the current. Orient the colony so that its broad side faces the direction of water movement. Some species more readily open their polyps in the aquarium than others. Fine reticulate colonies such as *Subergorgia* are particularly hard to maintain. If captive conditions are not in total accordance to those of their natural habitat, the polyps will not open. Constant nocturnal checks must be made to ensure polyps are opening.

Hardly any colorful gorgonian can be maintained under metal halide illumination due to their sensitivity to intense UV radiation. Even direct light from fluorescent tubes will prove to be excessive. However, keeping their requirements in mind, a suitable local can be created.

In reality, a specialized aquarium geared towards keeping photophobic reef organisms, such as colorful soft corals and sponges, is prescribed for successful care. Frequent, voluminous water exchanges using **mature** seawater are mandatory. A second aquarium (without animals) connected to the show tank will facilitate the process.

Algae are strongly contraindicated. As previously mentioned, the majority of gorgonians are incapable of cleaning themselves. Hence, colonies become overgrown with algae. Filamentous algae are the principal culprits, but smear algae do harm as well. Specialized aquaria containing only photophobic animals can be placed in total darkness for a few weeks to control the plague. Simultaneously increase the current.

Gorgonians are fragile organisms, and care must be executed during transport. The colony should be wholly submersed at all times. Any section exposed to air will quickly perish!

Unfortunately, while their beauty tempts everyone to add one of these anthozoans to their aquarium, maintenance challenges even experienced aquarists. Novices should refrain from purchasing gorgonians.

ORDER HELIOPORACEA

Before the discovery of *Heliopora coerulea*, the only living member of this order, blue corals were thought to be extinct, since the only signs of their existence were in fossil records. *Heliopora coerulea* is found in shallow, calm waters of coral reefs of the tropical Indo-Pacific. This hermatypic coral forms voluminous colonies with finger- to lobe-shaped appendages. Its multitude of pores, each of which contains a tiny, eight-tentacled polyp, give the colony a holey appearance. All polyps are interconnected by a densely branched canal system. The light blue calcareous skeleton, which consists of fused aragonite fibers, is covered by a 2–3 mm layer of blue-gray to gray-brown coenenchyme. The blue calcareous skeleton is only visible if a branch is broken off. The skeleton is so hard, it can be crafted into pearls for the jewelry trade.

BLUE CORALS IN THE AQUARIUM

Blue corals are very rarely kept in aquaria, mainly due to the fact that they are virtually never imported. Nevertheless, occasionally Indonesian imports do contain small colonies on reef rock, often with disc anemones of the genus *Actinodiscus*.
Care of the animals is relatively simple. There are several different growth forms, and the more filiform the colony, the less current they should be subjected to. The polyps open irregularly. In aquaria with appropriate illumination, all of *Heliopora caerulea*'s nutritional requirements are met by the metabolic products of its zooxanthellae, allowing feedings to be waived. Fluorescent illumination yields poor growth, i.e., metal halide illumination is more appropriate. Joint care with other anthozoans is possible, though stony corals should be given a generous buffer zone.

Taxonomy

Phylum: CNIDARIA	Cnidarians
Class: ANTHOZOA	Anthozoans
Subclass: OCTOCORALLIA	Octocorals

Order: GORGONARIA — Gorgonians

- Suborder: Scleraxonia
  - Family: Briareidae
    - Genus: *Briareum*
  - Family: Anthothelidae
    - Genera: *Diodogorgia*, *Iciligorgia*
  - Family: Subergorgiidae
    - Genus: *Subergorgia*
  - Family: Carallidae
    - Genus: *Corallium*
  - Family: Melithaeidae
    - Genera: *Acabaria*, *Clathraria*, *Melithaea*, *Mopsella*
- Suborder: Holaxonia
  - Family: Acanthogorgiidae
    - Genus: *Acalycigorgia*
  - Family: Paramuriceidae
    - Genera: *Muricella*, *Paramuricea*, *Villogorgia*
  - Family: Plexauridae
    - Genera: *Echinogorgia*, *Eunicea*, *Eunicella*, *Heterogorgia*, *Muricea*, *Plexaura*, *Plexaurella*, *Pseudoplexaura*, *Rumphella*

Family: Gorgoniidae
Genera: *Gorgonia*
Lophogorgia
Pseudopterogorgia
Pterogorgia
Family: Ellisellidae
Genera: *Ctenocella*
Ellisella
Junceella
Family: Chrysogorgiidae
Genus: *Trichogorgia*
Family: Isididae
Genus: *Isis*

Order: HELIOPORACEA — Blue corals
Family: Helioporidae
Genus: *Heliopora*

Plexaurella sp., Caribbean.

Order: GORGONARIA
Fam.: Briareidae

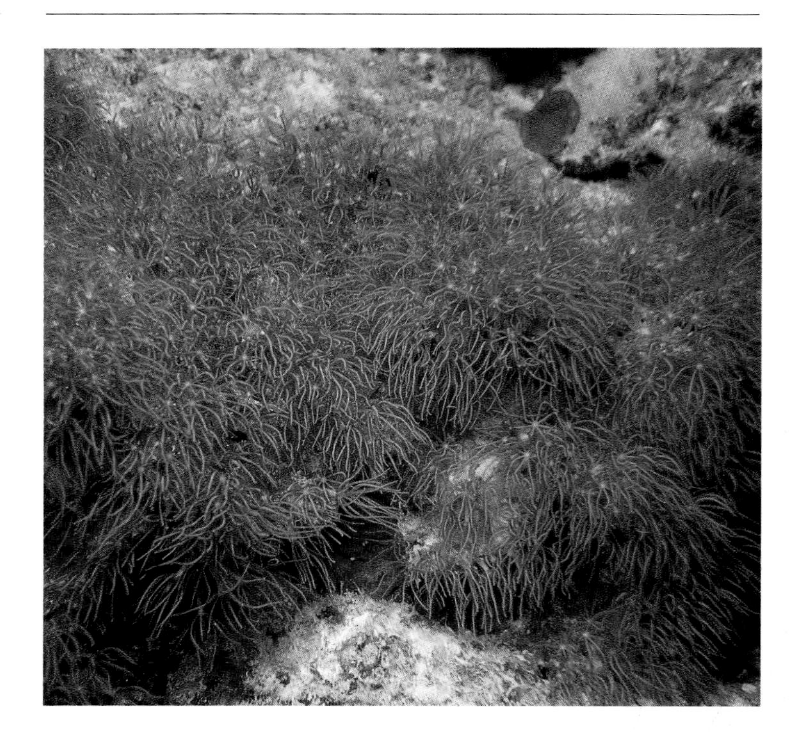

Briareum asbestinum (PALLAS, 1766)
Deadman's fingers, corky sea fingers

Hab.: Caribbean, Bahamas, southern Florida, Bonaire.

Sex.: None.

M.: The zooxanthellae make feedings unnecessary. Hence, this gorgonian is suitable for aquaria. Indo-Pacific soft corals, *Xenia* spp., and mildly toxic disc anemones make appropriate tankmates.

Light: Sunlight zone.

F.: *B. asbestinum* has lost its ability to capture particulate foods; therefore, foodstuffs are totally superfluous.

S.: An internal horny axis is lacking. While gorgonians generally form erect, branched colonies, *B. asbestinum* grows unbranched and finger-shaped or encrusting. The pictured colony is an encrusting form growing on dead corals. Greenish brown, exceedingly long-tentacled polyps protrude from the base. It is found from shallow water to moderate depths along almost all Caribbean reefs.

T: 18°–24°C, Ø: 80 cm, TL: from 100 cm, WM: m–s, WR: t, AV: 3–4, D: 2–3

Briareum asbestinum

Diodogorgia nodulifera HARGITT, 1901
Colorful sea rod

Hab.: Caribbean, Bahamas, southern Florida, Puerto Rico, Colombia.

Sex.: None.

M.: Unknown.

F.: Plankton.

S.: Colonies are erect with few branches, occasionally rodlike. The white, often transparent polyps are embedded in a red, red-orange, or orange coenenchyme and are diurnal as well as nocturnal. The rather stooped colonies settle on hard substrates at the reef bottom, underneath shaded overhangs, or on solitary boulders along deep, sandy sites. Since colorful sea rods live beyond common diving depths, they were not discovered until recently.

T: 18°–24°C, L: 30 cm, AV: 4, D: 4

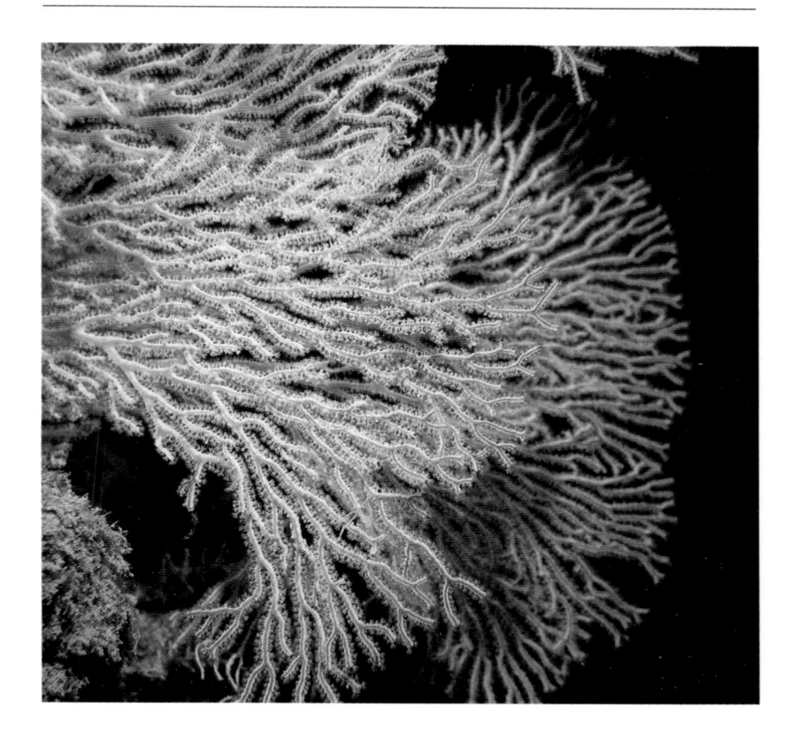

Iciligorgia schrammi DUCHASSAING, 1870
Deep water sea fan

Hab.: Caribbean, Florida, Bahamas, from Colombia to Brazil.

Sex.: None.

M.: Unknown.

F.: C; plankton.

S.: Growing to an impressive diameter of almost 2 m, these richly branched sea fans occur in an array of colors. Though occasionally found in shallow waters, they are only common in deep water. Current-swept reef and rock slopes are their habitat of choice. Depth is one factor that determines color—gray-brown to rust red in shallow water, lighter hues at deep depths, white at depth of more than 50 m. In deep water environments, their habitat coincides with that of black corals. The light polyps are arranged laterally along the branches.

T: 18°–24°C, **L**: over 100 cm, **AV**: 4, **D**: 3–4

Order: Gorgonaria
Fam.: Subergorgiidae

Subergorgia hicksoni STIASNY, 1940
Hickson's giant fan

Hab.: Red Sea, Indo-Pacific.

Sex.: None.

Soc.B./Assoc.: Some bivalves and feather and brittle stars live on *S. hicksoni.*

M.: Unknown.

F.: C; plankton.

S.: Under favorable conditions, *S. hicksoni* can grow to be more than 2 m in diameter. It predominantly inhabits rocky substrates of deep current-swept reef slopes. As twilight descends and plankton begins to rise from the depths, feather stars and other planktivores scale up these colonies, placing themselves at advantageous capture sites.

T: 20°–24°C, **L:** over 100 cm, **AV:** 4, **D:** 4

Subergorgia mollis (NUTTING, 1910)
Sea fan

Hab.: Tropical Indo-Pacific, Australia, the Philippines, Thailand, Japan to Taiwan.

Sex.: None.

M.: Unknown.

F.: C; plankton.

S.: These large, extensively branched reticulate fans always orient their broad side towards the prevalent current. Bivalves (*Pteria* sp.), brittle stars (*Ophiothela danae*), and feather stars (Colometridae) commonly use *S. mollis* as a substrate; small gobies are more infrequent visitors.

T: 20°–24°C, **L**: 80 cm, **AV**: 4, **D**: 4

Order: Gorgonaria
Fam.: Subergorgiidae

Subergorgia sp.
Sea fan

Hab.: Indo-Pacific, Maldives.

Sex.: None.

M.: Unknown.

F.: C; plankton.

S.: This large, erect sea fan has a complex reticulate network of secondary branches supported by stout primary branches. Using its wide "foot," *Subergorgia* sp. adheres to hard substrates. The broad side of the fan is always directed towards the prevailing current, thereby increasing the organism's filtering capacity. Plankton is captured at night, and the polyps remain retracted within the coenenchyme during the day. *Subergorgia* sp. occurs in a variety of colors. It inhabits exposed steep reef slopes and vertical walls in deep water environments.

T: 20°–24°C, L: over 100 cm, AV: 4, D: 4

Subergorgia suberosa (PALLAS, 1766)
Rough gorgonian, bushy gorgonian

Hab.: Tropical Indo-Pacific, Sri Lanka, Mauritius, Maldives, Thailand, the Philippines, Japan to Taiwan.

Sex.: None.

M.: Unknown.

F.: C; plankton.

S.: Colonies are planar or fascicular. Branches are long and cylindrical but have deep, well-defined grooves up their sides that make the branches appear oval. They grow on rocks and coral blocks in sandy areas along the bottom of reefs. Live colonies are brown to red-brown, bright red, or orange.

T: 20°–24°C, **L:** 70 cm, **AV:** 4, **D:** 4

Order: Gorgonaria
Fam.: Carallidae

Corallium rubrum (LINNAEUS, 1758)
Jewel coral

Hab.: Mediterranean to the Canary Islands.

Sex.: Dioecious.

M.: Maintenance is not recommended, as colonies have a short life span in aquaria.

F.: C; plankton.

S.: This richly branched, fascicular gorgonian has two classes of polyps—small polyps, whose purpose is water exchange, and feeding polyps. The former lack tentacles, while the latter, typical of this subclass, have eight pinnate tentacles. Light polyps are embedded in matt red coenenchyme. Along the center of the coenenchyme is a hard, calcareous axial rod. *C. rubrum* lives on hard dimly lit substrates down to depths beyond 100 m. Bright red, occasionally white, and very rarely black, the axial rods have been collected from the Mediterranean and crafted into jewelry since ancient times. Divers do not engage in collecting this coral today. It is gathered from the depths with specially designed drag nets, devastating entire underwater landscapes in the process.

T: 10°–18°C, L: 4–30 cm, AV: 4, D: 4

Acabaria biserialis KÜKENTHAL, 1908
Red reef crest gorgonian

Hab.: Red Sea.

Sex.: None.

M.: Unknown.

F.: C; plankton.

S.: Above is a detailed photo of *A. biserialis* in its natural habitat. Branches dichotomously divide along one plane, creating a fan-shaped colony. Nodes are enlarged, and internodes are of variable lengths. Polyps are pale and laterally arranged on the branches. This gorgonian lives in dim caves and under overhanging ledges of reef slopes at a wide range of depths.

T: 22°–26°C, L: 8 cm (pictured section), AV: 4, D: 4

Order: Gorgonaria
Fam.: Melithaeidae

Acabaria variabilis (HICKSON, 1905)
Delicate net fan

Hab.: Red Sea, Indian Ocean, Maldives.

Sex.: None.

M.: Unknown.

F.: C; plankton.

S.: *A. variabilis* has extraordinarily filiform branches, bright polyps, slightly enlarged nodes that are red, orange-red, or yellow, and light internodes. The branches are roughly arranged in a single plane. It lives on reef slopes in open water, along shaded areas beneath ledges, and within tiny caves where water movement is practically undetectable.

T: 20°–24°C, L: 40 cm, AV: 4, D: 4

Acabaria variabilis, Maldives.

Order: GORGONARIA
Fam.: Melithaeidae

Acabaria crosslandi STIASNY, 1938
Crossland's fan coral

Hab.: Red Sea, Indo-Pacific.

Sex.: None.

M.: Unknown.

F.: C; plankton.

S.: Using its broad "foot," *A. crosslandi* adheres to coral rocks along deep current-swept reef slopes. The multitudinous branches form a fan-shaped colony which faces the prevailing current. The coenenchyme is rust red to red-purple, and the expanded polyps are white.

T: 20°–24°C, L: 70 cm, AV: 4, D: 4

Clathraria maldivensis VAN OFWEGEN, 1987
Maldivian reticulated fan

Hab.: Indian Ocean, Maldives, Lakshadweep Islands.

Sex.: None.

M.: Unknown.

F.: C; plankton.

S.: Delicate, filiform branches grow within one plane to create a reticulate fan-shaped colony. The orange branches have light-colored nodes, parallel rows of small, white polyps along the sides, and forked tips. With its broad side always perpendicular to the prevailing current, *C. maldivensis* colonizes current-swept reef slopes both in the open and in shaded areas under ledges.

T: 20°–24°C, L: 35 cm, AV: 4, D: 4

Clathraria rubrinodis GRAY, 1859
Red knot corals

Hab.: Red Sea.

Sex.: None.

M.: Unknown.

F.: C; plankton.

S.: The scientific and common names were derived from the red axial rod. Numerous gnarled to antlerlike branches create a distinctive fascicular colony. The terminal segments of the branches are forked, and each of the tips is rounded. Coloration is variable. *C. rubrinodis* inhabits reef slopes, extending its branches far into the open water.

T: 20°–24°C, L: 60 cm, AV: 4, D: 4

Melithaea squamata (NUTTING, 1911)
Scaled giant sea fan

Hab.: Indian Ocean to the western Pacific.

Sex.: None.

M.: Unknown.

F.: C; plankton.

S.: This finely reticulate, fan-shaped colony can grow to extraordinary dimensions. Attached to oblique substrates such as rocks and corals by a broad "foot," the colony grows transverse to the current. The primary branches anastomose near the colony's base. *M. squamata* predominantly inhabits steep reef slopes down to 25 m.

T: 20°–27°C, L: 100 cm, AV: 4, D: 4

Order: GORGONARIA
Fam.: Melithaeidae

Mopsella sp.
Sea fan

Hab.: Tropical Indo-Pacific, Thailand.

Sex.: None.

M.: Unknown.

F.: C; plankton.

S.: These sea fans chiefly inhabit deep reef slopes, commonly in groups. They form fan-shaped, planar colonies which grow with their broad side transverse to the current. Main branches are clearly structured into nodes and internodes, and polyps are translucent and arranged in rows along two sides of each branch. Though species within this genus are at present simply identified by geographic origin, there are several autonomous species that inhabit Indo-Pacific reefs.

T: 20°–24°C, L: 100 cm, AV: 4, D: 4

Acalycigorgia sp.
Blue sea fan

Hab.: Tropical Indo-Pacific, Thailand.

Sex.: None.

M.: Unknown.

F.: C; plankton.

S.: Colonies branch in one plane. Bright blue polyps make these colonies especially notable, but unfortunately the noteworthy blue color is not retained in dried or preserved specimens. Shaded vertical substrates beneath ledges are their preferred habitat. There the voluminous, richly branched fans virtually fill the lumen.

T: 20°–24°C, L: 50 cm, AV: 4, D: 4

Order: Gorgonaria
Fam.: Paramuriceidae

Muricella sp.
Warty cave fan

Hab.: Red Sea.

Sex.: None.

M.: Unknown.

F.: C; plankton.

S.: Via a small base, colonies affix to ledges and hang into the shaded open space below the ledge. Branches extend in all directions and rarely anastomose. Polyps are the same color as the coenenchyme; each is seated in a pinlike warty calyx.

T: 20°–24°C, L: 40 cm, AV: 4, D: 4

Paramuricea clavata (RISSO, 1826)
Violescent sea whip, chameleon sea whip

Hab.: Mediterranean.

Sex.: None.

M.: If certain standards are maintained—good water quality, temperatures less than 18° C, and sufficient food—specimens with an undamaged coenenchyme can live for years in captive environments.

Light: Dim light.

F.: C; frozen substitute plankton, plankton, live *Artemia* nauplii, *Mysis*.

S.: In its natural habitat, *P. clavata* is encountered along rock walls at depths greater than 15 m. Sometimes it occurs in groups. The broad fan is richly branched and densely covered with polyps of the same color. Because part of the colony may be pale or yellow, its previous *nomen* was "chameleon."

T: 10°–18°C, L: 80 cm, TL: from 160 cm, WM: m–s, WR: b, AV: 3, D: 3

Paramuricea sp.
Sea fan

Hab.: Tropical Indo-Pacific, Thailand.

Sex.: None.

M.: Unknown.

F.: C; plankton.

S.: The slender base holds several large noninterlaced branches. Secondary and tertiary branches practically emerge at right angles from their supporting branches. The light polyps, which are located along two sides of the branches, are a distinct contrast to the red-orange coenenchyme. This erect gorgonian prefers to inhabit horizontal, hard substrates in deep water environments.

T: 20°–26°C, L: 30 cm, AV: 4, D: 4

Villogorgia sp.
Yellow sea fan

Hab.: Tropical Indo-Pacific, Thailand.

Sex.: None.

M.: Unknown.

F.: C; plankton.

S.: Colonies are small, delicate, filigree fans that grow transverse to the prevailing current. Polyps and coenenchyme are the same color. The majority inhabits shaded vertical walls or hangs from the underside of ledges. They are a favorite substrate for small brittle stars.

T: 20°–24°C, **L:** 30 cm, **AV:** 4, **D:** 4

Echinogorgia sp.
Sea fan

Hab.: Tropical Indo-Pacific, Thailand.

Sex.: None.

M.: Unknown.

S.: Many species of *Echinogorgia* are found on Indo-Pacific reefs. They have a very distinct growth pattern, and many grow to respectable proportions. To the best of our knowledge, individual species can only be classified by their geographic origin. Some species form several fans off a common base, all orientated in different directions. Species which live beneath ledges are usually small and filigree, but not reticulate.

T: 20°–24°C, L: 30–80 cm, AV: 4, D: 4

Eunicea calyculata (ELLIS & SOLANDER, 1786)
Warty sea rod

Hab.: Caribbean, Bermuda, West Indies, Gulf of Mexico, Puerto Rico, Colombia.

Sex.: None.

M.: Unknown.

F.: C; plankton.

S.: Supported by one stem, this erect colony has stout branches covered by rigid nubs. It chiefly occurs singly, never in great densities, on patch reefs or fore reefs in deep water environments. The polyps contain zooxanthellae.

T: 20°–24°C, **L**: 60 cm, **AV**: 4, **D**: 3–4

Eunicea tourneforti MILNE, EDWARDS & HAIME, 1857
Tournefort's knobby candelabrum

Hab.: Caribbean, Florida, Bahamas, Curaçao, Bonaire, Barbados.

Sex.: None.

M.: Unknown.

F.: C; plankton.

S.: Growth is bushy to chandelierlike. The sparse branches are tall, thick, and cylindrical. Polyps are diurnal as well as nocturnal and densely cover the rigid branches. *E. tourneforti* attaches to rocks, stones, and dead corals on clear water reefs exposed to a moderate current. Coloration is mainly gray or brown.

T: 20°–26°C, L: 60 cm, AV: 4, D: 4

Eunicella cavolinii (KOCH, 1887)
Yellow sea whip

Hab.: Mediterranean.

Sex.: None.

M.: Only small colonies can be maintained for extended periods of time in a captive environment. They require clean water, a shady site, and a medium to strong current.

Light: Moderate light.

F.: C; live plankton, e.g., *Artemia* nauplii and *Mysis*, and frozen substitute plankton.

S.: *E. cavolinii* is richly branched and fan-shaped. Growth is virtually planar, and the broad side of the colony is oriented transverse to the current. At depths of 15 m or greater, its slender base attaches to substrates along steep rock walls, spacious caves, and other shady sites. Polyps are diurnal as well as nocturnal. This species does not have zooxanthellae.

T: 10°–18°C, **L**: 30 cm, **TL**: from 160 cm, **WM**: m–s, **WR**: b, **AV**: 3, **D**: 3–4

Order: GORGONARIA
Fam.: Plexauridae

Eunicella singularis (ESPER, 1791)
White sea whip

Hab.: Mediterranean, Atlantic, English Channel.

Sex.: None.

Soc.B./Assoc.: Tunicates (*Clavelina*), calcareous tubeworms (*Salmacina*), and cowries (*Simnia* and *Cypraea*) are often found living among the branches of the white sea whip.

M.: Good water quality, a temperature of 18°C or less, shade, and a moderate to strong current are demanded for successful aquarium care.

Light: Moderate light.

F.: C; frozen substitute plankton, live *Artemia* nauplii, plankton.

S.: The sparse branches are upright and largely parallel. To a depth of 40 m, *E. singularis* attaches to rocks and sand and mud bottoms using a small adhesive disc. Polyps are yellow-brown to green, depending on the zooxanthellae.

T: 10°–18°C, **L**: 60 cm, **TL**: from 160 cm, **WM**: m–s, **WR**: m, **AV**: 2–3, **D**: 3–4

Eunicella verrucosa (PALLAS, 1766)
Sea fan, white sea whip

Hab.: Mediterranean, Atlantic.

Sex.: None.

M.: *E. verrucosa* requires shade and a moderate to strong current.

Light: Moderate light.

F.: C; frozen substitute plankton, live *Artemia* nauplii, *Mysis*, plankton.

S.: From 10 m depth, this slow-growing, erect species is most frequently found attached to hard substrates. It has a small base and sparse branches that extend in one plane. Rhythmic tidal currents provide the colony with sufficient food. Coloration ranges from white to yellow or pink.

T: 10°–18°C, L: 40 cm, TL: from 160 cm, WM: m–s, WR: m, AV: 2, D: 3–4

Heterogorgia natumani CASTRO, 1990
Golden sea whip

Hab.: Northwest Caribbean, Colombia to Brazil.

Sex.: None.

M.: Unknown.

F.: Plankton.

S.: *H. natumani* is a low-growing, highly branched coral. Growth is chiefly planar—some specimens might even produce fan-shaped colonies. Polyps are yellow to golden and diurnal as well as nocturnal, and the coenenchyme ranges from orange to brown. Most colonies are found along shaded locals, e.g., deep reef dropoffs or beneath ledges.

T: 18°–24°C, L: 40 cm, AV: 4, D: 4

Muricea muricata (PALLAS, 1766)
Spiny sea fan, spiny sea whip

Hab.: Caribbean, Florida, Gulf of Mexico, West Indies, Colombia.

Sex.: None.

M.: A long-lived, maintainable organism, though intense light is demanded.

Light: Sunlight zone; use metal halide illumination.

F.: C; plankton.

S.: When maintained under optimal conditions, the zooxanthellae meet all of this animal's nutritional needs, i.e., feeding is unnecessary. Growth is bushy or planar, and the branches bifurcate near the base. Because of the sharp, pointed calcium spicules in the coenenchyme, the long, rodlike branches are tough and thorny. Polyps are relatively large and diurnal as well as nocturnal. *M. muricata* occurs at various depths on coral fragments, sand, and hard substrates.

T: 20°–29°C, **L:** 60 cm, **TL:** from 150 cm, **WM:** m, **WR:** m, **AV:** 3–4, **D:** 3

Plexaura flexuosa LAMOUROUX, 1821
Bent sea rod

Hab.: Caribbean, Belize, Honduras, Colombia.

Sex.: None.

M.: Successful maintenance is contingent on metal halide illumination.

Light: Sunlight zone.

F.: C; plankton. Under optimal conditions, most of *P. flexuosa*'s nutritional needs are met by its symbiotic zooxanthellae.

S.: In its natural habitat, i.e., predominately patch reefs and outer reefs, *P. flexuosa* is represented by diverse growth forms and colors. Branches of small colonies are largely arranged in planar fashion, while large colonies are bushy and richly branched. The surface of the branches is rough and dry. This coral may be numerous, perhaps even the dominant species, in some regions.

T: 20°–29°C, **L:** 40 cm, **TL:** from 150 cm, **WM:** m, **WR:** m, **AV:** 3–4, **D:** 3

Plexaurella dichotomea (ESPER, 1791)
Double-forked sea rod

Hab.: Caribbean, Florida, Bahamas, West Indies, Bonaire, Curaçao, Gulf of Mexico, Colombia.

Sex.: None.

M.: When its requirement for intense illumination is fulfilled, it is a long-lived aquarium animal.

Light: Sunlight zone.

F.: C; plankton. Most of *P. dichotomea*'s nutritional requirements are met by its symbiotic zooxanthellae.

S.: The colony is erect and bushy with tall, dichotomously forked branches. When extended, the polyps give the colony a peltlike appearance. They retract into slitlike apertures in the coenenchyme if disturbed. Which growth form *P. dichotomea* assumes is greatly influenced by habitat. Hard substrates, outer and inner reefs, reef crests, and sandy areas of lagoons down to moderate depths are colonized.

T: 20°–29°C, L: 100 cm, TL: from 150 cm, WM: m, WR: m, AV: 3, D: 3–4

Plexaurella nutans KUNZE, 1916
Giant slit-pore sea rod

Hab.: Caribbean, Florida, Gulf of Mexico, West Indies, Colombia.

Sex.: None.

M.: *P. nutans* is suitable for aquaria. Even under unfavorable conditions, polyps will open. Colonies without a holdfast adhere within one week.

Light: Sunlight zone.

F.: C; plankton. Most of the organism's nutritional needs are met by its symbiotic zooxanthellae.

S.: The sparse branches are tall and stout; trumpetfishes often lurk for prey among the large branches. The long-tentacled, large polyps are diurnal as well as nocturnal, and when touched, retreat into the coenenchyme in an ever widening circle, demonstrating their interconnection. *P. nutans* lives on inner and outer reefs on fragmented coral as well as sandy substrates.

T: 20°–29°C, L: 100 cm, TL: 150 cm, WM: m, WR: m, AV: 3–4, D: 3

Pseudoplexaura porosa (HOUTTUYN, 1772)
Porous sea rod

Hab.: Caribbean, Florida, Bahamas, Bonaire.

Sex.: None.

M.: Unknown.

F.: C; plankton.

S.: Four species of *Pseudoplexaura* live in the western Atlantic. Definitive identification demands microscopic examination of the spicules. Colonies are predominately bushy with variable-length, dichotomously branched rods of variable length. The polyps, which are active day and nicht, cannot be completely retracted into the elliptical pores of the coenenchyme. The porous sea rod colonizes reef slopes, reef platforms, and reef bottoms; on reef bottoms, it will stand individually in the middle of an expansive surface. Coloration ranges from light brown to brown and purple-red.

T: 20°–26°C, L: 60 cm, AV: 4, D: 4

Order: Gorgonaria
Fam.: Plexauridae

Rumphella attenuata (NUTTING, 1910)
Bushy gorgonian

Hab.: Red Sea, Indo-Pacific, eastern Africa, Indonesia.

Sex.: None.

Soc.B./Assoc.: *R. attenuata* is compatible with other octocorals, but direct contact between this organism and stony corals or disc and colonial anemones should be avoided.

M.: Colonies can be maintained in a captive environment since they contain symbiotic zooxanthellae. Cautiously acclimate *R. attenuata* to metal halide illumination.

Light: Moderate light zone.

B./Rep.: Choose a shoot from a lateral branch, and sever it at the fork. From the same end, remove approximately 2 cm of the coenenchyme. Place the cleaned end into a hole in the decoration and anchor the shoot. The holdfast forms after about 2 weeks.

F.: C; plankton, substitute plankton. Daily meals are contraindicated.

S.: Bushy gorgonians have a fascicular growth form and numerous long whiplike, flexible branches. The polyps are extended day and night. Most colonies are olive green and occur on hard substrates of current-swept reefs from shallow to moderate depths. Note that the left corner of the pictured specimen has retracted polyps, while the right side has extended polyps.

T: 21°–29°C, L: 80 cm, TL: from 100 cm, WM: m, WR: b, AV: 3–4, D: 2

Rumphella sp., aquarium photo.

Rumphella sp.
Bushy gorgonian

Hab.: Red Sea, Indo-Pacific.

Sex.: None.

Soc.B./Assoc.: *Rumphella* sp. is compatible with other corals, but direct contact between this species and stony corals, disc anemones, or colonial anemones should be avoided. Do not place *Cladielle* or *Xenia* in the immediate vicinity.

M.: This species can be maintained in a captive environment. The current should not be overly strong or direct. Anchor the colony to a small outcropping so that the polyp-bearing branches hang freely into the water column.

Light: Sunlight. Carefully acclimate the colony to metal halide illumination.

B./Rep.: Choose a shoot from the lateral branches. Sever it at a bifurcation, and remove approximately 1 cm of coenenchyme from the shoot's end. Then place the end into a hole in the tank's decorative infrastructure and anchor the coral. A new holdfast forms in about 4 weeks.

F.: *Rumphella* sp. lives from its symbiotic zooxanthellae; therefore, it has scant need for foodstuffs.

S.: See *Rumphella attenuata* and the species on the following page.

T: 21°–29°C, **L**: 40 cm, **TL**: 100 cm, **WM**: m, **AV**: 3–4, **D**: 2

Rumphella sp.
Bushy gorgonian

Hab.: Red Sea.

Sex.: None.

M.: Unknown.

F.: C; plankton.

S.: These erect, bushy colonies have an extraordinary number of flexible, bifurcated branches. They are generally found on corals at the bottom of the reef and along reefs with a slight grade. Polyps are predominately nocturnal, and the coenenchyme is rough and dry to the touch.

T: 20°–26°C, L: 70 cm, AV: 4, D: 4

Gorgonia flabellum LINNAEUS, 1758
Venus sea fan

Hab.: Caribbean, Bermuda, Florida, Curaçao, Colombia.

Sex.: None.

Soc.B./Assoc.: *Cyphoma gibbosum*, a snail, preys upon the polyps of this coral, leaving characteristic purple feeding tracks on its surface. Some brittle stars use the colony as a substrate.

M.: Colonies usually reach pet stores as dried skeletons. These are almost exclusively used as decorative infrastructure.

F.: C; plankton.

S.: *G. flabellum* is the smaller of the two Caribbean species. Using a large, purple holdfast, the broad, fan-shaped colony anchors to corals and other hard substrates from the surf zone to a depth of 30 m. The fan, which consists of an intricate network of small branches, is held erect into the open water. Polyps are nocturnal. Because of its great flexibility, *G. flabellum* is capable of withstanding strong turbulence.

T: 21°–29°C, L: 40 cm, TL: from 150 cm, WM: m–s, WR: m–t, AV: 3–4, D: 4

Order: Gorgonaria
Fam.: Gorgoniidae

Gorgonia ventalina LINNAEUS, 1758
Common sea fan

Hab.: Caribbean, Bahamas, Bermuda, Florida, Colombia.

Sex.: None.

Soc.B./Assoc.: The flamingo tongue, *Cyphoma gibbosum,* feeds on the polyps of this coral, leaving typical purple feeding tracks on the surface of the colony.

M.: Pet stores sell life specimens and dried skeletons. The latter are used as decorative infrastructure for aquaria, where the intricate reticulate design quickly becomes coated with algae. With correct maintenance parameters, small specimens can be maintained in reef aquaria. Colonies that are well anchored and exposed to a strong current are capable of cleaning themselves of algae and dirt particles by "skinning."

Light: Metal halide illumination.

F.: C; plankton. However, colonies mainly subsist on the metabolites of their symbiotic zooxanthellae.

S.: Common sea fans use a large, broad, purple foot to affix itself to corals and other hard substrates. Stout primary branches are interconnected by an intricate mesh of secondary branches, giving rise to a reticulate fan. Growing to a diameter of over 2 m, these fan-shaped, planar colonies orient themselves transverse to the current. They occur from shallow to deep-water habitats. Exposed reefs often support groups of *G. ventalina*.

T: 21°–29°C, **L:** 120 cm, **WM:** m–s, **WR:** m–t, **AV:** 3–4, **D:** 4

Lophogorgia sp.
King sea fan

Hab.: Caribbean, Colombia.

Sex.: None.

M.: Unknown.

F.: Plankton.

S.: King sea fans have delicate branches of different heights, all originating from a common base. Polyps are arranged in alternating rows along the branches and are usually white or transparent, whereas the coenenchyme is orange, red, or purple. Oriented such that their broad side faces the prevailing current, the colonies settle along hard or open sandy substrates in deep water. Frequently their branches support a rich epifauna.

T: 18°–24°C, **L:** 60 cm, **AV:** 4, **D:** 4

Pseudopterogorgia americana (GMELIN, 1791)
Slimy sea plume, slimy sea feather

Hab.: Caribbean, Florida, Bahamas, Bonaire.

Sex.: None.

M.: Unknown.

F.: Plankton.

S.: Colonies are bushy with tall pinnate branches. As denoted by the sparse branches, the pictured colony is young. Full-grown colonies have exceptionally long (up to 2 m), purple-red to violet primary branches and pinnate secondary branches. The thick coat of mucus covering the colonies serves to capture food and protect the animals from sedimentation. They are found from shallow to deep water on hard substrates, largely on reef slopes, patch reefs, and underwater plateaus.

T: 20°–29°C, L: 40 cm, AV: 4, D: 4

Pseudopterogorgia bipinnata (VERRILL, 1864)
Bipinnate sea plume, forked sea feather

Hab.: Caribbean, Florida, Bahamas, San Andrés.

Sex.: None.

M.: The bipinnate sea plume is easily maintained in reef aquaria. Because of its size, regular pruning is required; attaching *P. bipinnata* to the bottom substrate allows for larger branches and a more natural growth form. Under fluorescent lighting, colors may pale; metal halide illuminations is tolerated well. The current should be strong, but indirect.

Light: Sunlight zone.

F.: C; plankton. Symbiotic zooxanthellae make daily feedings unnecessary.

S.: There are several species of *Pseudopterogorgia* in the Caribbean. Growth is characteristically immense and erect with numerous rough, pinnate branches. Polyps are nocturnal and arranged pairwise at right angles along the branches. The coenenchyme is light purple and does not secrete mucus. *P. bipinnata* lives on hard substrates from shallow water to depths of more than 30 m.

T: 21°–29°C, **L:** over 100 cm, **TL:** from 150 cm, **WM:** m–s, **WR:** b, **AV:** 2–3, **D:** 2–3

Order: Gorgonaria
Fam.: Gorgoniidae

Pterogorgia citrina (ESPER, 1792)
Yellow sea whip

Hab.: Caribbean, Florida, Bermuda, Colombia.

Sex.: None.

M.: Although it is occasionally sold in pet stores, *P. citrina* is considered very difficult to maintain in captive environments. It quickly attaches to offered substrates, but polyps only open under optimal conditions. With a timer and antagonistic powerheads, appropriate current rhythms can be provided.

Light: Under strong metal halide illumination, colonies eventually develop a bluish hue.

F.: C; plankton and substitute plankton. Feed once a week.

S.: The many branches of the fascicular colonies are flat and foliaceous. Slitlike apertures line the thin, sometimes red, edges. The coenenchyme is yellow to green or light brown, whereas the polyps are pale. Colonies are usually found in shallow water attached to hard substrates, mainly along current-swept reef sections.

T: 21°–29°C, L: 40 cm, TL: from 160 cm, WM: m–s, WR: b, AV: 2, D: 3–4

Ctenocella pectinata (PALLAS, 1766)
Comb gorgonian

Hab.: Red Sea, Indo-Pacific, Australia, Malaysia.

Sex.: None.

Soc.B./Assoc.: Brittle stars live on *C. pectinata*, and at night, feather stars scale its branches.

M.: Care is difficult. A strong current is needed for the polyps to open, and daily meals are required. Metal halide illumination is inappropriate.

Light: Moderate light.

F.: C; plankton, small substitute plankton.

S.: The vertical secondary branches typically emerge off the primary branches like teeth of a comb. Branches are of different lengths and arranged in one plane—transverse to the current. The coenenchyme is red to red-orange, and the numerous polyps are short and white. *C. pectinata* lives in deep water on reef sections away from direct current.

T: 21°–29°C, **L:** 80 cm, **TL:** from 150 cm, **WM:** m–s, **WR:** b, **AV:** 3, **D:** 4

Ellisella elongata (PALLAS, 1766)
Long sea whip

Hab.: Caribbean, Florida, Bermuda, Roatan, Gulf of Mexico.

Sex.: None.

M.: Unknown.

F.: C; plankton.

S.: *E. elongata* has long whiplike branches which fork close to the base and extend into the open water. The coenenchyme and polyps are red to red-orange; only the tentacles of the polyps are light. Shaded sites beneath rock ledges at depths greater than 250 m are where most specimens are encountered.

T: 20°–24°C, L: 100 cm, AV: 4, D: 4

Ellisella sp.
Fascicular sea whip

Hab.: Indian Ocean, Thailand.

Sex.: None.

M.: Unknown.

F.: C; plankton.

S.: *Ellisella* is a rare find along the reefs of Thailand. With its numerous elongated, slender switches emerging from a common base, the colony resembles a giant shaving brush. It anchors to dead corals along the reef bottom via a slender base.

T: 22°–26°C, L: 60 cm, AV: 4, D: 4

Ellisella sp.
Sea whip

Hab.: Tropical Indo-Pacific, Thailand.

Sex.: None.

M.: Unknown.

F.: C; plankton.

S.: Long, unbranched, whiplike branches are supported by a short base stalk. This nocturnal organism inhabits deep reef slopes and plateaus.

T: 20°–24°C, L: 60 cm, AV: 4, D: 4

Junceella juncea PALLAS, 1766
Sea whip

Hab.: Red Sea, Indo-Pacific.

Sex.: None.

Soc.B./Assoc.: *J. juncea* is colonized by brittle stars. At night, feather stars scale its branches.

M.: Unknown.

F.: C; plankton.

S.: Growth is erect and unbranched. While initially straight, the burden of their own weight and length and the influence of water currents result in misshapen stalks. When encountered, they are either coiled from their tip or upright like bent switches. Their growth pattern demonstrates extreme adaptation to current-swept reefs. They are normally found on protruding rock plateaus and other horizontal substrates.

T: 20°–24°C, L: over 100 cm, AV: 4, D: 4

Order: Gorgonaria
Fam.: Ellisellidae

Junceella sp. (= *rubra*?)
Sea whip

Hab.: Tropical Indo-Pacific, Thailand.

Sex.: None.

M.: Unknown.

F.: C; plankton.

S.: A colony consists of a long, erect, branched stalk. While the stalk normally has a curved tip, the entirety coils into shallow windings as it grows. Normally this organism lives on hard substrates of the lower reef slopes, generally in moderate current but occasionally exposed to strong water movement. Small colonies commonly inhabit dim roomy caves. At the commencement of twilight, large numbers of feather stars scale *Junceella* sp., taking advantage of the exposed site to capture plankton.

T: 20°–26°C, L: 80 cm, AV: 4, D: 4

Fam.: Chrysogorgiidae

Trichogorgia lyra BAYER & MUZIK, 1976
Lyra gorgonian

Hab.: Caribbean, Colombia.

Sex.: None.

M.: Unknown.

F.: C; plankton.

S.: The long, whiplike branches bifurcate near the base of the colony. Light, virtually transparent polyps march pinnately up the sides of the delicate branches. They and the coenenchyme are equal in color. At depths greater than 30 m, *T. lyra* settles on hard substrates of rubble zones and reef bottoms. This is a very uncommon species, which explains why it was not described until fairly recently.

T: 20°–24°C, **L**: 40 cm, **AV**: 4, **D**: 4

Isis hippuris LINNAEUS, 1758
Golden sea fan

Hab.: Indo-Pacific, Australia, Indonesia, Sri Lanka.

Sex.: None.

M.: For successful maintenance, this species must be fed 3–4 times a week. The current should be strong, yet indirect. It is a self-cleaning colony.

Light: Moderate light. Fluorescent tubes are considered to be a better source of illumination, as metal halides are not well tolerated.

F.: C; plankton and small substitute plankton.

S.: *I. hippuris* has a highly branched planar growth form. It is found on corals and other hard substrates from shallow to deep water, particularly in lagoons and along upper reef slopes. As denoted by the golden color, one half of the pictured specimen has extended polyps.

T: 21°–29°C, L: 40 cm, TL: from 150 cm, WM: m–s, WR: m–b, AV: 3, D: 3

Heliopora coerulea PALLAS, 1766
Blue coral

Hab.: Indo-Pacific, western Pacific, Chagos, Malaysia.

Sex.: None.

M.: *H. coerulea* generally considered an insignificant import.

F.: C; plankton.

S.: Blue corals are hermatypic. Because of their similar physical appearance and overlapping geographical distribution, fire corals and blue corals are difficult to distinguish in their natural habitat. Their identity only becomes obvious after a branch is broken off and the skeleton, blue in the case of *H. caerulea*, is exposed. Colonies vary in size, shape, and appearance. Erect forms can attain a height in excess of 2 m. Growing tips are light. Polyps are cream colored and, like soft corals, have 8 tentacles. Blue corals inhabit lagoons and shallow reefs, where in some areas they may be the dominant coral.

T: 22°–26°C, L: 30 cm, AV: 4 (protected species), D: 4

Subclass: Octocorallia
Order: Pennatulacea

Sea pens or sea pansies are octocorals that share certain characteristics with soft corals. The most significant difference between these two orders lies in the nature of the substrate they inhabit. Soft corals typically colonize hard substrates, whereas sea pens use a bulblike enlargement to anchor themselves in sand or mud bottoms. When caring for sea pens in a captive environment, at least 15–20 cm of sand will be needed. If the depth of the substrate falls short of the animals' requirements, they will curl up on the bottom of the aquarium!

Whether fleshy or filamentous, sea pens expand greatly in length and volume as siphonozooids pump water into the gastrovascular cavities. The expanded colony of *Sarcoptilus* sp. on the facing page has a typical arrangement of primary and secondary polyps—the stalk, or primary polyp, supporting two rows of secondary polyps.

Sea pens are dioecious. The entire colony is either male or female; there is a preponderance of female animals. Most simply release their gametes into the water where they are fertilized. A few are viviparous; that is, the eggs are fertilized and begin development internally. When they have reached the larval stage, they are released into the sea. For a few days the planula larvae are free-swimming. They then metamorphose into an elongated primary polyp and settle on the substrate. The inferior end becomes the stalk; it is frequently fortified by a horny, sometimes calcareous, axial rod. Secondary polyps emerge laterally and from the tip, creating the characteristic plume-shaped colony. Eight-rayed autozooids can easily be distinguished on expanded stalks, but the siphonozooids merely appear as small dots between the autozooids. Siphonozooids primarily take in and release water, allowing the colony to extend into the open water or deflate and disappear into the sand. About 40,000 polyps can be found on a single large sea pen.

Most sea pens are nocturnal, only emerging during the evening or at night to capture plankton. Daylight hours are passed buried in the sand. Inhabitants of the deep sea do not follow a diurnal cycle. Some species have colorful calcarious bodies sandwiched between their endoderm and ectoderm which make the colonies quite attractive. Many sea pens luminesce in the dark. This is ac-

Sarcoptilus sp., Malaysia.

complished through mucus secretions that emit light as they come in contact with water.

Sea pens are capable of locomotion. With the aid of the foot, sea pens burrow into the substrate to anchor themselves; likewise, they can also free themselves and move about at will. If their placement in the aquarium is not to their satisfaction, their ability to move will be called into play. At times sea pens may drift with the current, but this is a rare event in nature, since they normally adhere in locals with favorable conditions. Slight position adjustments are frequently undertaken using the pedal section.

There are about 300 species of sea pens found throughout the seas, from shallow waters to great depths (below 5,000 m). They range in size from a few centimeters to more than one meter. Few are suitable for captive environments, which explains why they are seldom encountered in pet stores. Deep-sea forms, which are also diurnal, are suitable for rather dark, coldwater tanks; nocturnal, shallow water forms do not make very interesting aquarium additions unless they can be enticed to a more diurnal lifestyle. Feeding solely during the day and abstaining from maintaining disturbing fishes may bring this about. Nevertheless, sea pens will never be well adapted, congenial additions to the aquarium.

As mentioned, most require about 20 cm sand (only a few of the small, broad species that are similar to soft corals are content with 5 cm). Since they consume plankton, a diet of crustacean larvae, small crustacea, or plankton substitutes (scraped mussel, crustacea, fishes, or worm meat as well as moistened dry diets) must be offered. A strong current capable of transporting the food to the polyps is a necessity.

Class: ANTHOZOA	Anthozoans
Subclass: OCTOCORALLIA	Octocorals
Order: PENNATULACEA	Sea Pens

Family: Pennatulidae — "True" sea pens
Genus: *Pennatula*

Family: Pteroeididae
Genera: *Pteroeides*
Sarcoptilus

Family: Veretillidae
Genera: *Cavernularia*
Veretillum

Family: Virgulariidae
Genus: *Virgularia*

Maldives, Indian Ocean.

Order: PENNATULACEA
Fam.: Pennatulidae

Pennatula sp.
Delicate sea pen

Hab.: West Indo-Pacific. *Pannatula* sp. is found at depths of 10–40 m.

Sex.: None.

Soc.B./Assoc.: *Pannatula* sp. normally lives singly on deep sand or mud substrates. Limit tankmates to tube anemones, soft corals, and gorgonians.

M.: To adequately anchor itself, *Pennatula* sp. requires a 20 cm substrate of white sand. Provide a current.

Light: Moderate to dim light.

B./Rep.: Separate sexes (dioecious). Gametes are simply released into the water. The resulting larvae are planktonic until they settle on a suitable site, where they transform into an elongated primary polyp and start a new colony. There is metagenesis or alternation of generations, i.e., *Pennatula* sp. reproduces sexually one generation then asexually the next.

F.: O; various types of plankton. Since these animals are nocturnal, feed them at night. The polyps face away from the current, and the tentacles capture small organisms that drift by in the countercurrent. A diet of crustacean larvae, small crustacea, and diverse plankton substitutes such as moistened commercial diets, scraped mussel, crustacean and fish meat, and minced worms is suitable.

S.: The yellow tip of the stalk signifies that the pictured colony is still growing. The virtually transparent lateral branches support prominent eight-tentacled polyps.

T: 20°–25°C, **L:** 10 cm, **TL:** from 50 cm, **WM:** m, **WR:** b, **AV:** 2, **D:** 3

Pennatula sp.
Stout sea pen

Hab.: Indian Ocean, primarily the Maldives. This species occurs at depths of 5–40 m.

Sex.: None.

Soc.B./Assoc.: The stout sea pen normally lives singly on sandy or muddy substrates. Only associate it with tube anemones, soft corals, and gorgonians.

M.: To adequately anchor itself, *Pennatula* sp. requires a 20 cm layer of fine, soft sand. A slight current is recommended.

Light: Sunlight to dim light.

B./Rep.: Dioecious. Eggs and sperm are simply released into the water. The resulting larvae are planktonic until they settle on a suitable substrate, where they transform into an elongated primary polyp and develop into a new colony. There is metagenesis or alternation of generations, i.e., the stout sea pen reproduces sexually one generation and asexually the next.

F.: O; various plankton. Since these animals are nocturnal, feed them at night. The polyps face away from the current, and the tentacles capture small organisms that drift by in the countercurrent. A diet of crustacean larvae, small crustacea, and diverse plankton substitutes such as moistened dry commercial diets, scraped mussel, crustacean and fish meat, and minced worms is suitable.

S.: This species can be recognized by its stout, almost round shape. Typically, parts of the "stalk" are pale to white or beige while the remainder is brown. While lateral branches and polyps are initially the color of the stalk, Their coloration can change. The Maldivian *Pennatula murrayi* is much taller and occurs at greater depths.

T: 20°–25°C, L: 10 cm, TL: from 50 cm, WM: m, WR: b, AV: 2, D: 3

Order: PENNATULACEA
Fam.: Pteroeididae

Pteroeides sp.
Dark sea pen

Hab.: West Indo-Pacific. *Pteroeides* sp. is found at depths of 5–50 m.

Sex.: None.

Soc.B./Assoc.: *Pteroeides* sp. normally lives singly on deep sand and mud substrates. Only tube anemones, soft corals, and gorgonians make suitable tankmates.

M.: To adequately anchor itself, *Pteroeides* sp. requires a 20 cm layer of fine, soft sand. A slight current should be present.

Light: Moderate to dim light.

B./Rep.: Dioecious. Eggs and sperm are simply released into the water. The resulting larvae are planktonic until they settle on a suitable substrate, where they transform into an elongated primary polyp and develop into a new colony. There is metagenesis or alternation of generations, i.e., *Pteroeides* sp. reproduces sexually one generation then asexually the next.

F.: O; various plankton. Since these animals are nocturnal, they should be fed at night. The polyps face away from the current, and the tentacles capture small organisms that drift by in the colony's countercurrent. A diet of crustacean larvae, small crustacea, and diverse plankton substitutes such as moistened dry commercial diets, scraped mussel, fish, and crustacean meat, and minced worms is suitable.

S.: The stalk is yellow to orange, the back of the lateral branches is largely white, and the polyps are dark green-brown.

T: 20°–25°C, **L**: 50 cm, **TL**: from 90 cm, **WM**: m, **WR**: b, **AV**: 2–3, **D**: 3

Sarcoptilus sp.
Light sea pansy

Hab.: West Indo-Pacific. Light sea pansies occur at depths of 15–50 m.

Sex.: None.

Soc.B./Assoc.: Light sea pansies are usually found singly on deep sand and mud substrates. Limit tankmates to tube anemones, soft corals, and gorgonians.

M.: To anchor properly, *Sarcoptilus* sp. requires a 20 cm layer of fine, soft sand. A slight current is recommended.

Light: Moderate to dim light.

B./Rep.: Dioecious. Eggs and sperm are simply released into the water. The resulting larvae are planktonic until they settle on a suitable substrate, where they transform into an elongated primary polyp and develop into a new colony. There is metagenesis or alternation of generations, i.e., *Sarcoptilus* sp. reproduces sexually one generation then asexually the next.

F.: O; various plankton. Since these animals are nocturnal, feed them at night. The polyps face away from the current, so the tentacles must capture small organisms from the colony's countercurrent. Offer crustacean larvae, small crustacea, and diverse plankton substitutes such as scraped fish, mussel and crustacean meat, minced worms, and moistened dry commercial diets.

S.: The shape—broadened towards the tip—and the uniform, almost white coloration are distinctive.

T: 20°–26°C, L: 20 cm, TL: from 50 cm, WM: m, WR: b, AV: 2, D: 3

Order: Pennatulacea
Fam.: Veretillidae

Cavernularia obesa (VALENCIENNES, 1831)
Obese sea pen

Hab.: Tropical and subtropical Indo-Pacific.

Sex.: None.

M.: Undamaged specimens are longevous. A layer of fine sand will permit obese sea pens to properly anchor themselves.

F.: Zooplankton and substitute plankton. Colonies must be individually fed.

S.: *C. obesa* inhabits sand or mud substrates in zones exposed to moderate currents. It is frequently found in quite bays and harbors. Some individuals are iridescent when exposed to long wave UV light.

T: 20°–24°C, **L:** 12–40 cm, **WM:** w, **AV:** 2–3, **D:** 2

Veretillum cynomorium (PALLAS, 1774)
Finger-shaped sea pen

Hab.: Northeastern Atlantic to the Mediterranean. It occurs at depths of 20-40 m.

Sex.: None.

Soc.B./Assoc.: This species is found singly on deep sand and mud substrates. It is a common species in some areas. Only tube anemones, soft corals, and gorgonians make suitable tankmates.

M.: To anchor itself properly, *V. cynomorium* needs a 20 cm layer of fine, soft sand. A slight current is recommended.

Light: Moderate to dim light.

B./Rep.: Dioecious. Eggs and sperm are simply released into the water. The resulting larvae are planktonic until they settle on a suitable substrate, where they transform into an elongated primary polyp and develop into a new colony. *C. cynomorium* passes through metagenesis, i.e., alternating sexual and asexual generations.

F.: O. Despite relatively large polyps, this sea pen demands tiny plankton. Offer crustacean larvae and diverse plankton substitutes. Since this is a nocturnal species, it should be fed at night.

S.: *V. cynomorium*'s cylindrical body is circumscribed by polyps; therefore, there is no particular orientation to the current. The animal's reluctance to anchor itself in captive habitats and the resulting tendency to fall over has given *V. cynomorium* a reputation of being a problematic species to care for. A close up of a polyp is pictured above. A very similar tropical species, *Cavernularia obesa*, is sometimes offered in pet stores.

T: 15°–18°C, **L:** 40 cm, **TL:** from 70 cm, **WM:** m, **WR:** b, **AV:** 2, **D:** 3

Virgularia sp. (probably *V. presbytes*, BAYER, 1955)

Whip sea pen

Hab.: Tropical western Atlantic, from the United States to Suriname. *Virgularia* sp. lives at depths of 20–40 m.

Sex.: None.

Soc.B./Assoc.: *Virgularia* sp. occurs singly on deep sand and mud substrates. Only invertebrate tanks housing fauna such as tube anemones, soft corals, and gorgonians make suitable captive habitats.

M.: This sea pansy requires 20 cm of fine, soft sand to properly anchor itself. A slight current is recommended.

Light: Moderate to dim light.

B./Rep.: Dioecious. Eggs and sperm are simply released into the water. The resulting larvae are planktonic until they settle on a suitable substrate. There they transform into an elongated primary polyp and develop into a new colony. There is alternation of generations, i.e., each generation alternates between sexual and asexual reproduction.

F.: O; various plankton. Since the whip sea pen is nocturnal, it should be fed at night. The polyps face away from the current; therefore the tentacles capture plankton from the countercurrent of the colony. Feed crustacean larvae, small crustacea, and diverse plankton substitutes such as scraped mussel, fish, and crustacean meat, minced worms, and moistened dry commercial diets.

S.: *Virgularia* sp. has a very tall stalk, short lateral branches, and relatively few polyps. Coloration varies from white and yellow. *Virgularia miralbilis* is a similar Caribbean species, but it is thinner and only grows to a height of 20 cm. There are additional similar species of *Stylatula* from the Caribbean and the Indo-Pacific.

T: 18°–24°C, **L:** 35 cm, **TL:** from 70 cm, **WM:** m, **WR:** b, **AV:** 2–3, **D:** 3

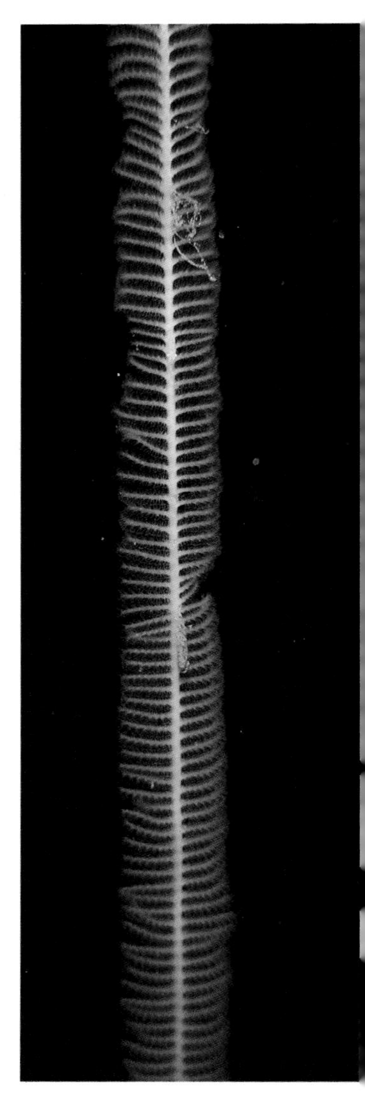

Plerogyra sinnosa

Coral Reefs

In ancient times the term "reef" was used by sailors to designate areas that were shallow and dangerous to sailing vessels. Just recently the word has come to have biological and geological significance.

Coral reefs are primarily calcareous structures. Though stony corals are the foremost contributors to reef formation, fire corals (*Millepora*) and two members of the subclass Octocorallia, *Heliopora coerulea* and *Tubipora musica* (blue coral and organ pipe coral, respectively), are also hermatypic.

In terms of productivity, complexity, and species diversity, coral reefs are second only to tropical rain forests. Due to their dependance on temperatures of at least 20°C and intense sunlight, reefs are restricted to clear, shallow, tropical and semitropical waters. Most coral reefs occur in coastal areas. The more than 700 coral species found in the Indo-Pacific give it the distinction of having the greatest number of coral species.

From a geologic standpoint, corals are of extraordinary importance because they gave birth to tall mountains and numerous mountain ranges over the course of geological time. Corals gave rise to dolomite; the Dachstein, Watzmann, and Eifel mountain ranges in Germany are evidence of this natural process. Many petroleum deposits originated from the organic remains of coral polyps. Even today, corals change the landscape as they form islands, creating new habitats for plants, animals, and man.

According to fossil records, the first coral reefs were formed about 350 million years ago. However, present day reefs only date back fifteen thousand years, a time frame that corresponds to the cessation of the latest ice age. With each ice age the polar ice caps grew; consequently the sea level dropped and coral reefs were exposed and died.

Based on formation, location, and shape in relation to their associated land masses, various reef types can be distinguished:

Fringing Reefs

Fringing reefs form parallel to coastlines, developing close to shore and growing seaward. Typically they develop in shallow water and are separated from the land mass by a narrow lagoon.
The extension of the fringing reef depends on the slope of the substrate—the steeper the substrate the narrower the fringing reef. Generally the upper elevations of these reefs correspond to the low tide mark. The reef flat and sections of the reef front are frequently intersected by deep channels and indentions, a result of wave action and water returning to the sea. This area of cuts and ridges is called the spur and groove zone. Beaches associated with fringing reefs are characteristically of sand and rubble. In tropical regions it is not uncommon for calcareous algae to consolidate the beach sand and rubble into beachrock—large, thick shinglelike plates that often cover kilometers of beach. The calcium from these plants is baked into the crevices of the beach material, leading to these rocklike structures. A fringing reef is basically subdivided into a reef flat and an outer reef slope. The reef flat frequently extends from the beach region to the edge of the reef. At that point the reef drops off into the depths of the sea, forming the outer reef slope or reef front. Most corals grow on the reef edge and outer reef slope.

Barrier Reefs

Forming parallel to the coastline but a great distance seaward are the barrier reefs. Structurally, barrier reefs are similar to fringing reefs, but the former typically has a deep, broad lagoon interspersed with patch reefs and is representative of a more mature reef. Barrier reefs are often so interrupted by channels that they resemble a fragmented ribbon (ribbon reef). Thanks to strong currents sweeping through the breaks, a rich coral garden is often found along the walls of the channels. Inarguably, the most familiar reef in this class is the Great Barrier Reef along Australia's eastern coast. This awe-inspiring sight is over 2,000 km long and exceeds 100 km at its widest points, making it one of the grandest if not **the** grandest animal construction on earth.

Atolls

Atolls fringe submarine volcanoes and mountain ranges, growing at the same rate that the substrate is collapsing. In the earliest stage of their evolution, atolls are fringing reefs of volcanic islands. As the island sinks, the fringing reef becomes a barrier reef, i.e., further from the coastline of the island with a respectable size lagoon in the intervening area. Eventually the island subsides until it is completely submersed, and a characteristic oval- to ring-shaped reef with an enclosed lagoon has developed. The lagoon may have a diameter of several kilometers. Atolls are common in the Indo-Pacific. Unlike the two previously mentioned reefs, atolls have rather distinct horizontal and vertical zones arising from such physical factors as the slope of the reef, amount of surge, depth, and the presence or absence of rivers.

Helmut Schuhmacher extensively defines and addresses these zones in his book "Korallenriffe" (Coral Reefs). This is an impressive reference for those interested in a more in depth look at reef zonation. Predictably, each of the above mentioned factors influences what zones are suitable for biological colonization. Delicate species are more commonly, if not necessarily, found on the inner reef slope or within lagoons where they are protected from the brunt of the sea's forces. A mind-boggling array of niches can be found in coral reefs, and animals and plants have risen to the challenge by adapting to what nature has provided.

Factors Influencing the Presence of Corals

Temperature

What governs the world's distribution of reefs? Since hermatypic corals depend on sunlight and a minimum temperature of 20°C, reefs are limited to shallow waters of tropical seas.
Within the narrow range of 26°–28°C, corals grow luxuriantly. Below 20°C corals, and therefore the entire reef, stop growing; this explains why areas with cold currents are bereft of reefs. Further,

temperatures above the listed range lead to cessation of growth or even death. Extreme temperatures, around 18°C or above 30°C, for short periods of time generally do not lead to permanent damage.

When maps of coral reefs are studied, it readily becomes apparent that the preponderance of reefs is found between the latitudes of 32°N and 32°S, corresponding to seas with an annual mean temperature of 23°C, the ideal temperature for coral reef formation.

Increased mass, i.e., growth, of a coral colony is achieved chiefly by asexual and/or sexual reproduction and polyp growth. Biotic and abiotic factors influence growth. As stressed above, water temperature is one of these factors. Since temperature is inversely related to depth—the deeper the water, the lower the temperature—hermatypic corals are not found in deep water. Hence, temperature is also a limiting factor for the vertical distribution of corals in tropical seas. Sixty meters, a depth corresponding to a temperature of 20°C, is considered the maximum depth at which hermatypic çorals can grow. Growth and reproduction are halted at temperatures below 20°C. Some corals inhabit deep water strata, but these corals not have symbiotic zooxanthellae; hence they are ahermatypic (non-reef-forming) and outside of this discussion.

Light

Light is an additional factor limiting vertical distribution. For reef-forming corals to grow and prosper, the requirements of their endosymbiotic zooxanthellae (unicellular algae) must be met. One of the main needs of zooxanthellae is light; light is essential to their photosynthetic processes. It follows that hermatypic corals are restricted to depths with adequate light penetration. It seems that 1% of the light intensity striking the water surface meets their needs. As depth increases, growth is drastically reduced. At 40 m, the light deficit is so great, growth is barely measurable. Illumination, or lack thereof, forces many corals to adapt. Planar and foliaceous growth forms predominate in deep waters because this shape exposes the greatest surface area to light, the limiting factor.

Oxygen

Dissolved oxygen is another important physical parameter. Only where there is sufficient available oxygen, i.e., the breaker zone and other shallow water areas, is growth assured. Under oxygen deficient conditions, coral polyps are able to use oxygen produced by their zooxanthellae; this oxygen is a by-product of the algae's photosynthetic process.

Nutrients

In some zones or biotopes, the lack of nutrients may restrict growth and even dictate the distribution of corals. Growth rate is directly proportional to the quantity of available nutrients. Corals are capable of utilizing a wide range of foodstuffs. Depending on species and their technique of food acquisition, the diet may range from zooplankton to dissolved inorganic substances, with zooplankton being the most prevalent. However, picture a food web or pyramid of productivity for a typical nitrate- and phosphate-poor tropical sea. In the first tier are the autotrophs or, in this case, phytoplanktonic organisms, the basis for most marine life. Unfortunately, the paucity of nitrates and phosphates holds phytoplankton populations in check. Hence, zooplankton and even corals are limited indirectly but very effectively by the lack of these nutritients. The symbiotic relationship with zooxanthellae is one way corals have adapted to life in nutrient-poor water. Physiologically, this is a very efficient alliance (See chapter on stony corals).

Water Movement

Water current greatly influences distribution and growth form. Only the most stalwart colonies are able to live along exposed regions near the surface, withstanding the force of the surf and current. As the surge diminishes at greater depths, delicate, highly branched forms become more predominant. Planar corals begin to appear as well, since light intensity is decreasing.

Virtually all hermatypic corals along any particular latitude live under equal conditions in regard to temperature, oxygen, salinity, solar radiation, and current; nevertheless, significantly different growth rates can be determined when species are compared. The fastest growing corals are the branched forms of *Acropora, Porites,* and *Pocillopora.* These colonies are found along brightly illuminated sections of the reef flat and upper reef slopes where there is ample current, solar radiation, warmth, and food. Other forms grow slower, and they may even stop growing altogether once they have achieved a certain dimension. Growth form and growth rate of the future colony is determined by where the planula larva settles.

Coral Growth and Development

Because corals could not be maintained alive and healthy over extended periods of time in captive environments, their life was largely a mystery. The question of how and at what rate corals grow has been the object of extensive studies. Thomas Goreau's innovative work answered many of these questions. In Jamaica he used radioactive calcium ions in the seawater and measured the rate that tracer ions were assimilated in the coral's skeleton. Since corals cannot distinguish between a stable and a radioactive ion, both isotopes were incorporated into the skeleton in the same relation as their concentrations in the seawater. Using these methods, it became possible to gather data on coral growth. Today corals are left in their natural biotope and growth rates are determined with underwater scales or by using pigments such as Alizarin Red S.

The majority of corals grow in close proximity to one another along the reef. Competition for food, light, and space is intense. To persevere, several survival mechanisms and strategies have developed. Nematocysts, growth inhibitors, and toxins are used to keep other corals at a distance, thereby ensuring that individuals have enough light, space, and unpolluted water. Sometimes these measures are inadequate to prevent more aggressive or faster-growing species from encroaching and even overgrowing more docile organisms. Interaction between corals and corals and their environment is intricate, and as of yet, little understood.

In recent decades, many marine biologists have attempted to unlock some of the secrets surrounding these primitive, mysterious animals. Emphasis has been placed on covering and defining the relationship that corals have with many other reef organisms. Certain terms have been adopted to explain these alliances. The terms endobiota and endosymbiosis identify the symbiotic relationship of unicellular algae living within the tissues of coral polyps (zooxanthellae). This is probably the most intimate association that corals have with another living entity. Boring organisms, i.e., boring bivalves and boring worms, are part of the cryptobiota. The boring activities these organisms engage in contributes to reef erosion. Plants and animals such as small snails, bivalves, and brittle stars that live in deep, dark crevices and at the base of dense coral branches are called hypobiota. Organisms that encrust corals and coral skeletons are called epibiota. These include epiphytic plants, sponges, molluscs, small anemones, and colonial anemones. Parabiota is a term to describe those organisms that occasionally seek shelter between the coral's branches, e.g., small crustacea and fishes. Most fishes are peribiota—they always live close to corals, but they are not intimate associates.

A coral reef, as long as it is still intact, usually represents a stable ecosystem. For continued stability, the sum of the primary production must be greater than all the combined forces that contribute to the reef's decay.

Fishes, bristleworms, sea urchins, starfishes, mussels, snails, sponges, plants and bacteria are some of the organisms that are harmful to the reef. Even the quality of the water, changing currents, municipal sedimentation (trash), and human population affect the equilibrium of this sensitive ecosystem, often deciding whether a reef flourishes or dies. Unlike most other animals, corals are sessile organisms. As such, they are at the mercy of their environment and predators.

In recent years, sections of the northeastern extension of Australia's Great Barrier Reef off the coast of Queensland have been decimated by *Acanthaster planci,* the crown-of-thorns starfish. This echinoderm underwent a population explosion. Because the crown-of-thorn starfish's diet is coral polyps—and startling amounts of them—increased numbers of this thorny, very large starfish have resulted in large reefs being grazed barren. To get an idea of the magnitude of the problem, consider that each full-grown starfish daily consumes about one square meter of coral polyps and as many as 14,000 starfish have been counted within one square kilometer.

Warming and pollution of the world's oceans seem to be weakening coral populations, making them susceptible to diseases. Black-band disease has become more prevalent in recent years. The causative agent of this disease is a blue-green alga that undergoes a sudden population explosion. The coral behind the black line is dead; hence its white color. Scientist are puzzled by the cause of this proliferation.

In modern times, man has significantly contributed to the irreparable damage of the delicate equilibrium of coral reefs of Australia, the Maldives, the Persian Gulf, and the tropical Pacific. Calcium rock is excavated to produce cement to construct streets and dwellings and to make fertilizer. With increasing industry, agriculture, and mining, the pollution of coastal regions has rapidly increased in recent years. The expanding tourism industry, fishing with dynamite, and the disposing of refuse into the sea all contribute to the declining health of the ocean's ecosystems. Perhaps releasing human and industrial sewage into the sea has the most damaging impact, however. It is loaded with pesticides, herbicides, and nutrients. Blue-green and green algae take advantage of this "fertilizer" by reproducing explosively. This leads to overgrown reefs, a condition that eventually results in the demise of the reef. Little imagination is required to picture coral reefs in the future if certain practises are not checked. Many reefs have succumbed to careless collecting and mining processes. They are now overgrown by algae and secondary colonizers. Slime algae drift in the current and smother all life in their wake. The tally has begun: if the corals die, the entire ecosystem dies! Man has unwittingly begun to destroy a unique natural creation.

Maldives

THE BEACH

Beaches are areas of shore that have a slight gradient and an unconsolidated sand and/or gravel substrate. Typically, beaches occur along peaceful bays, lagoons, or shorelines of small, sheltered coral islands. Sediment is brought by the surge and deposited along the beach. Waves continuously wash over the shore, keeping upper sediment layers in motion and further pulverizing them.

Maldives

Beachrock

Beachrock is a hard, rocklike substrate that forms in the tidal zone. Flat plates of sand and rubble mortared with calcareous algae cover the bottom. These plates may be as much as one meter thick and a kilometer in length. Everything the sea deposits is baked into one mass—bivalves, snails, corals, glass, beverage cans, and many other products of civilization. Why and how beachrock is formed has not been thoroughly investigated.

Maldives

THE LAGOON

Lagoons are bodies of water of variable depth and breadth between the beach and the reef moat. Their sand substrate is dotted with small patch reefs and coral skeletons that have been broken from the reef moat and carried into the lagoon by waves. This sheltered region is home to a large community of plants and animals that have successfully adapted to the sandy substrate. Green and brown algae commonly grow therein. Occasionally large seagrass meadows are found within the lagoon; these areas harbor a unique faunal community. Because of the high rate of sedimentation, few if any corals reside in this biotope. The small number that can survive are tolerant of a certain degree of sedimentation.

Barrier reef in the Maldives.

The Reef Moat

Because the reef moat falls dry during low tide, its population is limited to the most robust plants and animals. They must be able to withstand extreme biotic and abiotic factors such as desiccation, heavy surf, and severe fluctuations in salinity and temperature. Corals are generally intolerant of these conditions. Depressions in the reef moat fill with sand and coral fragments. During low tide these areas are tidepools that support reef organisms that failed to reach the open sea as the tide receded. Because fine sand and sediments are carried in during high tide, there are expansive sections of soft sandy substrate.

Australia's Great Barrier Reef.

The Reef Flat

The horizontal surface between the reef moat and the reef slope is called the reef flat. It has a variable extension, and it may occasionally fall dry under extreme water conditions. The reef flat is predominantly colonized by strong corals capable of surviving powerful surf. The upper sections of the coral colony that have been exposed are generally dead, while the lower, submersed parts continue to grow, creating "microatolls." As breakers hit, the water dispels its energy and begins to flow back to the sea. This runoff cuts channels into the reef flat. Algae and seagrasses will grow on the sand patches and in the deep rocky clefts that have resulted from erosion. The stout branched *Acropora* in the above photo are colonizing a reef flat along the Great Barrier Reef. Young reef sharks hunt within this coral labyrinth. A great number of small fishes, snails, and crustacea reside therein.

Maldives

The Reef Crest

Separating the reef from the sea is the reef crest, the pinnacle of the reef. This zone is hit with the full brunt of the surf. Seaward, the reef crest forms a heightened reef edge. Stalwart corals able to withstand incoming waves grow near the low tide line. Green algae and calcareous algae, the mortar of beachrock, are particularly prosperous along these sections of the reef. The shallow water translates into intense solar radiation and warm temperatures, two conditions favorable for growth. Fragments of calcareous algae baked into coral rock add to the reef's integrity and increase its ability to resist erosion.

Ribbon Reef, Australia.

The Reef Edge and the Spur and Groove Zone

The reef edge is a narrow region, a transitional zone between the horizontal reef flat and the steep outer reef slope. It has buttresses and grooves that follow the entire outline of the outer reef. Since it is below the intertidal zone, the reef edge is only exposed to air during extreme low tides. Growth is constant and directed seaward. The majority of the coral population is comprised of species and forms that thrive in current and strong surge.

Great Barrier Reef, Australia.

The Reef Slope

The reef slope is the seaward face of the reef that descends—abruptly or gradually—into the sea. It is divided into two distinct regions; the upper and lower fore reef. While rubble and debris are the predominant substrate at the foot of the slope, they gradually yield to sand expanses further away from the wall. On the upper reef slope, where light and fresh water are abundant, corals grow luxuriantly. Robust forms are prevalent due to the surge. More filigree corals grow along the lower reef slope, where the currents are weaker. The deeper the reef slope, the less solar radiation penetrates and, consequently, the sparser the coral population.

Great Barrier Reef, Australia.

THE REEF CHANNEL

Reef channels are the result of the force of the surf and incoming and outgoing tides, principally along the reef flat and the outer reef slope. Channels may even transect the reef, opening a passage for water to flow into and out of the lagoon unhindered. The intense light and strong currents bathe the corals in a bounty of nutrients and oxygen, fomenting exuberant growth.

Lizard Island, Australia.

The Fringing Reef

Fringing reefs are characteristically narrow extensions of the shoreline, or they may be separated from the coast by a shallow, coral-sand-and-rubble-bottom lagoon. The reef constantly grows seaward. Eventually, given correct conditions, a barrier reef develops from the fringing reef. Because of currents and surf, the reef front is sloped and grooved. The breadth of the fringing reef depends of the slope of the sea floor.

Painted coral pieces are commonly sold as souvenirs in markets of southeast Asia, the Maldives, and Australia. Harvested by divers, the natural coral is sold to small tourist shops that clean and paint the pieces. Entire reef sections are damaged by these activities, as the reef is incapable of regenerating at the same rate damage is incurred. A large proportion of the corals are broken during transport and are unsuitable for retail.

Unfortunately, coral rock is the only suitable building material found on many coral islands. Broken or blasted from reefs, the coral is used with mortar to construct edifications. Since the demand for housing is ever increasing and coral islands are often ideal sites for tourism, this activity significantly contributes to the demise of large sections of coral reefs.

Black-Band Disease

The causative agent of the black-band disease is *Phormidium corallyticum*, a blue-green alga. It attacks sick or weakened polyps of shallow water stony corals, slowly but insidiously conquering the entire colony. The border of the disease is visible as a dark band. As the band advances, a dead white coral skeleton is left in its wake. Rapid-growing secondary colonizers such as algae, sponges, and colonial anemones quickly establish themselves on the dead coral. It has been determined that water pollution does not foment the disease. The existence is a natural occurrence.

Calcareous Algae

Calcareous algae secrete calcium carbonate in the form of aragonite between their cell walls or on their thalli. While stony corals are indisputably the source of most of the reef's calcium carbonate, calcareous algae deposit significant quantities as well. But calcium carbonate deposition is not their only contribution to the reef; these algae mortar the reef together as they settle into cracks and crevices, thereby increasing the reef's overall integrity. The above photo is of a shinglelike calcareous red alga at 30 m depth covering a large area of dead coral rock. Many herbivores break and grind calcareous algae as they graze.

ORDER SCLERACTINIA

Anatomy

All stony corals excrete a calcium carbonate skeleton. These skeletons are the foundation and building blocks of coral reefs. A few scleractinian corals are solitary animals; however, the great majority are colonial.

Each animal, or polyp, is cylindrical and sac-shaped with two layers of tissue—the epidermis and the gastrodermis. A thin supportive mesogleal layer lies between the two. At the center of the cylindrical polyp is the gastrovascular cavity. The oral/anal orifice is on the superior end of the gastrovascular cavity, providing an opening to the environment. The gastrovascular cavity of every polyp is subdivided by at least 6 mesenteries, each of which is adhered to the basal plate and the oral disc. Their function is to increase the surface area of the gastrovascular cavity and secrete digestive juices to dissolve ingested food, rendering it absorbable. Long tubelike structures, mesenteric filaments, further contribute to this process. Foods that are too large to enter the gastrovascular cavity can be digested extragastrically by the mesenteric filaments. The mesenteric filaments are expelled through the mouth or the body wall for this purpose.

All polyps of the colony are covered by a thin tissue layer and interconnected by a richly branched vascular system. Hence, the nutrients captured by one polyp are food for the entire colony.

One or several rows of tentacles are arranged along the circumference of the circular oral disc. Occurring in multiples of 6, the tentacles are armed with batteries of nematocysts. Nematocysts, one of the most complex secretory products of the animal kingdom, are predation and defense organelles. Each species has characteristic nematocysts. Upon tactile stimulation, nematocysts discharge explosively from their capsule, stunning and capturing small planktonic organisms with adhesive threads. Spent nematocysts are quickly regenerated.

Reproduction

Scleractinians reproduce both sexually and asexually. Sexual reproduction involves the release of eggs and sperm and the formation and of a planula larva. Some species may be capable of producing both gametes within the same colony—hermaphrodites—while other species have both male and female colonies—dioecious. In hermaphroditic animals, egg and sperm cells mature at different times, precluding self-fertilization.

The gonads, the organs which produce the gametes, are found on the polyp's mesenteries. Depending on species, fertilization is external or internal. The former involves a simple release of sperm and eggs into the open water. In the case of internal fertilization, sperm cells must reach the gastrovascular cavity of a conspecific colony to fertilize the eggs. Once released into the environment through the oral opening, sperm actively swim and use currents to carry them to the mouth of a conspecific polyp.

A ciliated, free-swimming planula larva develops from the fertilized egg. From the time the larvae emerge from the mother polyp until they settle onto the substrate, they are plankters. The duration of this phase is highly variable—anywhere from days to weeks or months—and is crucial to the species' geographic distribution. Clearly, the longer the larvae drift, the further they are carried by currents and waves. Planula larvae are extremely vulnerable to predation and few survive.

The process by which larvae adhere to the substrate has only been observed and studied in the laboratory, never in their natural habitat. After the larva establishes itself, it forms a tiny primary polyp which expeditiously begins to secrete calcium carbonate and lay its first septa on the basal plate.

Sexual reproduction, which distributes the species, alternates with an asexual phase, which increases the population locally. There are two methods of asexual reproduction—fission and budding, or intratentacular and extratentacular division, respectively. Fission is by far the less common process, one almost exclusive to solitary species. The great majority of species produce colonies by budding. As the mother polyp buds, the resulting daughter polyps often remain attached by the coenosarc.

Nutrition

Tropical seas are poor in nutrients. Concentrations of dissolved organics vary from habitat to habitat and with the seasons, but normally only trace amounts of these substances, i.e., amino acids, carbohydrates, and fatty acids, are present.

In corals, like in other animals, nutrient intake is the key to species perpetuation. In view of the paucity of nutrients in their environment, it is obvious that corals had to develop many different mechanisms of food acquisition to compensate. Each of these methods are used autonomously or in seemingly random combinations.

Some animals are predatory and use nematocysts to capture their prey. Because corals are sessile animals, they depend on currents to supply sustenance and remove waste products. Zooplankton that drifts near corals is very efficiently seized by the nematocyst-bearing tentacles. Corals typically feed at night in accordance to the daily vertical migration of plankton. At night when plankton rises to the surface from the water's depths, polyps and tentacles extend and the organisms begin to feed. When polyps are not feeding, they are usually retracted into their corallites (depressions of the calyx). Divers that peruse the reef at night will see a much altered vista from that seen during the day. The size of the coral's prey is limited by the dimensions of the polyp and the extension of its tentacles; it follows that the diameter of the polyp is indicative of the size foods it is capable of dealing with. Only giant polyps of mushroom corals (*Fungia*) are able to capture small crustacea, bristleworms, and small fishes. The prey is introduced into the gastrovascular cavity through the mouth, absorbed by the cells of the gastrodermis, and digested in the feeding vacuoles. Indigestible items are expelled through the oral opening.

The entire body of the coral is capable of absorbing dissolved organic compounds, opening yet another avenue by which corals procure nutrients. Some of these compounds are ingested from the surrounding seawater, while others are by-products of the photosynthetic activity carried out by the coral's symbiotic zooxanthellae.

Zooxanthellae

Zooxanthellae, single-celled algae from the family Dinophyceae, help corals precipitate calcium carbonate, allowing them to form a skeleton at a substantially faster rate than corals that lack these symbiotic organisms. In fact, it has been calculated that corals with zooxanthellae deposit calcium carbonate ten times faster than corals without zooxanthellae. It has been hypothesized that the pH within the corals tissue increases as the zooxanthellae use available carbon dioxide during photosynthesis. An higher alkalinity is more favorable to skeletogenesis. Since zooxanthellae are plants, and therefore dependant on sunlight for photosynthesis, the advantages corals derive from this mutualistic beneficial relationship are relative to the quantity (and quality) of light available to the zooxanthellae. Significant numbers of hermatypic corals are solely encountered in shallow water, where light penetration is optimal.

Because zooxanthellae are so important for hermatypic corals, it is not surprising that they are passed from generation to generation through the planula larvae. These unicellular algae either adapt to one or several species. It has been determined that about 10 percent of the polyp's tissue mass is zooxanthellae. Almost one million of these tiny algae live in one square centimeter of tissue. The population is remarkably stable—a balance between less active algal cells being expelled and the rate of division within the polyp.

Zooxanthellae and the polyps constantly exchange nutrients to their mutual benefit. The unicellular algae use the sun's energy to produce carbohydrates and oxygen from carbon dioxide and water through photosynthesis. The oxygen is used by the polyps for respiration and metabolism. The algae can metabolize the corals' catabolites such as nitrate, phosphate, sulfate, and ammonia through short energetic pathways. It is obvious that this relationship evolved as a solution to the nutrient-poor environment of tropical seas. But it was the zooxanthellae's positive influence towards calcium carbonate deposition that provided for the successful evolution of corals; coral reefs are only present where destructive biological and chemical processes are outpaced by calcium deposition.

Under unfavorable, stressful conditions, e.g., long periods of darkness, abrupt temperature changes, or exposure to high concentrations of freshwater for extended periods of time, the zooxanthellae can be jettisoned from their host. When the algae leave their host, they grow a flagellum which enables them to swim. Under extreme situations of nutrient deficit, it seems that polyps convert from carnivores to herbivores and feed on their symbiont zooxanthellae.

Non-reef-forming corals (ahermatypic) lack algal symbionts; therefore, their habitat is extended beyond that of their hermatypic brothers to include areas without light. They are present in all seas and at virtually all depths, inhabiting shaded caves and areas beneath ledges. Since their rate of calcium assimilation is very low, they have insignificant skeletons and play no role in reef formation.

Corals in the Aquarium

by Dietrich Stüber, Berlin

The thought of successfully keeping a tiny piece of a coral reef in one's own living room has enticed many into the marine aquarium hobby. Using a few axioms, this art is within the grasp of all. The technical problems and their rather simple solutions were extensively addressed in the Marine Atlas Vol. 1. The following section will contain brief anecdotes concerning the particularities of the "reef aquarium hobby." Please take each account subjectively. It is very fulfilling when a particular technique leads to a healthy microcosm. Carefully select compatible tankmates and infrastructure that complements and fosters their lifestyle.
Every aquarium animal has a precious intrinsic value, meaning that each and every hobbyist has a moral responsibility to educate himself/herself well before making that first purchase. Many species are protected from extinction by laws. The ultimate goal of the aquarium hobby is not to stress the natural state of the environment, but to provide valuable information on species requirements and, when needed, a refuge for those animals whose natural biotope has become uninhabitable because of pollution or other environmental change, thereby preserving the species, at least in captivity.

Particularities of the Aquarium Hobby

For an encompassing understanding of the biological processes occurring in a coral reef and a reef aquarium, a specialized foundation of knowledge is needed. Supplying oneself with this information is easy in view of the abundance of literature on the subject. However, corals and their complex series of interactions within the community and the environment are so multifarious, even scientists are often befuddled. Because of the lack of long-term data, many aspects of coral reef biology are only theory. Reef aquarium keepers are in an uniquely advantageous situation; they can intimately observe a miniature coral reef biotope in the comfort of their own home, noting development, each species' demands for optimal growth, and the relationship they establish with their tankmates. The marine aquarium hobby should be guided by facts and science and motivated by more then mere entertainment and ephemeral novelty.

Gas Balance

The composition of dissolved gases in salt water, whether it be in the home aquarium or the open ocean, has the same relevance for marine animals that components of air have for us. The ratio of oxygen, nitrogen, and carbon dioxide is of particular importance. The balanced presence of these gasses in the aquarium is a precondition for successfully maintaining marine animals. It has been proven that the solubility of these gases depends on water temperature. As the temperature falls, solubility rises. Likewise, the solubility decreases with a rise in temperature. Temperatures exceeding 25°C may lower the dissolved oxygen level to such an extent, that oxygen becomes the limiting factor for stocking density. Cooler aquatic environments present better conditions for assimilation and biological processes. However, plants and animals are tolerant of just a narrow range of temperatures. When stocking an aquarium, especially with sessile invertebrates these criteria and water current are of paramount importance. In captive environments, plants can drastically alter the amounts of dissolved carbon dioxide and oxygen, thereby affecting some chemical processes, i.e., the carbonate/bicarbonate buffer system. Algae tend to grow vigorously under intense illumination. As the plants assimilate energy—photosynthesize—oxygen levels rise, possibly stabilizing dissolved oxygen. Unfortunately, photosynthesis uses free carbon dioxide, thereby lowering the concentration of dissolved carbon dioxide in the system. As a consequence, calcium solubility decreases and carbonate hardness is diminished. Corals and other animals with calcium carbonate skeletons rely heavily on carbonate hardness for continued growth. Good aeration is vital to bring carbon dioxide levels towards equilibrium, but carbon dioxide may have to be added as well. A protein skimmer and a good current further ensure stable oxygen levels for sessile invertebrates.

Because algae in coral reef tanks often overgrow sessile invertebrates and are especially difficult to control in this biotope, strictly regulate their presence.
Where marine organisms settle and how successfully they colonize particular sites is largely contingent on prevailing abiotic factors, or inorganic and physiographical conditions. In the sea, the most important conditions determining species distribution are the geological substrate, temperature, salinity, current, and nutrient levels. These parameters are particularly relevant for sessile animals because of their stationary lifestyle and their limited ability to physiologically adapt. Sessile animals that depend on currents to deliver foods such as plankton or dissolved nutrients are especially sensitive to environmental changes. Currents and salinity greatly influence shape, size, and distribution of the sea's inhabitants.
Whether animals taken from their natural environment thrive, merely survive, or parish largely depends on two factors: the animal's ability to adapt and how closely conditions in the aquarium follow those of its natural biotope. Many species, once they pass a certain state of maturity, have a reduced ability to adapt.

Illumination

Light, through plants, represents the basis of nutrition of many invertebrates, and it is the source of regulatory mechanisms that yield a state of equilibrium. If we consider the artificial illumination above the aquarium a substitute for the sun, then the essential nature of the light source becomes apparent. The optimal illumination intensity for reef aquaria corresponds to 30,000–35,000 lux at the water's surface (over the largest area possible). Corals and other light-dependant animals should be placed at least 10 cm below the surface, given the above conditions. Significantly higher or lower intensities are detrimental to higher algae and animals that live in symbiosis with algae.
At present, OSRAM's metal halide "D" is the only source on the market that satisfies the demands of corals in regard to spectrum and, most notably, intensity (although continued progress in illu-

mination technology will certainly offer additional alternatives by the time this book reaches the market). Intensity has particular importance because factors such as water depth and turbidity can dampen it, compromising the amount of energy reaching the plants (zooxanthellae) for photosynthesis. OSRAM's 67 actinic tubes or PHILIPS TLD 18 used concurrently with OSRAM's metal halide "D" may be advantageous, though I personally have never noted improved growth or survival.

Filtration

Filters are an irreplaceable aid for aquarists. The basic function of filters is to supply a hospitable environment for *Nitrobacter* and *Nitrosomonas*, genera of bacteria that are essential in converting toxic ammonia into less poisonous nitrate. Concentrations of the latter should be kept between 20 and 40 mg/l in the reef aquarium, an impossible task using filtration methods alone—periodic water exchanges will be needed. Other biological waste products that can accumulate problematically in the tank are phosphates. These ions can be rendered insoluble by adding calcium water. The calcium binds to the phosphate, and the resulting calcium phosphate falls out of solution. Nutrient levels in the aquarium should be kept on a par with concentrations in the sea.

Algae

Algae cannot be totally eradicated from the marine aquarium. It is illogical to combat them with either technical or chemical methods, since after all, they represent the basic dietary component for many microorganisms and filtering organisms. Additionally, they stabilize the biological equilibrium by using dissolved metabolites. Massive proliferation due to overfeeding, high stocking densities, lack of current, or excessive light must be avoided since the well-being of sessile animals as well as more complex algae (food competitors) could be hampered. One of the best methods to control algae is to keep the biotope in equilibrium—maintaining a balance of producers and higher tropic levels (consumers).

It is very difficult to imitate or harness many of the very complex natural cycles in captive environments. Therefore, all life and biological cycles established in our aquarium should be respected and disruptions avoided. Infrastructure should follow features from the natural biotope to the best of our capabilities. Following the above prescribed measures, a stable aquarium biotope can be achieved.

Basic Considerations for the Maintenance of Hermatypic Corals

Caring for hermatypic stony corals in the aquarium is very difficult. Some of the problems are technical others are rooted in keeping different species in a small, confined area. Providing a balanced diet with the proper amounts of trace elements presents yet another difficulty.

Caring for stony corals is admittedly not an easy endeavor, but it is well within the realm of the possible. Successful aquarists stand ready to attest to their achievements, but despite proof, there are still many skeptics.

Corals have ingeniously developed different feeding strategies. The methods each coral species has adopted depends on the environmental pressure, i.e., the means that yield the best survival within the coral's particular habitat. There are five methods corals use to acquire nutrients:

- Capturing planktonic organisms. Unless the tank is mature and has a rich community of microbes and a limited number of fishes, corals will be unable to utilize this method of food acquisition in aquaria.
- Consumption of large pieces of food. This feeding technique is primarily observed in solitary and bladder corals. The food is transported to the oral disc by the tentacles and digested in the gastrovascular cavity.
- Ingestion of microscopic food particles. As fishes eat, particulate leftovers fall from the water column onto the coral. The colony coats the particles with mucus and digests them extragastrically. Comparable situations may occur in the aquarium.

- Absorption of dissolved organic compounds. The entire surface of the organism is capable of absorbing dissolved organic compounds. It is assumed that corals utilize this feeding method in aquaria, though it has not been proven.
- Exchange of metabolic products between zooxanthellae and corals. A predominant proportion of a coral's nutritional requirements is fulfilled by their symbiotic zooxanthellae. Using light energy, carbon dioxide, and water, zooxanthellae produce oxygen and organic compounds, two useful products for corals. Because of the removal of carbon dioxide, the pH in the coral's tissue rises, creating a more favorable environment for calcium carbonate deposition. This mode of nourishment is arguably the most complex and the best known.

In years past, it was impossible to maintain stony corals in an aquarium. We now realize that the failure was due to inadequate light sources that did not provide the correct spectral composition. The coral's health is dependant on the well-being of its zooxanthellae. With the advent of bulbs that provide light comparable to that of sunlight, zooxanthellae, and therefore hard corals, can be successfully kept in captivity. The relationship between zooxanthellae and corals is not unique. Many invertebrates, i.e., the giant clam and colonial anemones, benefit from a like relationship with algae. What was not realized is that algae can meet all of their host's nutritional needs, making them virtually autotrophs. This greatly eases care in an aquarium. Water quality must be exemplary in an invertebrate tank, but not having to feed these animals lowers the quantity of nitrogenous compounds introduced into the captive biotope and largely reduces the load on the system's filters.

The entire dynamics of nutrient exchange and the biochemical adaptations that allow the algae to live within animal tissue are extremely developed and can be disrupted by imbalances in water chemistry, possibly resulting in death of both the algae and the host.

It is the tiny—microscopic, in fact—unicellular algae called zooxanthellae that grant corals their status as autotrophs. A large population of these algae reside in the tissues of reef-forming stony corals. Intense sunlight is one of their principal requisites. Zooxanthellae offer the delicate coral tissue effective protection against UV radiation. In turn, zooxanthellae are well protected from predators and have access to the host's metabolites, using them as nutrients. Carbon dioxide, a by-product of respiration, and nitrogen and phosphorus, wastes from digestion, are the chief contributions to the algae. Nitrogen and phosphorus are scarce in the nutrient-poor tropical seas, but an essential commodity for algae. Carbon dioxide in conjunction with light and water is converted into sugar and oxygen, two products of interest for the coral. As you can see, each benefits nutritionally by this interaction. Energy is passed efficiently from producers (zooxanthellae) to primary consumers (corals).

Special attention must be directed to providing corals with trace elements. For proper skeletogenesis, hard corals demand supplements of calcium, strontium, and iodine.

Fishes and other invertebrates receive sufficient quantities of these elements through water exchanges and their diet. Adding calcium through kalkwasser will help circumvent a cessation of growth.

Strontium and iodine are available in prepared formulas. Corals utilize strontium to keep aragonite in soluble form until it has been transported to the matrix, the site of calcium deposition. Hence, it is strontium that keeps the membrane pores leading to the matrix free of deposits. Dosage, which is primarily dependent on water volume and the population of calcareous algae and skeleton-building animals in the biotope, must be determined through trial and error.

Success or failure pivots on more than just physical and chemical conditions. Observation and recognition of interactions are of primary importance. Attentive care is a fundamental requisite, especially in regard to newly introduced animals. Is the coral adjusting to its new local? If the polyps are extended, then the surroundings are favorable, and the environment should be stabilized in such a state. Current, light, temperature, and other technical aspects of the hobby can be altered to accommodate the organism. Under no circumstance should the organism be asked

to adapt to the conditions. Damage incurred from improper current, light, etc., will prove fatal if corrective action is not promptly taken.
Sooner or later, the corals will begin to grow and force us to cull the invertebrate population to insure room for the flourishing reef. As their biomass increases, the aquarium's chemistry must be increasingly tailored to their needs.
Interactions among corals and other organisms also determine whether a coral thrives or wanes. How carefully tankmates are chosen and—where applicable—arranged will be expressed in the ultimate failure or success of the aquarium. A parrotfish or an aggressive tube anemone can be as fatal to corals as filter failure. Dead is dead.
Life in a tropical reef is often brutal. Space is at a premium. Corals have developed mechanisms to help them compete for and maintain their space. Being overgrown or stung to death is the consequence of failure. The number of nematocysts the animals deploy and the potency of the toxin delivered by the nematocyst are key to the battle. Aquarists cannot quantify the strength of the toxins; therefore, he/she is reduced to observation. Effects can only be gauged when tankmates begin to show a negative response to close proximity. The wisest course of action is to arrange the animals in groups with each group comprised of similar species. A large safety distance should be observed between the groups and around poisonous animals.

To Summarize

For the successful care of hermatypic corals, a mature aquarium that is at least 2 years old with relatively stable water chemistry is necessary. Stability chiefly refers to the nitrate concentration, salinity, °KH, and pH. The nitrate concentration should be between 20 and 40 mg/l. Avoid strong fluctuations. It is unnecessary to attempt to maintain the nitrate concentration near 0 mg/l, since experience has shown that hermatypic corals will not grow optimally under such conditions. The carbonate hardness must be kept as stable as possible, not falling below 6°–8°KH. Check the calcium concentration often. It ought to be approximately 400 mg/l. Always measure the calcium levels when carbonate hardness falls below 6°. Additionally, pH should be 7.8–8.3, temperature around 26°C, and specific gravity 1.024. Filamentous and smear algae should be absent or present in minimal quantities, and the fish density should be moderate. A protein skimmer, a current adapted to the animals, metal halide "D" illumination, and a 10 hr photoperiod are also imperative.

Nutrient Requirements of Hermatypic Stony Corals

Stony corals have developed several feeding strategies to better their survival in nutrient-poor tropical seas. Hermatypic corals occur in sunlight exposed regions of warm, oxygen-rich waters. There they subsist on nutrients found or produced *in situ*, since corals are sessile organisms and cannot search for more favorable locals. As explained earlier, many invertebrates harbor zooxanthellae in their tissue. In the case of stony corals, this symbiosis allows the corals to subsist on little more than sunlight, water, and oxygen, virtually making them autotrophic. This highly specialized form of nutrient acquisition plays a dominant role in the life processes of hard corals. Since this symbiosis is a very complex process, it is logical to assume that it would only be engaged during emergency situations, e.g., a shortage or absence of nutrients.
Today, I am convinced that stony corals can be successfully kept in aquaria only because of their zooxanthellae. True reefs can now be maintained in our living rooms. The digestibility of substitute foods has always be in question. Furthermore, the constant addition of feed in the form of zooplankton and other substitute foods of animal origin pollutes the water excessively. Even rigorous water exchanges, as repeatedly recommended in the past, are not a solution. Although important for other reasons, in the end they represent only a partial solution.
It is known that polyps of the genus *Xenia* execute pumping motions that have been interpreted as capturing food, but it has been

determined that these organisms are not capable of actively feeding. This misinterpretation has been to the animals' detriment, as aquarists "feed" the animal, thereby polluting the water. *Xenia* subsist quite well on light alone with the help of their symbionts.
To settle the question of whether stony corals can subsist on light alone, I maintained corals for an extended period of time without offering food. Another aspect of the trial was to determine if after 6 years the coral would again accept food once it was offered. For this experiment I chose corals with large polyps where feed intake can easily be determined visually, in this case a member of the genus *Pavona*. The coral had only been fed during the first year of its aquarium life. Since 1985, that is, for the last 6 years, the same coral has survived in my aquarium without receiving additional food. I later attempted to reactivate its feeding activity with various foodstuffs (*Mysis*, infusoria, dry foods, tablet foods, mussel meat). Although it was clear that its feeding reflex was normal, it had "forgotten" how to eat. The study conclusively proved that these corals can live for many years without foods from outside sources. But *Pavona* is not the only coral that I have maintained successfully without feeding; *Montipora monasteriata* is a hermatypic coral very different from *Pavona* in that it has small polyps and a different growth form, yet it, too, has survived for many years without foodstuffs. Therefore, I will clearly state that contrary to other opinions, it is unnecessary to feed significant amounts of foods to captive stony corals! Growth of corals in the aquarium is not primarily dependent on external nutrient availability. Space competition and water pollutants are likely to be of much greater influence.

Reproduction of Stony Corals

While corals can easily be observed growing, it is extremely rare to see a new colony develop from a planula larva. I have been fortunate enough to have witnessed the development of two corals colonies, *Seriatophora* and *Stylophora,* from the planula larval stage. The larvae of the *Seriatophora* coral adhere to the substrate at sites exposed to strong current. Until the coral has established a 10 mm diameter base, growth is encrusting. It takes 3 months for the coral to initiate noticeable vertical growth; at that time the hump practically divides in two. An additional three months are needed until the humps take the shape of two 15 mm branches. The base dies, but additional fine branches are produced, giving form to the new colony. Although the new colony is delicate and brittle, it is extremely adept at using its nematocysts to defend itself against space competitors. Once it has become established in the aquarium, it grows to a respectable height of 15 cm within 3 years. *Seriatophora* corals can be propagated by breaking branches off the parent stalk and securing them elsewhere in the aquarium. Ensure that the new site is exposed to a suitable current. Since illumination seems to be unimportant, we can deduce that this coral does not subsist solely on the metabolites of its symbiotic zooxanthellae. Experiences with other corals, namely *Acropora* and *Stylophora*, support this theory. Therefore, *Seriatophora* should be offered pulverized foodstuffs on occasion. Recently I anchored branches of *Seriatophora* to the bottom substrate of the aquarium and, in the process, reaffirmed the organism's scant need of light and discovered that the coral adapts well to different currents. On the bottom of the aquarium, the coral developed a more compact growth form. It even seems unimportant if the coral is pushed over by bottom-dwelling animals. Since they continued to grow as long as the current was sufficiently strong, it is suspected that on the bottom, a biotope rich in suspended mater, they have developed yet another feeding strategy in response to their environment. The viability of this type of propagation should be considered a positive sign. By all indications, *Seriatophora* is capable of successfully propagating under different environmental conditions, an especially advantageous trait as no two aquaria have identical chemical and biological conditions. The lifestyle and the different reproductive alternatives of corals still harbor many mysteries, and, the discovery of new, still unknown means of development and propagation are in store for us, thanks to our new found possibility to witness these events in an aquarium right in front of our very eyes.

Fungia corals are a particularity among stony corals, differing in shape as well as lifestyle from the majority. After separating from their stalk, *Fungia* corals live unattached on the substrate. Their ability to clean themselves of sedimentation is a clear adaptation to their habitat. In the past years, I have maintained and observed *Fungia* and *Herpolithon* species in my aquarium. In the aquarium they propagate by two different methods, both of which are unusual. After 3 years of aquarium life, one of the *Herpolithon* species slowly began to dissolve. The small tentacles were no longer extended and the skeleton became partially exposed. It appeared as if the coral was teetering on the throes of death. Filamentous algae immediately began to foul the exposed skeleton, a clear indication that the skeletal sections were not covered with live polyp tissue. During my rescue operation, which included removing the filamentous algae with a eyelash brush, I noted remnants of polyp tissue in the indentions of the septa. Afterwards, the polyp tissue grew until it protruded from the septa. Apparently, this was the site of a future coral. About 60 small individual polyps developed in this manner. After one year the remainder of the coral had divided into individual polyps as well. The protopolyps, each with a diameter of about 1 cm, formed new coral colonies. The very close neighboring "coral plates" united to form a new coral.
I had heard about the "normal" propagation of *Fungia* corals, but I had never witnessed the process in my aquarium until about a year ago when I noticed a small polyp on a stone. After 6 months the crown of the polyp rose from the substrate and a small coral plate became visible on the stalk. The plate broke from the stalk after about 8 months and has lead an independent life on the bottom of my aquarium ever since. Strangely, the stalk immediately formed a new plate. The site and type of development is definite proof that this coral developed from a larva.

Life Expectancy of Hermatypic Stony Corals

Today *Acropora, Seriatophora, Porites, Merulina, Pachyseris, Pectinia, Favia,* and *Fungia* are maintained and reproduced in captive environments. They are long lived and easy to keep, which is why every year more hobbyists add them to their aquaria. Knowledge regarding their maintenance is constantly increasing. Because these genera have been kept in aquaria for such a long time, they have little in common with their brothers in the sea. Their requirements, even their growth forms, are quite distinctive. Captive bred corals are more forgiving in regard to environmental mishaps than specimens newly collected from the wild. Perhaps the increased popularity of coral husbandry can be attributed to the more forgiving nature of these corals rather than to the betterment of the technical aspects of maintenance. In aquaria, a coral's response to different factors can be easily observed and gauged. Curious, scientific-minded aquarists find fertile ground here. By modifying such variables as light, current, placement, and perhaps nutrients, the best conditions for growth and propagation can be determined.

After steady growth for a period of time, one coral stalk started into an incessant decline. Contrary to previous suspicions, corals seem to have a definite life expectancy. Hence death may be a result of natural processes, not necessarily poor husbandry. Therefore, longevity cannot be predicted. Approximately 100 years? Perhaps more? Or less?

Branching corals from the genera *Acropora* and *Seriatophora* are particularly rapid growing, colonizing relatively large areas of the coral reef in a short time. It has been determined that *Acropora* grows approximately 16 cm annually in an aquarium. It is surely not difficult to calculate how long a branch would be if it would grow at the same rate for 100 years. But the fragile nature of *Acropora* prohibits these dimensions; it breaks long before this astounding size is ever attained. There are definitely many factors that influence a coral's lifespan. Space competition, strength of the current, colonizing substrate, overall water quality, and coral predators are probably all important modifiers. When a coral branch breaks from the colony, it generally continues to live if it does not fall prey to excess sedimentation and become covered on the bottom. New colonies originate from the fragments and develop independently. Does this method of propagation sound

familiar? Aquarists commonly duplicate this process in the aquarium. By breaking a coral branch from its base and securing it to a substrate, a new coral colony is established.
In general, corals can be kept for over 10 years in mature reef aquaria. Life expectancy depends on the species as well as all the abiotic and biotic factors that reign in the aquarium. The following example should explain:
While decorating a new aquarium it was impossible to remove an excessively large *Acropora* coral from the old aquarium in one piece; it had to be divided. Only a part of the divided coral stalk could be situated in an analogous site in the new aquarium. The remainder of the branches were placed in the immediate vicinity. After a few weeks it became apparent that the fragments had been placed in a more favorable site, since they developed better and grew faster than the old mother colony. In just one year, linear growth was 16 cm. The base of the mother colony, however, dissolved as it was shaded out by neighboring conspecific colonies. Note how just one factor, in this case shade, affected growth and lifespan.
A similar situation arose with a *Seriatophora*, a coral with a totally different growth form than *Acropora*. A 15 mm colony base grew from a planula larva. The base branched until the colony was arborescent. Over time the superior branches cast too much shade on the lower sections of the colony and as a consequence, the lower sections began to die.

Damaging Factors of Hermatypic Stony Corals

Space Competitors

Damages incurred from space competition due to different species of corals attempting to colonize a restricted space are the most frequent cause of impairment in an aquarium. The struggle for colonizing space is ruthless, and the most aggressive animals will prevail.

Parasites

Recently there seems to be an escalation in reports of parasite damage. External injuries on corals are readily apparent at the time of purchase. To increase the chances of recuperation, a mature reef aquarium is recommended. Because of stress associated with transport, newly introduced corals place a disproportional biological load on the filtration system. Ensure that this is taken into account to prevent the filter from becoming overloaded. Poor water quality combined with the presence of filamentous algae are two conditions a newly acquired coral will be unable to cope with in its weakened state, greatly diminishing its chances of survival.

Environmental Factors

Other types of injuries which are not as easily determined, manifest themselves in the following ways:
If newly introduced animals fail to open or grow, the coral may be subject to deficient illumination, insufficient current, an unfavorable placement, and/or toxins. Be aware of the latest information and techniques regarding reef aquaria. If the corals fail to adhere, then the animals are either too large or the base has died off. Avoiding large, bulky corals will circumvent the likelihood of this problem. If corals are to grow, they must be securely attached to the substrate. Even minimal movements, such as those induced by strong currents, prevent the coral from fastening to the substrate. If there is damage to the live tissue at the base, it is not always tragic; a mechanism springs into action to seal the wound. Some coral genera, for example, *Seriatophora*, are especially photophilic. When deciding where to affix cuttings, consider the illumination requirements of the species. Light and calcium deposition are directly related; hence, without sufficient light the coral cutting will not be capable of cementing itself to the substrate. If established, flourishing corals begin to pale and wane, the problem may be more elusive. The culprit is usually one of the various water chemistry parameters. Check the nitrate concentration

(should be less than 60 mg/l) and ensure that ample trace elements are present, particularly iodine and strontium. Overstocking the aquarium negatively influences growth. Beware of fishes that have high metabolic rates. An invertebrate population that fights among itself using various chemical and mechanical means also slows growth. Responsible aquarists take these aspects into account when selecting the aquarium's inhabitants and strive for a balanced environment.

It is vital to control nitrate concentrations by methods which do not result in excess fluctuations in the overall chemistry of the water. Phosphates, phenols, and bacterial toxins, products of metabolism, are further examples of chemicals that accumulate and diminish water quality. Though it may take years, without proper preventive measures there comes a time when an invisible threshold is crossed and aquarium conditions deteriorate without being noticed. When poor water quality begins to manifest itself through sickly animals, most aquarists respond by installing new filters which seem to solve the problem, but basically represent a transitory solution. A drastic reduction in the stocking density is a more appropriate solution, but this approach is rarely taken, since few aquarists want to rid themselves of part of their prized possessions. My years of experiences have shown that stony corals have little tolerance for new filters that drastically change their environment, a consequence of new treatment properties. It has been established that prolonged use of activated carbon may remove vital trace elements to deficient levels. Although certain losses of trace elements are accrued through protein skimmers, the benefits clearly outweigh the disadvantages. Trace elements can be replaced through monthly water exchanges. Do not exchange more than 5% of the total volume at any given time. Deficiencies of trace elements in the water can also be influenced by the substrate. The composition of the substrate of every aquarium is an important factor in regard to water quality. In this context, animals that colonize the substrate, especially those that loosen and clean it, are influential.

Tankmates

New information concerning the source of damage to hermatypic corals has come to light, a particularly insidious, treacherous etiology. We are talking about injuries inflicted by the tank's own inhabitants. Although these organisms also damage corals in the sea, they usually do not present a danger to the population because of the sheer number of corals. But in the confines of an aquarium, these predators can have a catastrophic effect. The greatest danger comes from feeding specialists, particularly nudibranchs and some starfishes. It is preferable to remove these animals from reef tanks than to wait until the damage has been done. According to my observations, two species of snails have been identified as true coral eaters. Both are nocturnal and hardly visible despite their conspicuous coloration. They predominantly attack branching corals and are capable of eating a 5 cm branch in the course of one night. One of the species—which can only be found with a magnifying glass—grazes on *Galaxea* corals and colonial anemones. When coral polyps refuse to open and coat themselves in mucus, this gastropod is likely the irritant. Unfortunately it is difficult, if not impossible, to find and eliminate this predator.

Bacteria

Additional, hardly quantifiable or avoidable damage is caused by bacteria. It is an undisputable fact that pathogenic bacteria exist in the aquarium, but presently there is little recourse. Bacterial infections manifest themselves in a characteristic fashion—starting at the base then proceeding up the branches, the coral begins to dissolve. Remove the diseased sections outside the aquarium and reestablish the healthy remnants. Always strive to discover the source of the coral's injuries. Preventive action can only be taken when the etiology is known. By pooling our experiences and drawing logical conclusions, the reef aquarium hobby can make strides forward and illness and death can be successfully combatted. Once this level is reached, the care of corals will become a fascinating adventure. However, progress is only possible through collaboration and information exchanges among aquarists.

Taxonomy

Phylum: CNIDARIA	Cnidarians
Class: ANTHOZOA	Anthozoans
Subclass: HEXACORALLIA	Six-Tentacled Anthozoans
Order: SCLERACTINIA	Stony Corals

Suborder: Astrocoeniina
Family: Astrocoeniidae
Genus: *Stephanocoenia*
Family: Pocilloporidae
Genera: *Madracis*
Pocillopora
Seriatopora
Stylophora
Family: Acroporidae
Genera: *Acropora*
Anacropora
Astreopora
Montipora

Suborder: Fungiina
Family: Agariciidae
Genera: *Agaricia*
Leptoseris
Pachyseris
Pavona
Family: Siderastreidae
Genus: *Siderastrea*
Family: Poritidae
Genera: *Alveopora*
Goniopora
Porites
Family: Fungiidae
Genera: *Fungia*
Heliofungia
Herpolitha
Sandalolitha

Suborder: Faviina
- Family: Faviidae
  - Genera: *Caulastrea*
    - *Cladocora*
    - *Colpophyllia*
    - *Diploria*
    - *Echinopora*
    - *Favia*
    - *Favites*
    - *Manicina*
    - *Montastrea*
    - *Moseleya*
    - *Platygyra*
    - *Solenastrea*
- Family: Trachyphylliidae
  - Genus: *Trachyphyllia*
- Family: Oculinidae
  - Genera: *Galaxea*
    - *Oculina*
- Family: Meandrinidae
  - Genera: *Dendrogyra*
    - *Dichocoenia*
    - *Meandrina*
- Family: Mussidae
  - Genera: *Cynarina*
    - *Isophyllia*
    - *Lobophyllia*
    - *Mussa*
    - *Mycetophyllia*
    - *Scolymia*
    - *Symphyllia*
- Family: Pectinidae
  - Genera: *Mycedium*
    - *Pectinia*
- Family: Merulinidae
  - Genera: *Hydnophora*
    - *Merulina*

Suborder: Caryophylliina
Family: Caryophylliidae
Genera: *Caryophyllia*
Euphyllia
Eusmilia
Physogyra
Plerogyra
Polycyathus

Suborder: Dendrophyllina
Family: Dendrophylliidae
Genera: *Astroides*
Balanophyllia
Leptopsammia
Tubastrea
Turbinaria

Seriatopora hystrix

Acropora sp.

Favia sp.

Porites sp.

Pavona sp.

Scolymia sp.

Euphyllia ancora

Caulastrea fracata

Stephanocoenia michelinii EDWARDS & HAIME, 1848
Blushing star coral

Hab.: Caribbean, Florida, Bermuda, Bahamas, Bonaire, Curaçao.

Sex.: None.

M.: Unknown.

F.: C; plankton.

S.: Colonies are massive, hemispherical, or encrusting. The calyxes have a diameter of 2–3 mm and 24 septae arranged in 3 cycles. Blushing star corals principally occur on reef slopes and underwater platforms from shallow water to depth of more than 80 m. Polyps are nocturnal and predominantly brown.

T: 23°–27°C, Ø: 50 cm, AV: 4 (protected genus)

Madracis decactis LYMANN, 1859
Ten-ray star coral

Hab.: Caribbean, Florida, Bahamas, Bermuda.

Sex.: None.

M.: Unknown.

F.: C; plankton.

S.: Short, thick knobs project outward from a crusty coral mass. The knobs are densely covered with polyps which are extended during the day as well as at night. The low growth form is an adaptation to strong currents. Vertical, hard substrates from shallow waters to moderate depths, including reef sections with strong water turbulence, are colonized. The color of the polyps is variable.

T: 23°–27°C, Ø: 20 cm, AV: 4

Madracis mirabilis (DUCHASSAING & MICHELOTTI, 1860)
Yellow pencil coral

Hab.: Caribbean, Florida, Bahamas, Bermuda.

Sex.: None.

M.: Unknown.

F.: Plankton.

S.: The colony is a very brittle clump of slender, fingerlike branches. As the polyps are extended during the day as well as at night, they impart a bushy appearance. *M. mirabilis* is predominantly found in shallow waters, but it occasionally inhabits sandy areas of outer reefs. Extensive lawns may be encountered in static regions of calm bays just below the water surface.

T: 23°–27°C, Ø: 20 cm, AV: 4

Pocillopora damicornis (LINNAEUS, 1858)
Rasberry coral, birdnest coral

Hab.: Red Sea, eastern Africa, Indo-Pacific.

Sex.: None.

M.: *P. damicornis* can be maintained in aquaria, and with appropriate illumination, it growths relatively fast.

B./Rep.: This coral has been propagated both sexually (larvae) and asexually (fragmentation).

F.: C; plankton.

S.: Colonies are erect with numerous blunt-tipped branches. The cluster of small calyxes at the ends of the branches is reminiscent of a raspberry. Sunny sites close to the water surface are *P. damicornis*'s preferred habitat. Coloration is variable.

T: 23°–27°C, Ø: 10 cm, AV: 4 (protected genus)

Pocillopora eydouxi EDWARDS & HAIME, 1860
Cauliflower coral

Hab.: Red Sea, Indo-Pacific to Galapagos and Gorgona.

Sex.: None.

M.: Unknown.

F.: C; plankton.

S.: The erect branches are stout, short, and laterally compressed. These branches can either grow separate or form a compact clump. In turbulent areas, the growth form is particularly stalwart. *P. eydouxi* can be found to just beneath the water surface. Coloration is variable.

T: 23°–27°C, Ø: 60 cm, AV: 4 (protected genus)

Seriatopora caliendrum EHRENBERG, 1834
Needle coral

Hab.: Red Sea, eastern Africa, Indo-Pacific, Malaysia, Borneo.

Sex.: None.

M.: Satisfactory growth can be expected in aquaria. Because this is a strongly stinging scleractinian, there should always be a respectable distance between it and other corals.

Light: Moderate light.

F.: C; plankton.

S.: *S. caliendrum* inhabits calm, upper reef slopes. The colony is made up of numerous delicate branches. Branch tips are light-colored and needlelike. Calyxes are arranged in rows along the branches. It is nocturnal.

T: 23°–28°C, **Ø:** 15 cm, **TL:** from 150 cm, **WM:** m, **WR:** m, **AV:** 3–4, **D:** 4

Seriatopora hystrix DANA, 1846
Thorns of Christ coral

Hab.: Red Sea, eastern Africa, Indo-Pacific.

Sex.: None.

Soc.B./Assoc.: Fishes and shrimp are suitable associates.

M.: Successful maintenance depends on illumination and current. *S. hystrix* is intolerant of filamentous algae.

F.: Substitute plankton, liquid foods.

S.: Numerous extremely delicate, thorny branches emerge from a slender base. Because of their fragility, colonies most often occur in sheltered areas. The small calyxes are arranged in parallel rows along the branches. Coloration is extremely variable.

T: 23°–27°C, **Ø:** 25 cm, **TL:** from 150 cm, **WM:** s, **WR:** t–m, **AV:** 4 (protected genus), **D:** 4

Stylophora pistillata ESPER, 1797
Pistil coral, cluster coral

Hab.: Red Sea, eastern Africa, Madagascar, Indo-Pacific.

Sex.: None.

M.: *S. pistillata* can be kept in intensely illuminated aquaria. For maximum exposure, situate the colony close to the water surface. The attractive pink coloration is retained in the aquarium. Skeletons of this species are used as decorative infrastructure.

Light: Sunlight zone.

F.: C; plankton.

S.: Though colonies are usually branched and columnar or digitate, placement and substrate may bring about other growth forms. These organisms are very prevalent on Indo-Pacific reefs. From shallow to deep water, pistil corals grow on hard substrates on fore reefs where there is good water circulation. Coloration is variable.

T: 23°–27°C, Ø: 20 cm, AV: 4 (protected genus)

Stylophora subseriola DANA, 1846
Finger coral

Hab.: Red Sea, Indo-Pacific.

Sex.: None.

M.: Although these corals can be maintained in an aquarium, they are very slow-growing. They require a lot of light and usually lose against space competitors.

Light: Sunlight zone.

F.: C; plankton.

S.: These colonies have numerous terminal, digital branches, each of which has a light tip. Colonies are diurnal as well as nocturnal. Finger corals generally reside on hard substrates of fore reefs that are exposed to a moderate current. Coloration is variable.

T: 23°–27°C, Ø: 10 cm, AV: 4 (protected genus)

Acropora abrolhosensis VERON, 1985
Staghorn coral

Hab.: Australia to Houtmann Rocks.

Sex.: None.

M.: Unknown.

F.: C; plankton.

S.: Colonies are bushlike with straight branches. Axial calyxes are large; in contrast, radial calyxes are significantly smaller and round. They inhabit lagoons and areas subjected to a slight current.

T: 23°–27°C, Ø: 40 cm, AV: 4 (protected genus)

Acropora cervicornis (LAMARCK, 1816)
Staghorn coral

Hab.: Caribbean, Florida, Bahamas, Bermuda.

Sex.: None.

M.: Care in captive environments is problematic. Proper growth depends on kalkwasser and strontium supplements and the absence of filamentous algae. *A. cervicornis* is unsociable towards other cnidarians, often displacing them.

Light: Sunlight zone.

F.: Substitute plankton.

S.: *A. cervicornis* inhabits shallow lagoons, reef crests, and the upper zone of the fore reef. The numerous branches have pointed tips and small, laterally protruding tubular calyxes covering their surface. The dead base is occupied by many small invertebrates and algae. Fishes, crustacea, and feather and brittle stars commonly seek shelter among the labyrinth of branches.

T: 23°–27°C, Ø: 100 cm, AV: 4 (protected genus)

Order: Scleractinia
Fam.: Acroporidae

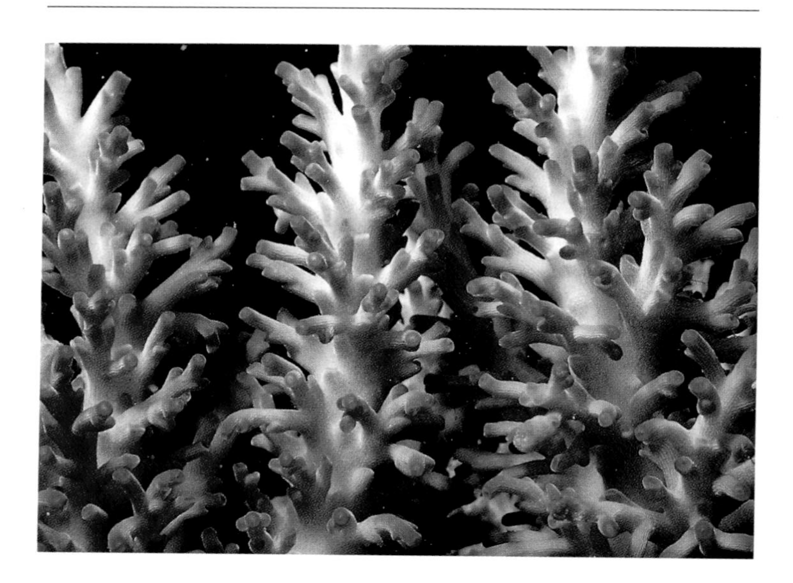

Acropora echinata (DANA, 1848)
Thorny staghorn coral

Hab.: Indo-Pacific, Australia, Marshall Islands to Samoa.

Sex.: None.

M.: Unknown.

F.: C; plankton.

S.: The thorny antler coral has a multitude of branches. As seen in the close-up photo, the branches, especially the secondary branches, are covered in uniform, delicate calyxes, giving them the appearance of bottle brushes. *A. echinata* inhabits shallow, sunlight-bathed clear water reefs that are sheltered from all but a slight current. Coloration is variable.

T: 23°–27°C, Ø: 40 cm, AV: 4 (protected genus)

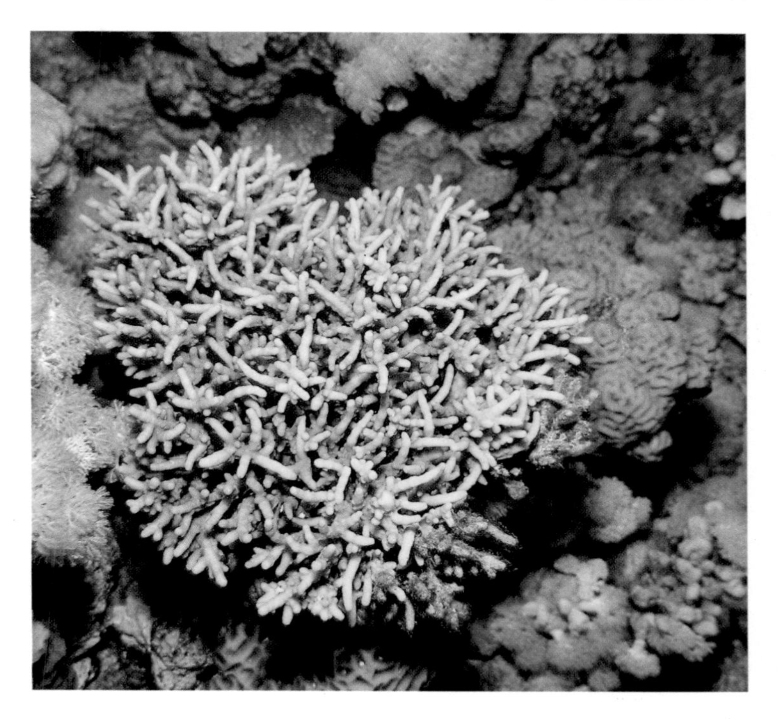

Acropora granulosa (EDWARDS & HAIME, 1860)
Rough antler coral

Hab.: Red Sea, Indo-Pacific to Tahiti.

Sex.: None.

M.: Unknown.

F.: C; plankton.

S.: Colonies are low-growing, elliptical to round, flat, and almost horizontal. The short, upward-pointing branches are covered with tubular calyxes. Rough antler corals are chiefly found in lagoons and on sheltered reef slopes. Color variations are rare; colonies are usually gray or cream-colored.

T: 23°–27°C, Ø: 60 cm, AV: 4 (protected genus)

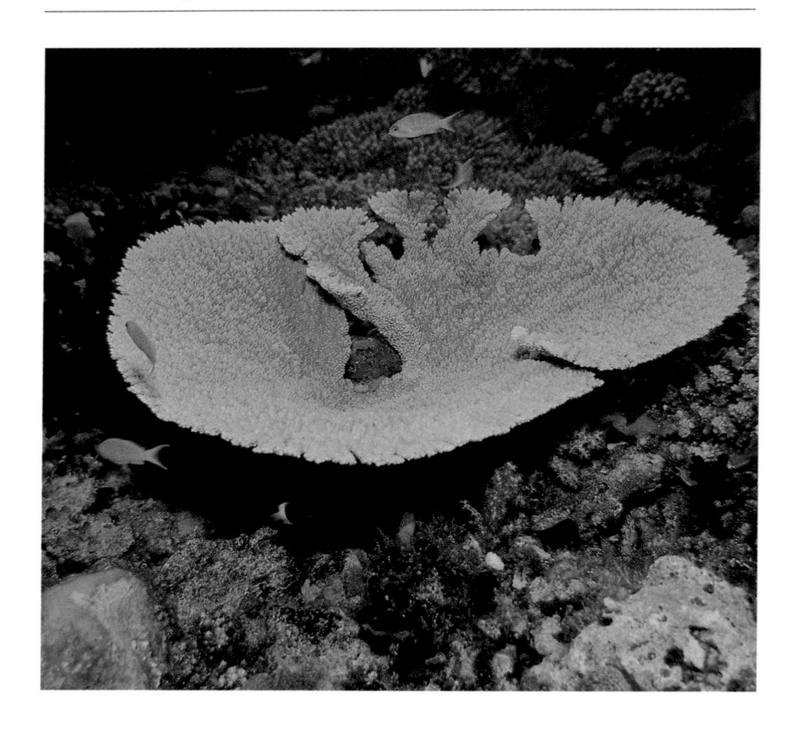

Acropora hyacinthus (DANA, 1846)
Table coral

Hab.: Red Sea, Indo-Pacific.

Sex.: None.

M.: Unknown.

F.: C; plankton.

S.: This large, broad, flat tabular coral has one or more whorls supported by a stout stem. Delicate, erect branches create spiral, rosettelike arrangements on the superior side of the colony. Although it breaks easily, the pieces will continue to grow. It usually occurs along reef tops and reef slopes. Coloration is variable.

T: 23°–27°C, Ø: 100 cm, AV: 4 (protected genus)

Acropora palifera (LAMARCK, 1816)
Finger coral

Hab.: Red Sea, Indo-Pacific.

Sex.: None.

M.: Unknown.

F.: C; plankton.

S.: *A. palifera* forms strong columnar to plate-shaped colonies. The few branches are short and stout with blunt tips. It is most commonly found along shallow, current-swept reef slopes.

T: 23°–27°C, Ø: 25 cm, AV: 4 (protected genus)

Acropora palmata (LAMARCK, 1816)
Elkhorn coral

Hab.: Caribbean, Florida to Venezuela.

Sex.: None.

M.: Unknown.

F.: C; plankton.

S.: *A. palmata* is an exceptionally strong coral that grows in the reef's surge zone, breaking the waves along the upper reef slopes. Its broad, shovellike branches radiate outward from a central stem. New growth along the edge of the branches is white because of the lack of zooxanthellae, which have yet to establish themselves in these areas. This species plays a major role in reef-building and is the predominant coral of many Caribbean reefs. Broken pieces will continue to grow, forming new colonies.

T: 23°–27°C, Ø: 100 cm, AV: 4 (protected genus)

Acropora squarrosa (EHRENBERG, 1834)
Staghorn coral

Hab.: Red Sea, Indo-Pacific, Malaysia.

Sex.: None.

M.: Unknown.

F.: C; plankton.

S.: Colonies are richly branched and dense. Although shallow waters of calm bays and reef lagoons are where these corals are most commonly encountered, upper reef slopes occasionally support growth as well. Some areas have lawnlike extensions of *A. squarrosa*.

T: 23°–27°C, Ø: 100 cm, AV: 4 (protected genus)

Acropora sp.
Staghorn coral

Hab.: Indo-Pacific, Malaysia.

Sex.: None.

M.: Unknown.

F.: C; plankton.

S.: *Acropora* sp. forms massive colonies. The short, thick branches have an extremely broad base and bright tips. It is found along shallow, current-swept reefs.

T: 23°–27°C, Ø: 60 cm, AV: 4 (protected genus)

Acropora sp.
Staghorn coral

Hab.: Indo-Pacific, Thailand.

Sex.: None.

M.: Unknown.

F.: C; plankton.

S.: Colonies have an extraordinary number of branches and many small uniform calyxes. Lateral branches are fine and delicate. Almost all biotopes along the reef are colonized except sandy lagoons and areas exposed to strong currents. These low-growing colonies of are capable of covering several square meters of substrate.

T: 22°–27°C, **Ø:** 200 cm, **AV:** 4

Order: Scleractinia
Fam.: Acroporidae

Anacropora sp.
Staghorn coral

Hab.: Indo-Pacific, Australia.

Sex.: None.

M.: Unknown.

F.: C; plankton.

S.: Colonies have numerous branches. *Anacropora* sp. is a very aggressive coral that is intolerant of other corals in its immediate vicinity. Because of its cryptic habitats, e.g., niches beneath overhangs and around cave openings, even attentive divers often overlook this coral.

T: 23°–27°C, Ø: 40 cm, AV: 4 (protected genus)

Astreopora gracilis BERNARD, 1896
Gracious star coral

Hab.: Red Sea, Indo-Pacific.

Sex.: None.

M.: Unknown.

F.: C; plankton.

S.: Growth is massive and generally hemispherical. The cylindrical calyxes protrude in all directions. Gracious star corals occur at all depths, particularly on calm reef sections.

T: 23°–27°C, **Ø**: 60 cm, **AV**: 4 (protected genus)

Montipora foliosa (PALLAS, 1766)
Foliaceous micropore coral

Hab.: Red Sea, Indo-Pacific to New Hebrides and Fiji.

Sex.: None.

M.: Unknown.

F.: C; plankton.

S.: Colonies consist of broad, flat, partially interlocked plates. Calyxes are arranged in rows. Since actively growing zones initially lack zooxanthellae, the outer edges of the plates are often white.These corals are found on protected reef slopes. Coloration is variable.

T: 23°–27°C, Ø: 100 cm, AV: 4 (protected genus)

Montipora monasteriata (FORSSKÅL, 1775)
Star coral

Hab.: Red Sea, Indo-Pacific, Hawaii, Thailand, Australia.

Sex.: None.

M.: This coral grows well in aquaria. Avoid placing stony corals, particularly *Acropora*, in its immediate vicinity, as *M. Monasteriata* is subordinate in front of virtually all other hermatypic corals.

F.: C; plankton.

S.: *M. monasteriata* forms massive, encrusting, or thick platelike colonies that have a granular to lumpy surface. The edge of the colony is frequently lighter. Reef slopes and deep reef plateaus exposed to strong currents are *M. monasteriata*'s habitat. Coloration ranges from brown to pink.

T: 23°–28°C, **L**: 60 cm, **AV**: 4

Montipora tuberculosa (LAMARCK, 1816)
Micropore coral

Hab.: Red Sea, Indo-Pacific to Japan and the Marquesas Islands.

Sex.: None.

M.: Unknown.

F.: C; plankton.

S.: Colonies are platelike and slightly folded. Calyxes are small and segregated by knots. Although nocturnal, the polyps are sometimes extended during the day. Coloration is variable. *M. tuberculosa* is distributed worldwide along all reef zones.

T: 23°–27°C, Ø: 70 cm, AV: 4 (protected genus)

Agaricia agaricites (LINNAEUS, 1758)
Tan lettuce leaf coral, lettuce coral

Hab.: Caribbean, Florida to Venezuela.

Sex.: None.

M.: Unknown. Skeletons, because they turn exceptionally white after being washed, are marketed as decorative items.

F.: C; plankton.

S.: *A. agaricites* forms massive hemispherical or encrusting colonies. As light diminishes with depth, this coral compensates by growing more discoid. The surface of the colony is covered by irregular ridges and valleys. Edges are extremely thin, white, dentate, and bent outward. From shallow water down to great depths, *A. agaricites* colonizes hard substrates. Coloration is variable.

T: 23°–27°C, Ø: 60 cm, AV: 4

Agaricia lamarcki EDWARDS & HAIME, 1851
Lamarck's lettuce leaf coral, Lamarck's sheet coral

Hab.: Caribbean, West Indies, Florida, Bermuda.

Sex.: None.

M.: Unknown.

F.: C; plankton.

S.: The thin, flat, foliaceous colonies are attached to the substrate by a small stem. These corals are solely found at depths greater than 20 m, where the broad folia are directed to catch the greatest amount of light. The small, nocturnal polyps are only found on the superior side of the folia.

T: 23°–27°C, Ø: 80 cm, AV: 4

Agaricia tenuifolia DANA, 1848
Thin leaf lettuce coral

Hab.: Caribbean, Florida to Colombia.

Sex.: None.

M.: Unknown.

F.: C; plankton

S.: The erect, foliaceous growth of this coral resembles a lettuce leaf. Folia emerge from a common base which is often colonized by calcareous algae and small invertebrates. In shallow, tranquil bays, *A. tenuifolia* is capable of covering several square meters of substrate. Small colonies can be found along the edge of seaweed meadows.

T: 23°–27°C, Ø: 100 cm, AV: 4

Leptoseris cucullata (ELLIS & SOLANDER, 1786)
Sunray lettuce coral

Hab.: Caribbean, Florida, Bermuda, Bahamas.

Sex.: None.

M.: With bright illumination and a directed current, this species can be maintained in aquaria. As soon as the colony begins to grow cup-shaped, the water current must bathe the cup to keep it clean. Do not permit the small cup to become overgrown.

Light: Sunlight zone.

F.: C; plankton.

S.: Young colonies are cup-shaped and circular. In contrast, large colonies are exceedingly thin, irregular plates. The tiny, nocturnal polyps are only located on the upper surface of the colony. Sunray lettuce corals occur on reef slopes and beneath overhangs to a depth of 90 m; occasionally they grow on sandy substrates.

T: 23°–27°C, Ø: 40 cm, AV: 4

Leptoseris scabra VAUGHAN, 1907
Wave coral

Hab.: Indo-Pacific, Hawaii, Tahiti, Malaysia.

Sex.: None.

M.: Unknown.

F.: C; plankton.

S.: Colonies are usually composed of encrusting, flat, foliaceous branches and vertical pillars or cylinders. The colonies' edges are white. *L. scabra* lives in moderate currents along reef slopes.

T: 23°–27°C, Ø: 20 cm, AV: 4

Pachyseris speciosa (DANA, 1846)
Groove coral

Hab.: Red Sea, Indo-Pacific.

Sex.: None.

M.: Unknown.

F.: C; plankton.

S.: *P. speciosa* forms thin discoid to shingle-shaped sheets. The sheets are predominantly horizontal and may occasionally overlap on large colonies. The dense concentric ridges give the coral a washboard surface. Calyxes are arranged along the grooves. This species grows on hard substrates and crushed coral. The inferior side of the folia is not completely adhered to the substrate. Coloration is variable.

T: 23°–27°C, Ø: 70 cm, AV: 4 (protected genus)

Pavona cactus (FORSSKÅL, 1775)
Cactus coral

Hab.: Red Sea, Indo-Pacific to the Marshall Islands.

Sex.: None.

M.: *P. cactus* is one of the few foliaceous corals suitable for aquarium maintenance. Due to its delicate nature, contact with all colonial anemones, disc anemones, and sea anemones must be avoided.

F.: C; plankton.

S.: Colonies are foliaceous. The thin, brittle folia are erect, sinuate, and lobed. *P. cactus* inhabits lagoons and reef slopes that have a slight current.

T: 23°–27°C, Ø: 10 cm, AV: 4 (protected genus)

Order: Scleractinia
Fam.: Agariciidae

Pavona clavus (DANA, 1848)
Star column coral

Hab.: Red Sea, Indo-Pacific to the Galapagos Islands.

Sex.: None.

M.: Unknown.

F.: C; plankton.

S.: *P. clavus* typically forms large, voluminous colonies. In shallow water, colonies consist of a multitude of individual columns. Growth is planar to discoid in deep water. When star column corals inhabit areas exposed to strong currents, i.e., the tidal zone of the upper reef slopes, they assume a strong growth form. Coloration is variable.

T: 23°–27°C, Ø: 60 cm, AV: 4 (protected genus)

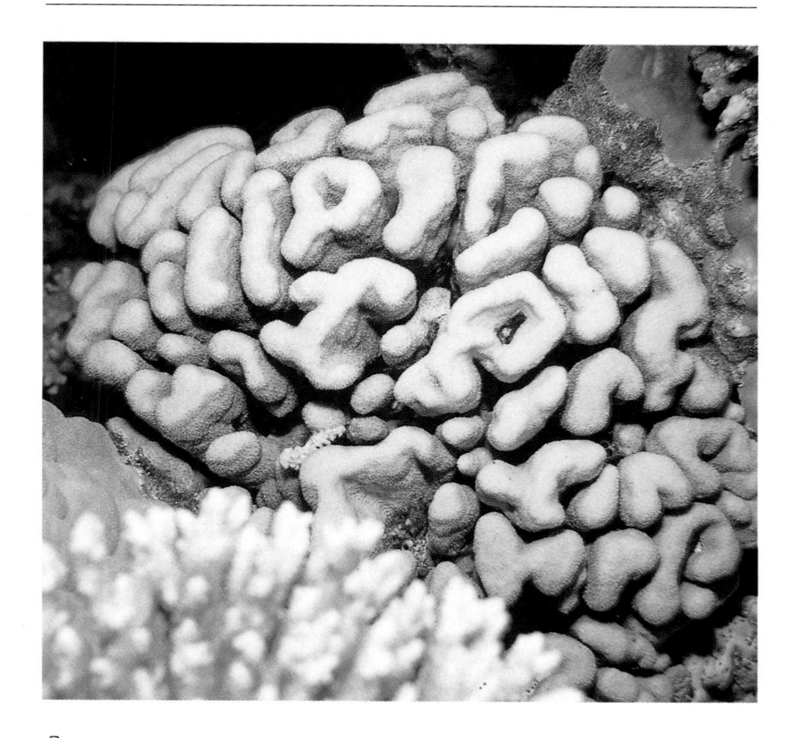

Pavona sp.
Star column coral

Hab.: Red Sea.

Sex.: None.

M.: Unknown.

F.: C; plankton.

S.: Growth is massive. The strong, parallel or irregular columns are separated by wide trenches. Star column corals grow along current-swept upper reef slopes.

T: 23°–27°C, **Ø:** 50 cm, **AV:** 4 (protected genus)

Order: SCLERACTINIA
Fam.: Siderastreidae

Siderastrea siderea (ELLIS & SOLANDER, 1786)
Round starlet coral, massive starlet coral

Hab.: Caribbean, Florida to Venezuela.

Sex.: None.

M.: Unknown.

F.: C; plankton.

S.: Colonies are usually hemispherical. Calyxes are 4–5 mm in diameter with over 48 septae. Polyps are nocturnal and slightly reddish brown. Although round starlet corals generally settle along shallow upper reef slopes, they occasionally occur at greater depths.

T: 23°–27°C, Ø: 30 cm, AV: 4

Alveopora sp.
Porous coral, daisy coral

Hab.: Indo-Pacific, Malaysia.

Sex.: Colonies are dioecious (separate sexes).

M.: Unknown.

F.: C; plankton.

S.: Branches are digital with round tips. Usually calcareous algae and invertebrates are found growing on the colony's dead base. Polyps are both diurnal and nocturnal. Extensive areas along shallow reefs are often covered by *Alveopora* sp.

T: 23°–27°C, Ø: 100 cm, AV: 4 (protected genus)

Goniopora lobata EDWARDS & HAIME, 1860
Coarse porous coral, flowerpot coral

Hab.: Red Sea, eastern Africa, Indo-Pacific to Fiji and Samoa.

Sex.: Dioecious (separate sexes).

M.: Aquarium care is very difficult. The required intense illumination fosters the growth of *Ostreobium*, a boring green alga. This alga enters the calcium skeleton, and the polyps become progressively smaller and blisterlike until they die.

F.: Substitute plankton, *Artemia* nauplii, liquid foods, microplankton.

S.: Growth is massive and hemispherical. Polyps are very long and tubular and always extended. At the terminal end of the polyp, there is a round oral disc which has 24 tentacles along its edge and a mouth, usually of contrasting color, at its center. The color of the zooxanthellae that live in the polyps' tissues determines the color of the colony. *G. lobata* lives on reef platforms and reef slopes that are exposed to strong to very strong currents.

T: 23°–27°C, Ø: 20 cm, AV: 4 (protected genus)

Goniopora planulata (EHRENBERG, 1834)
Anemone coral

Hab.: Red Sea, Indo-Pacific to Fiji and Samoa.

Sex.: Dioecious (separate sexes).

M.: Unknown.

F.: C; plankton.

S.: This species has a club-shaped form, rather like a lump on a dead stem. Only the surface of the lump is covered by live tissue. Polyps are exceptionally long and always extended. The tentacles are reminiscent of those found on anemones. Coloration is variable.

T: 23°–27°C, **Ø:** 60 cm, **AV:** 4 (protected genus)

Goniopora sp.
Porous coral

Hab.: Red Sea, Indo-Pacific, Malaysia.

Sex.: Dioecious (separate sexes).

M.: Unknown.

F.: C; plankton.

S.: Colonies are massive and hemispherical. The long, fleshy polyps have 24 tentacles which are extended day and night. The photo above was taken on an upper reef slope. Most porous corals colonize hard substrates.

T: 23°–27°C, Ø: 30 cm, AV: 4 (protected genus)

Porites astreoides LAMARCK, 1816
Yellow porous coral, mustard hill coral

Hab.: Caribbean, Florida to Venezuela.

Sex.: None.

M.: Unknown.

F.: C; plankton.

S.: Growth is massive, hemispherical, or encrusting, depending on depth and current. Shallow water forms are predominantly mounds. In deep water, yellow porous corals flatten into planar colonies. *P. asteroides* occurs from shallow water to a depth of 70 m. Coloration is variable.

T: 23°–27°C, **Ø:** 30 cm, **AV:** 4 (protected genus)

Porites cylindrica DANA, 1846
Cylindrical porous coral

Hab.: Red Sea, Indo-Pacific to Tonga and the Marshall Islands.

Sex.: None.

M.: Unknown.

F.: C; plankton.

S.: *P. cylindrica* is richly branched and generally affixed to the substrate with a broad, encrusting base. Its branches are cylindrical with white, rounded tips. Calyxes are very shallow, and the polyps are nocturnal. *P. cylindrica* is often the dominant species in shallow lagoons. Coloration is variable.

T: 23°–27°C, Ø: 30 cm, AV: 4 (protected genus)

Porites cylindrica

Porites lobata DANA, 1846
Lobed porous coral

Hab.: Red Sea, Indo-Pacific, Australia.

Sex.: None.

M.: Unknown.

F.: C; plankton.

S.: *P. lobata* forms a large hemispherical dome just beneath the water surface in lagoons and along rock reefs and reef platforms. During low tide, the upper part of the colony is exposed and dies. It is then eroded by sun, wind, and waves until only the water-covered lateral edges are alive and growing. "Microatolls," substrates for a multitude of sessile and boring organisms, are thus created.

T: 23°–27°C, Ø: 100 cm, AV: 4 (protected genus)

Porites lutea EDWARDS & HAIME, 1860
Yellow porous coral

Hab.: Indo-Pacific, Australia.

Sex.: None.

M.: Unknown.

F.: C; plankton.

S.: The above photo shows an unusually massive hemispherical specimen that has a diameter of several meters. Use the diver in the background as a size reference. The surface of the dome-shaped colony is riddled with recesses and crevices. *P. lutea* is found along reef tops just beneath the water surface.

T: 23°–27°C, Ø: 100 cm, AV: 4 (protected genus)

Porites porites (PALLAS, 1766)
Thick finger coral, finger coral

Hab.: Caribbean, Florida to Venezuela.

Sex.: None.

M.: Thick finger corals are very sensitive and therefore difficult to maintain in captivity. Live specimens are rarely seen in pet stores; dried skeletons are more commonly sold.

F.: Substitute plankton.

S.: In shallow bays and calm lagoons, *P. porites* is reef-forming and capable of covering extensive areas with dense growth. The multitude of branches are digital with blunt tips. Sponges, algae, and small invertebrates grow on the colony's dead base. Polyps are small and diurnal as well as nocturnal.

T: 23°–27°C, Ø: 100 cm, AV: 4 (protected genus)

Porites porites var. *divaricata* LESUEUR, 1821
Thin finger coral

Hab.: Caribbean, Colombia.

Sex.: None.

M.: Unknown.

F.: C; plankton.

S.: *P. porites* var. *divaricata* is a growth form of *P. porites*. Colonies are small, and the branches are thin, 4–5 mm in diameter, and bifurcated near the tip. The base of the colonies is dead and buried in the sediment. To a depth of 50 m, thin finger corals are found on coral sand and among coral pieces.

T: 23°–27°C, Ø: 7 cm, AV: 4 (protected genus)

Order: Scleractinia
Fam.: Poritidae

Porites sp.
Porous coral

Hab.: Great Barrier Reef, Australia.

Sex.: None.

M.: Unknown.

F.: C; plankton.

S.: This *Porites* sp. is clearly under attack by a red boring sponge of the genus *Cliona,* a hostile habitat competitor. The polyps in the vanquished territory have been destroyed. Because of its greater aggressiveness, the sponge will prevail.

T: 23°–27°C, Ø: 20 cm, AV: 4 (protected genus)

Fungia fungites (LINNAEUS, 1758)
Mushroom coral

Hab.: Red Sea, eastern Africa, Indo-Pacific.

Sex.: None.

M.: Since mushroom corals cannot be maintained in aquaria for extended periods of time, they should not be purchased. Skeletons are frequently sold as decorative items.

F.: Substitute plankton, *Mysis*, small shrimp.

S.: Mushroom corals are not colonial—each disc represents one polyp. Polyps of young specimens are supported by a stem and resemble a mushroom (hence the name!). Adults live unattached on sand and rubble substrates. Because these organisms are able to clean themselves, they are occasionally encountered in sediment-rich waters. The long, radial, dentate septa are easily seen. Mushroom corals have a great capacity of regeneration.

T: 23°–27°C, Ø: 15 cm, AV: 4 (protected genus)

Fungia klunzingeri DÖDERLEIN, 1901
Klunzinger's mushroom coral

Hab.: Red Sea, Indo-Pacific, Australia.

Sex.: None.

M.: Unknown.

F.: C; plankton.

S.: In comparison to other species of mushroom corals, Klunzinger's mushroom coral is small, flat, and round (rarely oval). It occurs on any and all substrates. Septa are fine, dense, dentate, and predominantly brown, though there are a few interspersed white septa.

T: 23°–27°C, Ø: 5 cm, AV: 4 (protected genus)

Fungia scruposa KLUNZINGER, 1879
Mushroom coral

Hab.: Red Sea, Indo-Pacific.

Sex.: None.

M.: Unknown.

F.: C; plankton.

S.: *F. scruposa* is a large, predominantly light brown mushroom coral that has a thick, heavy, round to oval skeleton and strong, dentate septa. It is found on corals and coral fragments of reef tops and slopes. A rare species in the Red Sea.

T: 23°–27°C, **Ø:** 25 cm, **AV:** 4 (protected genus)

Heliofungia actiniformis (QUOY & GAIMARD, 1833)
Plate coral

Hab.: Red Sea, Indo-Pacific, Philippines, Micronesia to New Caledonia.

Sex.: None.

M.: *H. actiniformis* does not fare well in captive environments over extended periods of time. Fish and crustacean tankmates cause undue stress, and other cnidarians must be kept at a distance. Provide a sand and/or rubble substrate.

F.: Substitute plankton, *Mysis*, small crustacea.

S.: This solitary, unattached coral has a flat, discoid skeleton. Because of its long tentacles (the longest of any coral), *H. actiniformis* looks like an open anemone. The tentacles have pale, spherical tips and are extended day and night. The oral disc is striped. It is generally found on rocks, corals, and rubble substrates. Novices may easily confuse this species with sea anemones.

T: 23°–27°C, Ø: 15 cm, TL: from 120 cm, WM: s, WR: t–m, AV: 4 (protected genus), D: 4

Herpolitha limax HOUTTUYN, 1772
Slipper coral, hedgehog coral

Hab.: Red Sea, Indo-Pacific, eastern Africa.

Sex.: None.

Soc.B./Assoc.: This species can be housed with other cnidarians; however, it cannot be kept with fishes or crustacea.

M.: Slipper corals are occasionally sold in pet stores. They are considerably easier to maintain than mushroom corals, but longevity in aquaria is still short. Dim light, a strong current, and a substrate of coarse coral debris are required.

Light: Dim light.

B./Rep.: Propagation is through fragmentation.

F.: Feed substitute plankton and *Mysis* twice a week.

S.: *H. limax* is an unattached, tongue-shaped coral that is made up of several polyps (colonial). Overlapping septa extend off the central fossa to the rounded lateral edges and continue on the underside of the colony. Generally, the slipper coral is encountered on protected reefs, lying upon the substrate. It occasionally becomes wedged between corals.

T: 23°–27°C, **L:** 15–20 cm, **AV:** 4 (protected genus)

Order: SCLERACTINIA
Fam.: Fungiidae

Sandalolitha robusta QUELCH, 1886
Robust hat coral, robust basket coral

Hab.: Indo-Pacific, Thailand, eastern Pacific to Tahiti.

Sex.: None.

M.: Unknown.

F.: C; plankton.

S.: *S. robusta* is a colony of many polyps. Structurally it is large, very heavy, and round to oval. The surface is rough and spiny. This coral lies detached on crushed coral substrates of submarine terraces and slightly sloped reefs. There it may be flipped over by waves and, occasionally, by fishes. Nocturnal. Small fishes use the underside of the colony as a spawning substrate. Colonies are predominantly light to dark brown.

T: 23°–28°C, L: to 75 cm, AV: 4

Sandalolitha robusta

Order: SCLERACTINIA
Fam.: Faviidae

Caulastrea curvata WIJSMANN-BEST, 1972
Curved finger coral, torch coral

Hab.: Indo-Pacific, Australia, Malaysia.

Sex.: None.

M.: Unknown.

F.: C; plankton.

S.: Colonies are low-growing, elongated, twisted, and branched. Calyxes are distinct with a diameter of approximately 8 mm. These organisms occur along protected reef slopes.

T: 23°–27°C, Ø: 20 cm, AV: 4

Caulastrea furcata DANA, 1846
Finger coral, torch coral

Hab.: Indo-Pacific to Tonga and Samoa.

Sex.: None.

M.: Unknown.

F.: C; plankton.

S.: Finger corals are mostly small and low-growing. Polyps are fleshy, variable-colored, and nocturnal. Unless they are about to divide—in which case they may have an irregular shape—calyxes are round or oval and about 10 mm in diameter. Calyxes appear independent and separate, but they are joined to a common base. The numerous septa are clearly visible.These corals prefer rocky reefs that are exposed to strong currents.

T: 23°–27°C, **Ø**: 8 cm, **AV**: 4

Order: Scleractinia
Fam.: Faviidae

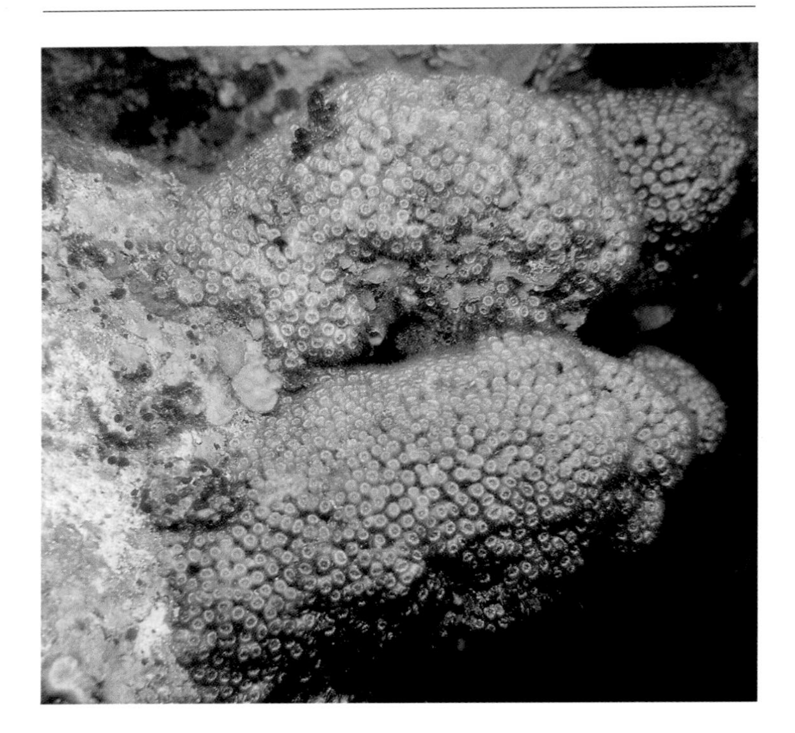

Cladocora cespitosa (LINNAEUS, 1767)
Lawn coral, tube coral

Hab.: Mediterranean, Balearic Islands to Asia Minor.

Sex.: None.

M.: With a strong current, good aeration, and appropriate quantities of very minute foods, *C. cespitosa* can be maintained in aquaria. Over time the zooxanthellae leave their host and the polyps pale, eventually turning transparent.

F.: Smallest plankton.

S.: Growth form and size are dependent on habitat and current. Along the upper rocky littoral, *C. cespitosa* forms low-growing mats or fascicular colonies. Calyxes are cylindrical and round in cross-section. Polyps are brown, and the tips of the tentacles are transparent. This species has zooxanthellae.

T: 10°–20°C, Ø: 60 cm, AV: 3

Colpophyllia amaranthus (HOUTTUYN, 1772)
Coarse Atlantic brain coral

Hab.: Caribbean, Florida to Colombia.

Sex.: None.

M.: Unknown.

F.: C; plankton.

S.: Growth is generally flat, but an occasional hemispherical colony is not unusual. The meandering valleys are a different color in comparison to the ridges and broader than those of *C. natans*. It occurs from shallow waters down to great depths.

T: 23°–27°C, Ø: 60 cm, AV: 4 (protected genus)

Colpophyllia natans (HOUTTUYN, 1772)
Large grooved brain coral, boulder brain coral

Hab.: Caribbean, Florida, Bahamas, Colombia.

Sex.: None.

M.: Unknown.

F.: C; plankton.

S.: Depending on depth, placement, and water current, *C. natans* assumes different growth forms. Growth is normally hemispherical with deep meandering valleys and high crests. Each crest has two parallel crest lines and 8–9 septa per centimeter. Valleys are about 1–2 cm wide and 1 cm deep. *C. natans* occurs from shallow water to 50 m depth. Coloration is variable.

T: 23°–27°C, **Ø**: 20 cm, **AV**: 4

Diploria strigosa (DANA, 1848)
Common brain coral, symmetrical brain coral

Hab.: Caribbean, Florida to Venezuela.

Sex.: None.

Soc.B./Assoc.: *D. strigosa* is frequently colonized by the calcareous tubeworm *Spirobranchus giganteus*.

M.: Unknown.

F.: C; plankton.

S.: *D. strigosa* forms massive, hemispherical, or flattened colonies. Valleys are meandering, and there are 15–20 septa per centimeter of wall. This is probably the most prevalent brain coral of the Caribbean. Coloration is variable.

T: 23°–27°C, Ø: 50 cm, AV: 4 (protected genus)

Order: Scleractinia
Fam.: Faviidae

Diploria labyrinthiformis (LINNAEUS, 1758)
Depressed brain coral, grooved brain coral

Hab.: Caribbean, Florida to Colombia.

Sex.: None.

M.: Unknown.

F.: C; plankton.

S.: Colonies are large, massive, and predominantly hemispherical in form. The double, labyrinthine ridges make the colonies quite distinct. The specimen on the facing page is atypical in that the ridges and fossa predominantly run in straight parallels. Sinuous valleys only appear in the course of additional growth. These corals occur from shallow to deep water. Colonies are nocturnal.

T: 23°–27°C, Ø: 20 cm, AV: 4 (protected genus)

Diploria labyrinthiformis

Order: SCLERACTINIA
Fam.: Faviidae

Echinopora lamellosa (ESPER, 1795)

Hab.: Red Sea, Indo-Pacific to Fiji and Samoa.

Sex.: None.

M.: Its fragility aside, this species fares exceptionally well in aquaria.

Light: Sunlight zone.

B./Rep.: Because they are very brittle, large colonies should be subdivided into smaller colonies.

F.: Zooxanthellae eliminate their need for food in the aquarium.

S.: Usually the folia are horizontal or vertically arranged in an overlapping spiral pattern, but occasionally they form cylindrical tubes. Calyxes are separate, round, and cylindrical with 3–4 septa per centimeter. Reef slopes exposed to a moderate current are where *E. lamellosa* resides. Coloration is variable.

T: 23°–27°C, **Ø:** 80 cm, **TL:** from 150 cm, **WM:** m, **WR** b, **AV:** 3–4 (protected genus), **D:** 3–4

Favia favus (FORSSKÅL, 1775)
Honeycomb coral, pineapple coral

Hab.: Red Sea, Indo-Pacific to Samoa.

Sex.: None.

M.: Unknown.

F.: C; plankton.

S.: *F. favus* forms massive, hemispherical, or encrusting colonies. Calyxes are round to oval, 1–2 cm in diameter, and slightly protuberant. Colonies are nocturnal. Coloration is variable; some colonies may fluoresce.

T: 23°–27°C, Ø: 60 cm, AV: 4 (protected genus)

Favia fragum (ESPER, 1797)

Star coral, golf ball coral

Hab.: Caribbean, Florida to Venezuela.

Sex.: None.

M.: Unknown.

F.: C; plankton.

S.: Colonies are relatively small and either hemispherical or encrusting. Calyxes are round to oval or asymmetrical, 5–6 mm in diameter, with 36–40 septa. Coloration is predominantly yellowish. Stones, coral skeletons, and shipwrecks are colonized in shallow lagoons and along reef slopes to a depth of 30 m.

T: 23°–27°C, **Ø:** 15 cm, **AV:** 4 (protected genus)

Favia maxima VERON, PICHON & WIJSMAN-BEST, 1977
Large star coral

Hab.: Indo-Pacific, Thailand, Malaysia, Australia, New Caledonia.

Sex.: None.

M.: Unknown.

F.: C; plankton.

S.: While *F. maxima*'s massive colonies are not common, they are very conspicuous along reef slopes, their chief habitat. Calyxes are roughly circular and 2–3 cm in diameter. Septa are irregular but systematically aligned. Coloration ranges from light brown to yellow-brown or green.

T: 23°–26°C, **Ø:** 80 cm, **AV:** 4 (protected genus)

Favia speciosa (DANA, 1846)
Star coral, pineapple coral

Hab.: Red Sea, Indo-Pacific, Malaysia.

Sex.: None

M.: Unknown.

F.: C; plankton.

S.: Twelve millimeter calyxes densely cover the surface of these massive, hemispherical, or almost spherical colonies. The septa are fine, numerous, and uniform. Star corals generally inhabit reef slopes and other regions exposed to moderately strong currents.

T: 23°–27°C, Ø: 35 cm, AV: 4 (protected genus)

Favia sp.
Star coral, pineapple coral

Hab.: Indo-Pacific, Malaysia.

Sex.: None.

M.: Unknown.

F.: C; plankton.

S.: Growth is hemispherical, encrusting, or discoid. The polyps are relatively large and nocturnal. Hard substrates of lower reef slopes are where this *Favia* species is most often found.

T: 23°–27°C, Ø: 30 cm, AV: 4 (protected genus)

Favia stelligera (DANA, 1846)
Star coral, pineapple coral

Hab.: Red Sea, Indo-Pacific to Hawaii.

Sex.: None.

M.: Unknown.

F.: C; plankton.

S.: Colonies are hemispherical, columnar, or flat. Some colonies form large individual pillars. They are found along every reef zone at virtually all depths. Coloration is variable.

T: 23°–27°C, Ø: 80 cm, AV: 4 (protected genus)

Favites peresi FAURE & PICON, 1972
Peres star coral

Hab.: Red Sea.

Sex.: None.

M.: Unknown.

F.: C; plankton.

S.: Colonies are massive, hemispherical, or encrusting. Calyxes are fused. On reef slopes exposed to strong currents, *F. peresi* colonizes hard substrates.

T: 23°–27°C, **Ø:** 20 cm, **AV:** 4 (protected genus)

Favites sp.
Moon coral

Hab.: Red Sea.

Sex.: None.

M.: Unknown.

F.: C; plankton.

S.: Colonies are hemispherical or planar. Calyxes are regular, dense, and polygonal in cross-section; they are separated from each other by high walls. Septa are serrated. These corals are normally found on current-swept sites along reef slopes. Coloration is variable.

T: 23°–27°C, Ø: ?, AV: 4 (protected genus)

Manicina areolata var. *mayori* (LINNAEUS, 1758)
Common rose coral

Hab.: Caribbean, Florida to Colombia.

Sex.: None.

M.: Unknown.

F.: C; plankton.

S.: This is a growth form of *Manicina areolata*; however, the variety *mayori* is larger, and its broad base is affixed to the substrate. The common rose coral is hemispherical with meandering fossa. Because the common rose coral can clean itself, it occurs in sediment-rich as well as clear waters. Coloration is variable.

T: 23°–27°C, Ø: 15 cm, AV: 4

Montastrea annularis (ELLIS & SOLANDER, 1786)
Boulder coral, boulder star coral

Hab.: Caribbean, Florida to Colombia.

Sex.: None.

M.: Unknown.

F.: C; plankton.

S.: Boulder corals are massive and hemispherical in shallow water, but flat and broad and either encrusting or shingle-like in deep water. These forms are ecomorphs, i.e., intrapecific variations as a result of certain environmental factors, in this case current and light (depth). Calyxes are star-shaped, and polyps are extended at night. Coloration is variable. *M. annularis* is one of the most important reef-forming corals of the Caribbean.

T: 23°–27°C, Ø: 70 cm, AV: 4

Montastrea cavernosa (LINNAEUS, 1767)
Great star coral, large-cupped boulder coral

Hab.: Caribbean, Florida to Brazil and western Africa.

Sex.: None.

M.: Unknown.

F.: C; plankton.

S.: Along many clear water Caribbean reefs, this species plays a very important role in reef-building. The growth form adopted depends on the depth at which the colony lives. The 10 mm calyxes densely cover the colony's surface. Polyps are anemonelike, variably colored, and nocturnal. The great star coral inhabits shallow to deep water habitats along reef slopes and on coral heads of fore reefs.

T: 23°–27°C, **Ø**: 60 cm, **AV**: 4

Moseleya latistellata QUELCH, 1884
Large star coral

Hab.: Indo-Pacific, Malaysia, Australia.

Sex.: None.

M.: Unknown.

F.: C; plankton.

S.: Colonies are flat or massive and frequently unattached. As a result of extratentacular reproduction (budding), *M. latistellata* has a large central calyx with several small calyxes along its periphery. Septal teeth are fine and numerous, and polyps are very fleshy. Colonies are nocturnal. Reef slopes exposed to strong currents are preferred. Coloration is variable.

T: 23°–27°C, **Ø:** 50 cm, **AV:** 4

Platygyra lamellina (EHRENBERG, 1834)
Indo-Pacific brain coral

Hab.: Red Sea, Indo-Pacific to Polynesia.

Sex.: None.

M.: Unknown.

F.: C; plankton.

S.: Colonies are massive and hemispherical or flat. Valleys and walls are long and narrow. The rounded septa are densely arranged. Most *P. lamellina* colonies occur on reef platforms and within the tidal zone of upper reef slopes.

T: 23°–27°C, Ø: ca. 100 cm, AV: 4 (protected genus)

Solenastrea hyades (DANA, 1848)
Knobby star coral

Hab.: Caribbean, Florida to Venezuela.

Sex.: None.

M.: Unknown.

F.: C; plankton.

S.: Knobby star corals are extremely rare, solitary ahermatypic corals. They are massive and predominantly hemispherical with an irregular, lumpy surface. From shallow water to a depth of 20 m, colonies live on mucky substrates and close to harbors, even in sediment-rich waters.

T: 23°–27°C, Ø: 20 cm, AV: 4

Trachyphyllia geoffroyi (aquarium photo).

Trachyphyllia geoffroyi (AUDOUIN, 1826)
Rose coral

Hab.: Red Sea, tropical Indo-Pacific.

Sex.: None.

M.: This species can be maintained in an aquarium for years. Moderate light, current, and a hard substrate should be provided.

F.: C; plankton.

Light: Moderate light zone.

S.: *T. geoffroyi* is a secondary free-living coral. Initially, the colony is attached to a stone or a shell of a bivalve or gastropod. As it grows, the weight increases accordingly and the colony breaks from its point of attachment. Adults are small, flat, and detached. Polyps are large and fleshy. Colonies occur in a variety of colors such as green, pink, blue, and brown, and some fluoresce. Tentacles are only extended at night. A few colonies are found on crushed coral substrates, but the great majority occur on sand- or mud-bottomed lagoons or reef slopes.

T: 22°–29°C, Ø: 12 cm, TL: from 100 cm, WM: m, WR: m–b, AV: 3–4, D: 2–3

Galaxea fascicularis (LINNAEUS, 1767)
Star coral, scalpel coral

Hab.: Red Sea, Indo-Pacific to Fiji and Samoa.

Sex.: None.

Soc.B./Assoc.: *G. fascicularis* can easily be associated with small fishes and invertebrates.

M.: Aquarium maintenance is not difficult. Intense illumination and a strong current are required. High nitrate concentrations are detrimental.

Light: Sunlight zone.

F.: Substitute plankton, *Artemia* nauplii and adults, and *Mysis*.

S.: Colonies are flat, encrusting, hemispherical, and unbranched. Calyxes are round to oval with sharp, spike-shaped septa. The white-tipped tentacles are only extended at night. Star corals occur from shallow to deep waters along calm reef slopes which nevertheless have good circulation. Coral skeletons are their substrate of choice. Coloration is variable.

T: 23°–27°C, Ø: 20 cm, TL: from 150 cm, WM: s, WR: t, AV: 4 (protected genus), D: 3–4

Oculina diffusa LAMARCK, 1810
Ivory bush coral, diffuse ivory bush coral

Hab.: Caribbean, Florida to Venezuela.

Sex.: None.

M.: Unknown.

F.: C; plankton.

S.: Ivory bush corals are found in lagoons, seaweed beds, and shallow waters that have high rates of sedimentation. Most lie detached on the sediment; a few adhere to a hard substrate. Colonies are fascicular and richly branched. Corallites are small, knoblike, and clearly separate.

T: 23°–27°C, Ø: 10 cm, AV: 4 (protected genus)

Order: SCLERACTINIA
Fam.: Meandrinidae

Dendrogyra cylindrus EHRENBERG, 1834
Pillar coral

Hab.: Caribbean, Florida, Bahamas, San Andrés, Providencia.

Sex.: None.

M.: Unknown.

F.: C; plankton.

S.: *D. cylindricus* has a broad base which supports several massive, very tall, bristly pillars. Clusters of pillars are frequent. Polyps are almost always extended. This species settles on coral rocks in shallow, calm waters; it does not grow in the surge zone.

T: 23°–27°C, Ø: 100 cm, AV: 4 (protected genus)

Dichocoenia stokesi EDWARDS & HAIME, 1848
Stokes' starlet coral, elliptical star coral

Hab.: Caribbean, Florida, Bahamas to Brazil.

Sex.: None.

M.: Unknown.

F.: C; plankton.

S.: Stokes' starlet corals are hemispherical or flat with round, oval, or elongated calyxes and star-shaped polyps that are only extended at night. From shallow water to a depth of 70 m, colonies inhabits current-swept reef slopes. They are predominantly yellow.

T: 23°–27°C, Ø: 50 cm, AV: 4 (protected genus)

Meandrina meandrites (LINNAEUS, 1758)
Tan brain coral, maze coral

Hab.: Caribbean, Florida, Bahamas, Bermuda.

Sex.: None.

M.: Unknown.

F.: C; plankton.

S.: Colonies are irregular, massive, flat, hemispherical or, in deep water, shinglelike. Valleys are meandering, long, and sinuous in sections. Polyps are only extended at night. Tan brain corals are very common on Caribbean reefs, where they inhabit virtually all depths and reef zones.

T: 23°–27°C, Ø: 30 cm, AV: 4 (protected genus)

Meandrina meandrites var. *danai* (LINNAEUS, 1758)
Butterprint rose coral

Hab.: Caribbean, Florida to Colombia.

Sex.: None.

M.: Unknown.

F.: C; plankton.

S.: This relatively small colonial coral lives detached on sand and rock substrates. The colony's base is conical. By inflating its tissue, *M. m.* var. *danai* can right itself when it is flipped over by fishes. It is also capable of ridding itself of heavy sedimentation, but it will continue to grow even if part of the colony becomes covered. The butterprint rose coral inhabits unstable substrates and thick seaweed beds to depths of more than 50 m.

T: 23°–27°C, **L:** 12 cm, **AV:** 4 (protected genus)

Cynarina lacrymalis (EDWARDS & HAIME, 1848)
Teary star coral

Hab.: Southern Red Sea, Indo-Pacific to New Caledonia.

Sex.: None.

M.: This species does well in captive environments, largely because of its tolerance of a wide range of biotic and abiotic factors. Brown animals should be placed in dimly lit sections.

Light: Moderate light zone.

F.: Substitute plankton, *Artemia* nauplii.

S.: This solitary species is either free-living or attached to a hard substrate by a slender base. The polyp is round or oval, very large, fleshy, and fluorescent under actinic light. Its surface has blisterlike protuberances. Septa are radial and sharply dentate. *C. lacrymalis*'s tentacles are predominantly extended at night.

T: 23°–27°C, Ø: 10 cm, AV: 3

Isophyllia multiflora VERRILL, 1901
Rose coral

Hab.: Caribbean, Florida to Colombia.

Sex.: None.

M.: Unknown.

F.: C; plankton.

S.: *I. multiflora* is massive and hemispherical with meandering valleys. Ridges are broad with 8–12 large, dentate septa per centimeter. From shallow to deep water, this coral inhabits outer reef slopes. Coloration is variable.

T: 23°–27°C, Ø: 10 cm, AV: 4

Isophyllia sinuosa (ELLIS & SOLANDER, 1786)
Sinuous cactus coral

Hab.: Caribbean, Florida to Colombia.

Sex.: None.

Soc.B./Assoc.: Small fishes and shrimp are suitable tankmates.

M.: This species grows best in reef aquaria that have low nitrate and phosphate concentrations. A strong negative reaction is demonstrated towards filamentous algae and excessive numbers of colonial anemones.

F.: Substitute plankton, liquid foods.

S.: Without exceeding a diameter of 20 cm, *I. sinuosa* forms massive, hemispherical, and meandroid colonies. Crests are long and sinuous with 7–9 septa per centimeter. Septal margins have sharp teeth. Sinuous cactus corals are capable of establishing colonies in lagoons and on reef slopes to a depth of 10 m. Some are affixed to hard substrates with a small base, while others are detached. Coloration is very variable.

T: 23°–27°C, **Ø**: 15 cm, **TL**: from 100 cm, **WM**: m, **WR**: m, **AV**: 4, **D**: 4

Lobophyllia corymbosa (FORSSKÅL, 1775)
Lobed cup coral, folded cup coral, meat coral

Hab.: Red Sea, eastern Africa, Indo-Pacific.

Sex.: None.

M.: See *L. hemprichii.*

F.: Substitute plankton.

S.: Colonies are massive and hemispherical. The 2–3 cm, round to oval corallites sit on the terminal ends of club-shaped branched columns. Coloration is variable, depending on the rugose polyp tissue. Lobed cup corals colonize current-swept reef slopes.

T: 23°–27°C, Ø: 50 cm, AV: 4 (protected genus)

Lobophyllia costata (DANA, 1848)
Dentate flower coral, meat coral

Hab.: Red Sea, Indo-Pacific to Tahiti.

Sex.: None.

M.: See *Lobophylllia hemprichii.*

F.: Substitute plankton.

S.: This massive coral attaches to coral rock via a small base. The fossa are long and broad. Septal margins are armed with long, sharp spines. Polyps are fleshy and nocturnal. Coloration is variable.

T: 23°–27°C, Ø: 15 cm, AV: 4 (protected genus)

Lobophyllia hemprichii (EHRENBERG, 1834)
Flower coral, meat coral

Hab.: Red Sea, eastern Africa, Indo-Pacific to the Marshall Islands.

Sex.: None.

M.: All *Lobophyllia* species can be maintained in aquaria, and with good care, they live for years. Direct contact with disc or colonial anemones should be avoided.

F.: At night, feed substitute plankton, plankton, fish eggs, and liquid foods. Maintain a moderate current.

S.: Colonies are massive and flat to hemispherical. Corallites are either individual or merged into meanders and separated from their neighboring corallites or meanders. Polyps are thick, fleshy, and nocturnal. *L. hemprichii* may be the dominant coral species along lower reef slopes, its principal habitat.

T: 23°–27°C, **Ø**: 40 cm, **TL**: from 150 cm, **WM**: s, **WR**: m, **AV**: 4 (protected genus)

Order: SCLERACTINIA
Fam.: Mussidae

Mussa angulosa (PALLAS, 1766)
Large flower coral, spiny flower coral

Hab.: Caribbean, Florida, Bahamas, Colombia.

Sex.: None.

M.: Unknown.

F.: C; plankton.

S.: *M. angulosa*'s large corallites are on the tips of pillar-shaped branches. The large, fleshy polyps have numerous septa, each of which has sharp, pointed serrations along its margin. Coloration is variable; under ultraviolet light, *M. angulosa* will fluoresce. It is most commonly encountered on current-swept reef slopes from shallow water down to depths of more than 60 m. The large flower coral is an extremely aggressive space competitor.

T: 23°–27°C, Ø: 60 cm, AV: 4

Mycetophyllia ferox WELLS, 1973
Rough cactus coral

Hab.: Caribbean, Florida, Jamaica, West Indies, Colombia.

Sex.: None.

M.: Unknown.

F.: C; plankton.

S.: Rough cactus corals are flat, almost round, and meandering. By a small base, they attach to hard substrates at depths of 10–40 m. The long, broad valleys are interconnected at the center of the colony, and some neighboring walls are fused. *M. ferox*, one of the most aggressive corals, is totally intolerant of space competitors. Coloration is variable.

T: 23°–27°C, Ø: 25 cm, AV: 4

Mycetophyllia lamarckiana EDWARDS & HAIME, 1848
Large cactus coral, ridged cactus coral

Hab.: Caribbean, Florida to Colombia.

Sex.: None.

M.: Unknown.

F.: C; plankton.

S.: Colonies are relatively flat with an upright rim. Crests are dentate and have 8–10 septa per centimeter. These relatively small and uncommon corals are found along reef slopes to a depth of 60 cm.

T: 23°–27°C, Ø: 15 cm, AV: 4

Scolymia australis (EDWARDS & HAIME, 1849)
Australian fungus coral

Hab.: Indo-Pacific, Australia.

Sex.: None.

M.: Unknown.

F.: C; plankton.

S.: This small, solitary coral attaches to hard substrates along outer reefs and within bays and harbors to a depth of 50 m, usually on shaded sites under ledges. Larvae are not broadly dispersed; sometimes daughter polyps settle within reach of the mother polyp. The septa are systematically arranged, and each septal margin is sharply dentate. Coloration is variable.

T: 23°–27°C, Ø: 3–4 cm, AV: 4 (protected genus)

Scolymia lacera (PALLAS, 1766)

Large-cupped fungus coral, Atlantic mushroom coral

Hab.: Caribbean, Florida, Bahamas, Bermuda, Colombia.

Sex.: None.

M.: All *Scolymia* enjoy longevity in aquaria. As long as the large-cupped fungus coral is placed in the shade away from direct light, it can be expected to live for 5–6 years in the aquarium.

F.: Feed zooplankton, *Mysis*, or fish fry 2–3 times a week.

S.: The large-cupped fungus coral is the largest solitary coral of the Caribbean. The polyp is exceptionally large, fleshy, and colorful. It divides frequently. Septa are strong and radial with sharp, pointed teeth of variable length. This species is found at moderate to great depths along reef slopes. Coloration is variable.

T: 23°–27°C, Ø: 15 cm, AV: 4

Symphyllia recta (DANA, 1846)
Wrinkle coral

Hab.: Indo-Pacific, Maldives, Thailand, Australia to the Marshall Islands.

Sex.: None.

M.: Unknown.

F.: C; plankton.

S.: Colonies are dome-shaped to hemispherical. Valleys are 1–2 cm wide and normally a distinct color in comparison to the walls. Under close examination, the many small oral openings can be seen in the valleys. The tentacles are retracted during the day. These corals inhabit shallow waters of protected upper reef slopes and fringing reefs.

T: 23°–28°C, Ø: 70 cm, AV: 4 (protected genus)

Symphyllia sp.
Undulate cup coral

Hab.: Indo-Pacific, Malaysia.

Sex.: None.

M.: Unknown.

F.: C; plankton.

S.: Colonies are massive and meandroid and flat to hemispherical. Neighboring walls are fused. The polyps, located in the valleys, only extend their tentacles at night. At moderate depths, *Symphyllia* sp. lives along reef slopes that are exposed to moderate water movement.

T: 23°–27°C, Ø: 30 cm, AV: 4 (protected genus)

Mycedium elephantotus (PALLAS, 1766)
Elephant ear coral

Hab.: Red Sea, Madagascar, Indo-Pacific to Tahiti and the Marshall Islands.

Sex.: None.

M.: Unknown.

F.: C; plankton.

S.: *M. elephantotus* forms flat, almost discoid colonies. Calyxes project from the surface like noses; they are 3–4 mm high and separate from each other. Fishes commonly damage the very thin, uneven edge of the colony. This species is usually found on hard substrates of reef slopes that are subjected to some current. Polyps are nocturnal. Coloration is variable.

T: 23°–27°C, Ø: 40 cm, AV: 4 (protected genus)

Pectinia lactuca (PALLAS, 1766)
Lettuce coral

Hab.: Southern Red Sea, Indo-Pacific to Japan.

Sex.: None.

M.: See *Pectinia paeonia*.

F.: C; plankton.

S.: These flat, foliaceous colonies look like a piece of crumbled-up paper. They have high, sharply ascending, irregular walls which are basically consistent in height. Valleys are of different widths. At a wide range of depths, *P. lactuca* inhabits reef slopes that are exposed to little current. Colonies are nocturnal. Coloration is variable.

T: 23°–27°C, Ø: 25 cm, AV: 4 (protected genus)

Pectinia paeonia (DANA, 1846)
Lettuce leaf coral

Hab.: Southern Red Sea, Indo-Pacific to Japan.

Sex.: None.

M.: The lettuce leaf coral can easily be kept in aquaria, but do not allow sediments to settle in the folds. Avoid placing aggressive stony corals in its immediate vicinity, as *P. paeonia* does not fare well in front of space competitors. It has remarkable regenerative powers.

Light: Dim light.

F.: Feeding is unnecessary, since the polyps have zooxanthellae.

S.: Colonies are flat and broad with irregular, erect projections. They live at moderate depths along fringing reefs that are exposed to strong currents. Coloration is variable.

T: 23°–27°C, Ø: 20 cm, AV: 4 (protected genus)

Order: Scleractinia
Fam.: Merulinidae

Hydnophora rigida (DANA, 1846)
Thorny coral, branch coral

Hab.: Red Sea, Indo-Pacific, Micronesia to New Caledonia.

Sex.: None.

M.: Unknown.

F.: C; plankton.

S.: Colonies are arborescent, richly branched, and lack a massive base. Branch diameter is variable. Small polyps, which are sometimes extended during the day, densely cover the colony's surface. *H. rigida* settles on various reef sections; they are occasionally found in lagoons, but the majority grow on protected reef slopes. Coloration is variable.

T: 23°–27°C, Ø: 15 cm, AV: 4 (protected genus)

Merulina ampliata (ELLIS & SOLANDER, 1786)
Curly encrusting coral

Hab.: Red Sea, eastern Africa, Madagascar, Indo-Pacific to the Philippines and Samoa.

Sex.: None.

M.: Unknown.

F.: C; plankton.

S.: Growth is encrusting, discoid or cabbagelike. *M. ampliata* seems to pass through several distinct growth forms. The colony is initially flat and encrusting with a concentric growth line. Later, small protrusions are produced. These develop into vertical, branched plates. The periphery of the colony is always pale and very thin. Reef slopes and lagoons are its preferred habitat. It is nocturnal.

T: 23°–27°C, **Ø:** 20 cm, **AV:** 4 (protected genus)

Caryophyllia inormata (DUNCAN, 1873)
Round carnation coral

Hab.: Mediterranean.

Sex.: Polyps are viviparous.

M.: With sufficient food, a strong current, oxygen-rich water, and temperatures that are markedly less than 20°C, *C. inormata* can be maintained in an aquarium.

Light: Dim light.

F.: C; plankton.

S.: *C. inormata* is a solitary coral. They are frequently arranged in small groups. Never, however, are they organized like a colony. Generally the calyx is round in cross-section and the basal section is unconstricted. Costa are shallow and vertically sinuate. Polyps are usually transparent, and the tentacles have enlarged tips. Round carnation corals are found within dim caves and crevices, beneath overhangs, and on rock walls in deep water environments.

T: 7°–18°C, Ø: 1 cm, TL: 70 cm, WM: m–s, WR: m–b, AV: 3, D: 2–3

Euphyllia ancora (VERON & PICHON, 1980)
Bean coral, hammer coral, anchor coral

Hab.: Indo-Pacific, Australia, Malaysia.

Sex.: None.

M.: All specimens are broken during transit. Because of its toxicity, a species tank is best. *E. ancora* needs calcium-rich water and an attachment point at the bottom of the aquarium.

Light: Dim light.

F.: Substitute zooplankton.

S.: Colonies can reach immense proportions. Polyps cannot be fully retracted, but unless they are contracted, the skeletal structure is obscured. The tentacles are especially long, arranged in rows, and have bulbous tips that look like beans. THese corals colonize reef slopes and reef bottoms, predominantly on horizontal substrates. Colonies are diurnal as well as nocturnal. Coloration is variable.

T: 23°–27°C, Ø: 100 cm, AV: 4 (protected genus)

Eusmilia fastigiata (PALLAS, 1766)
Smooth flower coral

Hab.: Caribbean, Florida to Colombia.

Sex.: None.

M.: Unknown.

F.: C; plankton.

S.: *E. fastigiata* has short, thick branches. Assorted small invertebrates and algae utilize the dead base as a substrate. Corallites may be separate or 2 or 3 may fuse. The margins of the 72 septa are sharp and smooth. It is found from shallow to deep water environments, often covering extensive areas. Most colonies are yellow.

T: 23°–27°C, Ø: 10 cm, AV: 4

Eusmilia fastigiata var. *flabellata* (PALLAS, 1766)
Elongate smooth flower coral

Hab.: Caribbean, Bahamas, Florida, Jamaica, Bonaire.

Sex.: None.

M.: Unknown.

F.: C; plankton.

S.: Elongate smooth flower corals are a growth form of *E. fastigiata*. Corallites are elongated, twisted, and widely-spaced. Polyps are nocturnal and virtually transparent. Various small invertebrates, predominantly sponges, occupy the base of the corallites and the interspaces. Colonies may cover extensive tracts of lower reef slopes.

T: 20°–27°C, Ø: 15 cm, AV: 4

Physogyra lichtensteini EDWARDS & HAIME, 1851
Pointed bladder coral

Hab.: Indo-Pacific, Australia, the Coral Sea.

Sex.: None.

M.: Because of their size, only broken pieces reach importers. These pieces quickly expire. The skeletons are very attractive.

F.: Plankton, *Artemia* nauplii.

S.: *P. lichtensteini* is massive and meandroid with numerous broad valleys and long, pointed septa. It is covered by a multitude of finger- to grape-shaped vesicles during the day; when disturbed, the vesicles are immediately withdrawn. The polyps, which are not totally retractable, are only extended at night. *P. lichtensteini* inhabits protected reefs, frequently on vertical walls. Coloration is variable.

T: 23°–27°C, Ø: 30 cm, AV: 4 (protected genus)

Plerogyra sinuosa (DANA, 1846)
Bubble coral, bladder coral, pearl coral

Hab.: Red Sea, Indo-Pacific to Samoa.

Sex.: None.

M.: See *Physogyra lichtensteini*.

F.: C; plankton, *Artemia* nauplii.

S.: Colonies are meandroid, occasionally gigantic, with blister- to grape-shaped vesicles covering their surface. White lines on the surface of the vesicles mark high densities of nematocysts. Polyps are nocturnal. Valleys are deep, and the walls are warty. Bubble corals live on exposed reef slopes.

T: 23°–27°C, Ø: 100 cm, AV: 4 (protected genus)

Polycyathus muellerae (ABEL, 1959)
Mueller's coral

Hab.: Mediterranean to Portugal and the Canary Islands.

Sex.: None.

M.: Unknown.

F.: C; plankton.

S.: This ahermatypic coral mainly occurs in lawnlike stands in shallow-water caves and grottos and beneath overhangs (<30 m). Polyps are by and large transparent, and the tentacles have small spherical tips.

T: 10°–18°C, Ø: 30 cm, AV: 4

Fam.: Dendrophylliidae

Astroides calycularis (PALLAS, 1766)
Star coral, cup coral

Hab.: Southern Mediterranean, Morocco, Algeria, Tunisia, Sicily.

Sex.: Polyps are viviparous.

M.: *A. calycularis* is best kept in a dark aquarium. Algae are detrimental.

F.: Colonies must be fed individually with *Artemia*, *Mysis*, and substitute plankton.

S.: This species is colonial. Growth is either flat and dense or tall and sparse. The star coral forms mats on rocks and stones within caves and under overhangs. Generally the upper edges of the calyxes are fused. Septa are thin and straight. Polyps are yellow to yellow-orange and lack zooxanthellae.

T: 12°–20°C, L: 2–4 cm, AV: 4

Order: Scleractinia
Fam.: Dendrophylliidae

Balanopyllia europaea (RISSO, 1826)
Wart coral

Hab.: Mediterranean, Atlantic.

Sex.: None.

M.: This species can be kept successfully for many years in an aquarium. However, *B. europaea* must be introduced with its substrate.

Light: Sunlight zone.

F.: C; *Cyclops*, *Daphnia*, and small planktonic crustacea.

S.: *B. europaea* is a solitary coral of the upper sublittoral zone of rocky coasts. It is well adapted to the surge zone of shallow waters; there it cements itself to stones, rocks, or other hard substrates. The polyps are retracted during the day, only opening at night. There are wartlike growths on the transparent tentacles of the polyps (name!). Calyxes are broader than they are tall with a thickened, porous rim. Septal margins are granular to finely acicular.

T: 10°–24°C, Ø: 1–1.5 cm, TL: 80 cm, WM: m–s, WR: t, AV: 3, D: 2

Leptopsammia pruvoti LACZE-DUTHIERS, 1897
Yellow carnation coral

Hab.: Mediterranean, uncommon in the Adriatic Sea.

Sex.: None.

M.: It can be kept in dimly lit aquaria.

Light: Dim light.

F.: Substitute plankton.

S.: *L. pruvoti* may be solitary or colonial. Colonies are low-growing and lawnlike. They are generally encountered at depths of 10–50 m on secondary hard substrates and rocks, particularly on cave walls and beneath overhanging ledges. Sometimes there are great numbers of *L. pruvoti* at these sites. Calyxes are irregular with straight to slightly curved septa and costal ridges. Polyps do not have zooxanthellae.

T: 23°–27°C, **L**: 2 cm, **AV**: 3

Order: Scleractinia
Fam.: Dendrophylliidae

Tubastrea coccinea LESSON, 1834
Red coral, orange coral, orange cup coral

Hab.: Cosmopolitan, Red Sea, Caribbean, Indo-Pacific.

Sex.: None.

Soc.B./Assoc.: Small fishes and invertebrates make appropriate tankmates. If possible, avoid aufwuchs feeders.

M.: Red corals have high feeding requirements. Place colonies on a shaded site on a slanted or vertical substrate. They thrive best when they share the tank with numerous calcareous algae. Algae cover, nitrates, and phosphates are detrimental to the colonies. Since the animals are crepuscular, actinic lighting is recommended.

F.: Stimulate the animals to feed with very fine plankton. Afterwards, *Artemia* nauplii can be introduced. Polyps also feed on small fishes in nature.

S.: These animals prefer to live on shaded sites along vertical walls. They cover large areas of harbor installations and shipwrecks and small spaces under overhangs. Polyps are a bright orange with yellow tentacles; they lack zooxanthellae. In locals where illumination is minimal during the day, the tentacles are extended.

T: 23°–27°C, Ø: 30 cm, TL: from 150 cm, WM: m, WR: m–b, AV: 4, D: 4

Tubastrea micrantha EHRENBERG, 1834
Black cup coral, black tube coral

Hab.: Red Sea, Indo-Pacific.

Sex.: None.

M.: See *T. aurea*.

F.: Substitute plankton, *Artemia* nauplii.

S.: Colonies are erect, richly branched, and arborescent. Black cup corals are typically found growing on coral skeletons in the surf zone of the fore reef. There they orient themselves perpendicular to the current. Most small colonies occur at cave entrances and under overhanging ledges. The polyps are twilight and night active and lack zooxanthellae. Coloration is variable, but the polyps are never black or dark brown like the colony.

T: 23°–27°C, L: 80 cm, AV: 3

Turbinaria frondens (DANA, 1846)
Cup coral

Hab.: Indo-Pacific, Thailand, Japan, Fiji, Samoa, Australia.

Sex.: None.

M.: Unknown.

F.: C; plankton.

S.: *T. frondens* lives in deep waters on all types of substrates. Growth is horizontal or vertical and predominantly cup-shaped. Cup-shaped colonies have a slender base and a large circular opening. Coloration ranges from greenish brown to yellowish or gray.

T: 20°–24°C, L: 80 cm, AV: 4

Turbinaria mesenterina LAMARCK, 1816
Folded lettuce coral, scroll coral

Hab.: Red Sea, eastern Africa, Indo-Pacific.

Sex.: None.

Soc.B./Assoc.: This coral can be associated with other corals, cleaner shrimp, and fishes. It thrives when housed with numerous red algae.

M.: It is relatively maintainable in aquaria. Unfortunately, only broken fragments reach stores.

F.: Colonies should be fed specifically 1–2 times per week with substitute plankton.

S.: *Turbinaria* is the only genus of reef-forming corals within the family Dendrophylliidae. Colonies are vase-shaped, massive, or lobed folia and plates. Folded lettuce corals are chiefly encountered on clear water reef slopes that have good water circulation; there they cover up to several square meters of coral substrate. The polyps are pale, and possess zooxanthellae. Colonies are nocturnal.

T: 23°–27°C, Ø: 50 cm, AV: 2

The coiled wire coral, *Cirrhipathes anguinea*, Maldives.

Order Antipatharia

Antipatharians are commonly called black corals because of their brown to black skeleton. In the jewelry trade, the skeletons are fashioned into beads, necklaces, amulets, etc. Each polyp of these colonial corals has 6 tentacles. Colonies generally grow on hard substrates.

Unlike stony corals, the skeleton of black corals does not contain calcium. It is thorny and made of an elastic, hornlike substance not unlike the skeletal material of gorgonians. While there are parallels between gorgonian and antipatharian skeletons, there are differences as well.

The axial skeleton is covered by a thin, soft, living layer, the coenenchyme. Embedded within and interconnected through the common basal substance of the coenenchyme are the cylindrical, small—just a few millimeters in size—polyps. Though rare, some polyps will have eight instead of six tentacles encircling their oral opening. Their short, thick, unbranched tentacles are non-retractable.

Antipatharians feed on minute phytoplankton which is whorled in by the ciliated epithelium.

Colonies have a wide variety of growth forms, e.g., long and whiplike, brush- and fan-shaped, fascicular, and arborescent. Fans and feathers are formed when branches emerge from a single level of the stem. Neighboring branches may fuse together, creating reticulate growth patterns. Three or more branches may grow radially from the main axis. Colonies with a series of whorled branches look like a bottle brush.

When unfurled, whiplike forms may reach an impressive length of 6 m. Some species never exceed a few centimeters.

Very little is known about the biology of black corals. This paucity of information is primarily due to the fact that the great majority of species grow at depths beyond 100 m. Few are distributed from 0 to 50 m depth. Species descriptions are often done on specimens that have been collected from dredged materials. It is a rare privilege for marine biologists to observe black corals in their natural habitat.

Order: ANTIPATHARIA

Most known species live in warm tropical seas. A few exceptions occur along temperate latitudes such as the central and western Atlantic and the Mediterranean.
Likewise, very little is known about their reproductive biology. It is known that the polyps of one colony are either male or female. Hermaphroditic colonies seem to be the minority.
So far there are no known predators, although numerous animals are frequently observed living on or among their branches, e.g., crustacea, small fishes, brittle stars, bivalves, snails, and crinoids. The exact interactions between black corals and these organisms is unknown.
Most black corals are protected by the Washington Endangered Species Act. They play an insignificant role in the marine aquarium hobby.

Taxonomy

Phylum: CNIDARIA	Cnidarians
Class: ANTHOZOA	Anthozoans
Subclass: HEXACORALLIA	Six-Tentacled Anthozoans
Order: ANTIPATHARIA	Black Corals

Family: Antipathidae
Genera: *Antipathes*
Stichopathes
Cirrhipathes

Antipathes atlantica GRAY, 1857
Gray sea fan, black coral

Hab.: Caribbean, Trinidad, Jamaica, Colombia.

Sex.: Colonies are dioecious (separate sexes).

M.: Unknown.

F.: H; plankton.

S.: Colonies are erect, richly branched, and attached with a small holdfast. The delicate branches emerge from a short stalk. Polyps are extended during the day as well as at night. Fan-shaped colonies are usually positioned transverse to the prevailing current. Gray sea fans are found in deep water environments.

T: 18°–22°C, L: 80 cm, AV: 4 (protected genus), D: 4

Order: ANTIPATHARIA
Fam.: Antipathidae

Antipathes dichotoma PALLAS, 1766
Black coral

Hab.: Red Sea, Indo-Pacific to the Galapagos Islands.

Sex.: Dioecious, i.e., male and female gametes are produced by different colonies.

M.: Unknown.

F.: H; plankton.

S.: The richly branched, fascicular or fan-shaped colonies are oriented transverse to the current. Numerous thin, thorny branches radiate from the stem. The largest specimens hail from tropical seas. *G. dichotoma* generally occurs on hard substrates of reef slopes and terraces in deep water. Jewlery is fashioned from the black skeletons.

T: 22°–26°C, **L:** 100 cm, **AV:** 4 (protected genus), **D:** 4

Antipathes furcata GRAY, 1857
Black coral

Hab.: Caribbean, Bahamas, Bermuda, Jamaica, Barbados, Colombia.

Sex.: Dioecious.

M.: Unknown.

F.: H; plankton.

S.: Supported by a single short stem, the erect, delicate, long branches grow within one plane. This species does not have pinnules. It is found in deep water environments on all hard substrates.

T: 18°–22°C, L: 60 cm, AV: 4 (protected genus), D: 4

Antipathes hirta GRAY, 1857
Bottle-brush black coral

Hab.: Caribbean, Trinidad, Barbados, Colombia.

Sex.: Male and female gametes are produced on separate colonies.

M.: Unknown.

F.: H; plankton.

S.: Above is a detail photo of a stalk of *A. hirta*. Because of the arrangement of the pinnules around the stalk, the scant branches look like bottle brushes. Both sides of the stalk have 4–6 rows of pinnules. This species is found on deep reef slopes down to a depth of 80 m.

T: 18°–22°C, L: 50 cm, AV: 4 (protected genus), D: 4

Antipathes thamnea WARNER, 1981
Black coral

Hab.: Caribbean, Trinidad, Barbados, Colombia.

Sex.: Colonies are dioecious.

M.: Unknown.

F.: H; plankton.

S.: *A. thamnea* chiefly colonizes hard substrates in deep water biotopes. Colonies are erect with numerous branches which extend in a single plane. Pinnules are arranged in long rows along the branches. Depending on age, the coenenchyme is either rust red (young) or brown (old). The polyps look like little anemones.

T: 18°–22°C, L: 60 cm, AV: 4 (protected genus), D: 4

Order: ANTIPATHARIA
Fam.: Antipathidae

Antipathes sp.

Hab.: Indo-Pacific, Malaysia.

Sex.: Dioecious.

M.: Unknown.

F.: H; plankton.

S.: Colonies are fascicular. Primary branches are long and whiplike, whereas secondary branches are delicate. *Antipathes* sp. grows on corals and coral debris, only projecting slightly above the substrate. Polyps are diurnal and nocturnal.

T: 22°–26°C, **L:** 40 cm, **AV:** 4 (protected genus), **D:** 4

Antipathes sp.

Hab.: Indo-Pacific, Malaysia.

Sex.: Colonies have separate sexes (dioecious).

M.: Unknown.

F.: H; plankton.

S.: These pinnate colonies grow on dead wire corals (*Cirrhipathes*) along current-swept lower reef slopes. Branches are covered with small, 6-tentacled polyps.

T: 22°–26°C, **L**: 8 cm, **AV**: 4 (protected genus), **D**: 4

Antipathes sp., Malaysia.

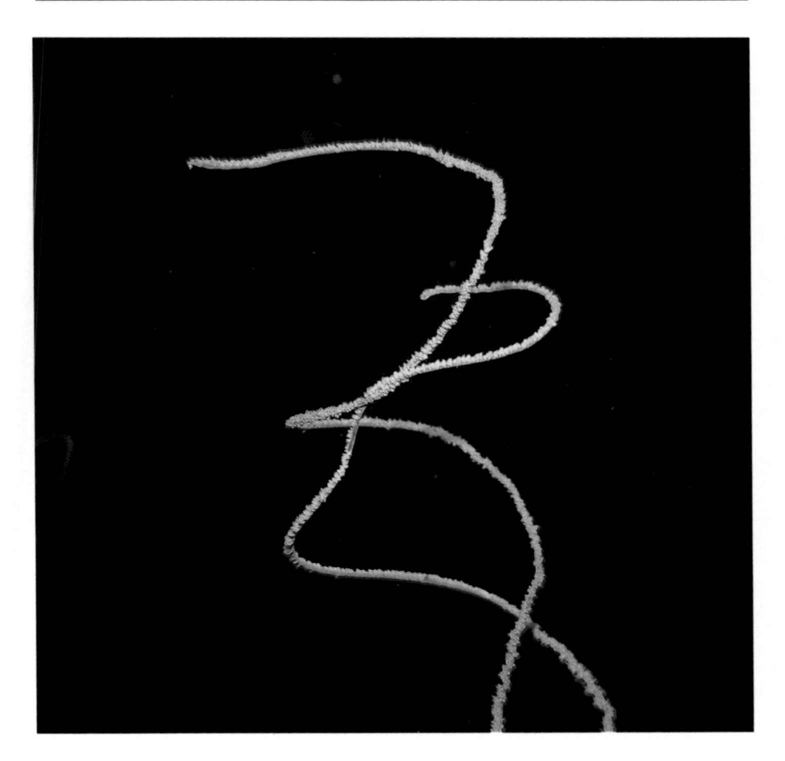

Stichopathes leutkeni BROOK, 1889
Spiraling wire coral

Hab.: Caribbean, Jamaica, Curaçao, Colombia.

Sex.: Dioecious, i.e., male and female gametes are produced by different colonies.

M.: Unknown.

F.: H; plankton.

S.: Capable of attaining a length in excess of 2 m, this species is erect, unbranched, and wirelike. If the colony is large, the terminal third will always be coiled. The polyps, which are found along the stalk, are always extended. It does not have pinnules. *S. leutkeni* is generally found on deep vertical walls, submarine platforms, and beneath ledges.

T: 18°–22°C, L: 150 cm, AV: 4 (protected genus), D: 4

Cirrhipathes anguina DANA, 1846
Coiled wire coral

Hab.: Indo-Pacific, Maldives, Australia, New Caledonia, Hawaii.

Sex.: Dioecious.

M.: Unknown.

F.: H; plankton.

S.: Attached by a small holdfast, *C. anguina* lives on reef slopes and terraces at depths of 15–60 m. It is erect, unbranched, and coiled. Coloration of the coenenchyme varies, but it is usually olive green to orange-red. Polyps are generally light, almost transparent, and expanded during the day.

T: 22°–26°C, **L:** 70 cm, **AV:** 4 (protected genus), **D:** 4

Cirrhipathes sp. (*spiralis* ?)
Spiraling wire coral

Hab.: Indo-Pacific, Maldives.

Sex.: Colonies have separate sexes.

M.: Unknown.

F.: H; plankton.

S.: The terminal third of these erect, whiplike colonies is coiled. With their small holdfast, they attach to deep, steep drop-offs and terraces. The light polyps, which are arranged radially around the stalks, are extended during the day as well as at night. The coenenchyme is rust red. They are very common along Indo-Pacific reefs.

T: 22°–26°C, L: 300 cm, AV: 4 (protected genus), D: 4

Phylum: CTENOPHORA

Comb jellies or sea walnuts are asymmetrical, transparent, delicate organisms that have a simple morphology. Represented by about 80 species, these largely pelagic organisms have a worldwide distribution. Because of their superficial similarity to medusoid cnidarians, comb jellies were initially classified into the phylum Cnidaria. Today it has become clear that comb jellies are very divergent from cnidaria. They have been accordingly placed in a separate phylum, the Ctenophora. Comb jellies do not have a polyploid stage, one major differentiating characteristic between Ctenophora and Cnidaria.

Anatomy

The majority of comb jellies are spherical to ovoid, but even within these shapes there are distinct features that differentiate species. The eight comb rows, characteristic of this phylum, are formed from a series of ciliated plates that are interconnected through basally fused cilia.

The cilia along the comb rows beat in sequence, propelling the comb jellies "head first" through their aqueous environment. As the cilia move, the organisms luminesce. Guidance and coordination of movements are accomplished by an uncentralized nervous system. The nerves are located just beneath the epidermis, achieving their highest density under the eight comb rows.

Besides numerous sensory cells distributed over the entire body, comb jellies have statocysts which function as an equilibrium organ, conveying the animal's relative position in the water. With this information, the organism directs its movements. Each statocyst is covered by a transparent dome and contains statoliths resting on tufts of cilia.

Even though comb jellies are composed of 99% water, their anatomy is elaborate. From the outside inward, the body wall of ctenophores is composed of the epidermis, mesoglea, and the endodermis. The central layer, the mesoglea, lends support to the organism. It is composed of proteinaceous material and longitudinal and circular strands of muscles. Numerous amebocytes circulate within.

Nutrition

Food is ingested by the ctenophores' extendable mouth. From the mouth the food passes through the long pharynx into the infundibulum, or central stomach. The infundibulum has a branched canal system throughout the mesoglea. The intestinal canal divides into four excretory canals, two of which void to the outside through small pores.
The canal system transports nutrients within the body of the comb jelly. For this purpose, it is lined with a ciliated epithelium.

Reproduction

All comb jellies are hermaphrodites, and reproduction is sexual. Fertilization of the gametes predominantly occurs in the open water; a few species incubate their young in a brood pouch. The development of comb jellies differs fundamentally from that of cnidaria. Each meridional canal has two gonads in its thickened wall, an ovary and a testis. Eggs and sperm are released into the sea through the mouth. The fertilized egg develops into a free-swimming cydippid larva. Comb jellies are dissogenic, i.e., twice during their life-span they are sexually mature—once in their larval stage and again as an adult. Shortly following hatching, comb jellies reproduce for the first time. Subsequently, the gonads degenerate and ctenophores attain their full adult size. Once again they mature sexually and reproduce.

Lifestyle

Pelagic species are largely at the mercy of currents. In calm waters, great numbers of ctenophores commonly congregate close to the water surface. They migrate into deep water when seas become turbulent. There are some benthic species of comb jellies. When they were first discovered, these animals were believed to be a transitional form between Cnidaria and Plathelminthes. This remains unconfirmed, and till today, all theories on the origins of comb jellies remain speculation.

Phylum: CTENOPHORA

Members of the class Tentaculata sport two long, retractable tentacles. The tentacles and their pouches, which contain the tentacles in their retracted state, are found on the organism's aboral pole. Extended and hauled through the water as the comb jelly swims, the tentacles catch a myriad of prey. Since Ctenophora lack nematocysts, they have an alternate capture mechanism, the colloblasts. These are tentacular cells which produce an adhesive. Colloblasts can be used repeatedly. From time to time the tentacles are cleaned by the mouth.
Comb jellies that lack tentacles (class Atentaculata) are predominantly predators. By extending their body and greatly stretching the oral opening, they engulf their prey.

Taxonomy

Phylum: CTENOPHORA	Comb Jellies
Class: TENTACULATA	Tentacled comb jellies
Order: CYDIPPIDA	
Family: Pleurobrachiidae	Sea gooseberries
Genus: *Pleurobrachia*	
Order: LOBATA	Warty comb jellies
Family: Bolinopsidae	
Genera: *Leucothea*	
Mnemiopsis	
Order: CESTIDA	Venus girdle
Family: Cestidae	
Genus: *Cestum*	
Class: ATENTACULATA	Nontentacled comb jellies
Order: BEROIDA	Beroe's comb jellies
Family: Beroidae	
Genus: *Beroe*	

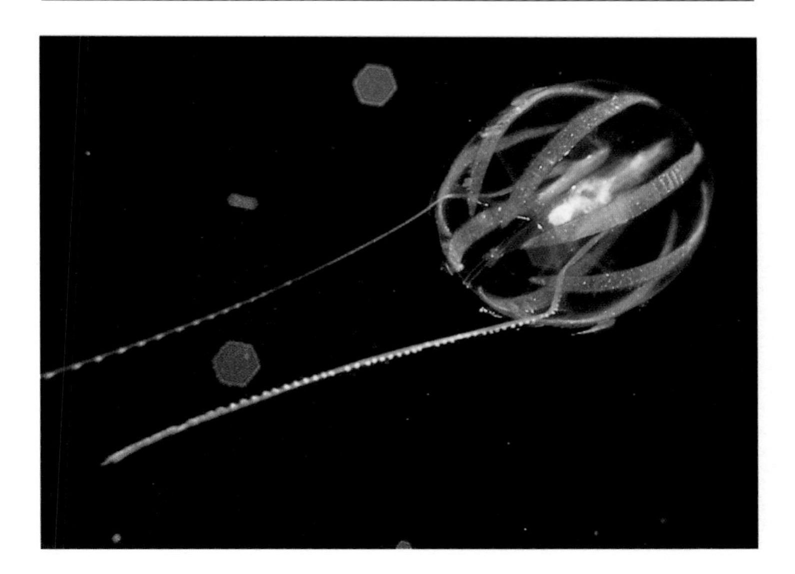

Pleurobrachia pileus (= *rhodophis*) (MÜLLER, 1776)
Sea gooseberry

Hab.: The genus is cosmopolitan; it is found in all European seas.

Sex.: Hermaphrodites.

M.: Unknown.

F.: C; plankton.

S.: Sea gooseberries have spherical to ovoid bodies. The 8 broad comb rows extend from the anal pore to the oral pole and are covered by dense cilia. Locomotion is by whiplike cilia. There are two tentacle pouches, one on each side of the organism. Each pouch holds a long pinnate tentacle. The tentacles, which are 20 times longer than the jelly's cross section, can be extended and dragged as capture threads. Sea gooseberries are pelagic organisms that are highly dependent on current. They are often found in schools. Generally these animals tolerate decreases in salinity and temperature.

T: 0°–26°C, L: 2–3 cm, AV: 4, D: 4

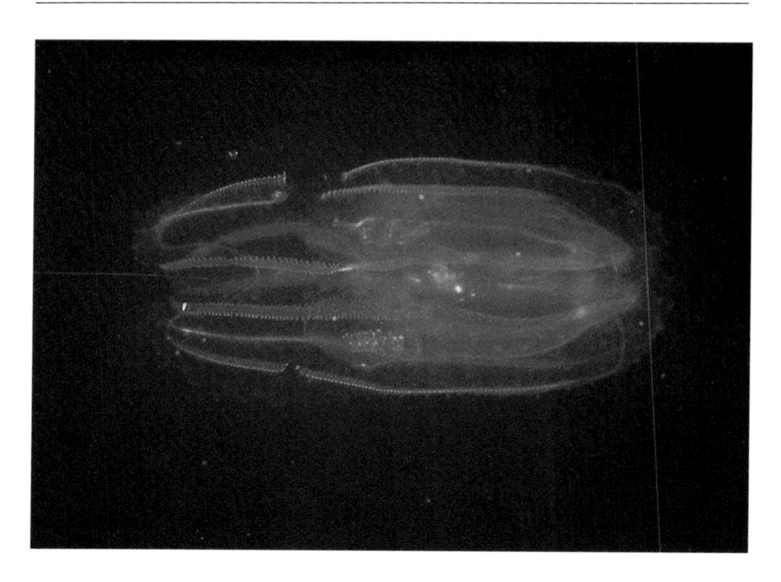

Leucothea multicornis (QUOY & GAIMARD, 1824)
Warty comb jelly

Hab.: Mediterranean, Atlantic, North Sea, western Baltic Sea, Black Sea.

Sex.: Hermaphrodites.

M.: Unknown.

S.: This pelagic species has an elongated ovoid body that is laterally compressed and largely transparent. The animal's oral lobes are the same size as its body. The protrusive corporal papillae have poison glands to aid in food procurement. Two worm-like appendages are located on the tentacles. During the spring and summer, *L. multicornis* forms large schools near coastlines.

T: 8°–18°C, L: 10 cm, AV: 4, D: 4

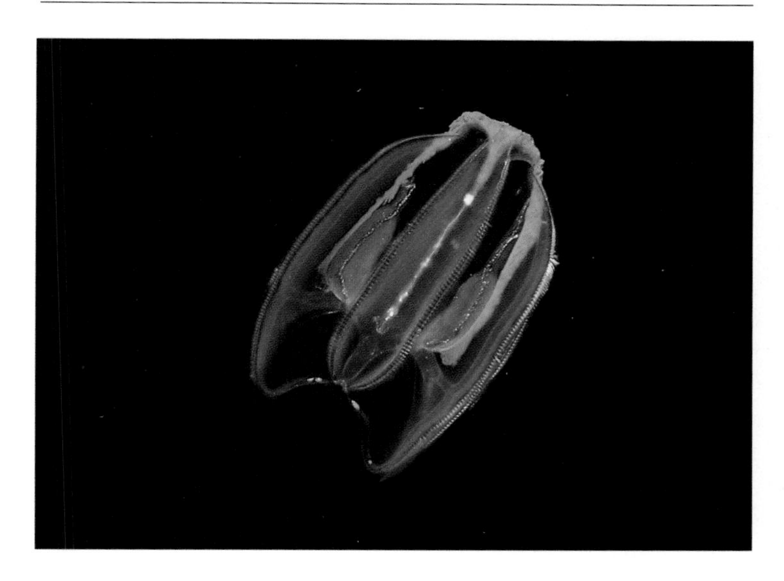

Mnemiopsis leidyi AGASSIZ, 1865
Leidy's comb jelly, sea walnut

Hab.: Western Atlantic, South Carolina, New England, Sea of Asov.

Sex.: Hermaphrodites.

M.: Unknown.

F.: C; plankton.

S.: *M.leidyi* is predominantly pear-shaped with lobate appendages elongating its body wall. Eight bands (combs) lie on a series of plates and continue onto the lobes. Light is emitted along the combs. When this species undergoes a population explosion, oyster beds are at risk, as it consumes oyster larvae. Leidy's comb jelly is pronouncedly tolerant of large temperature and salinity fluctuations.

T: 5°–20°C, L: 10 cm, AV: 4, D: 4

Order: Cestida
Fam.: Cestidae

Venus Girdle

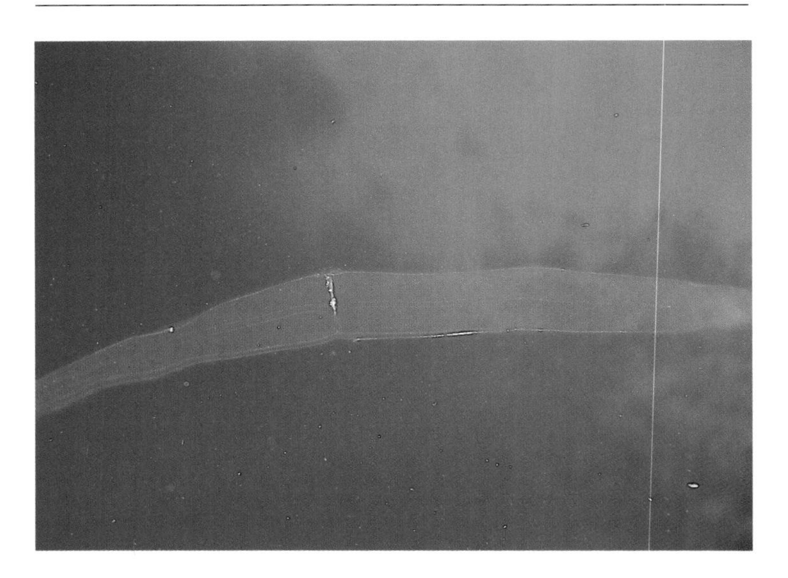

Cestum veneris LESUEUR, 1813
Venus girdle

Hab.: In all warm seas, Mediterranean, tropical Atlantic.

Sex.: None.

M.: Unknown.

F.: Plankton.

S.: The body is greatly compressed, flat, and transparent. In its larval stage, *C. veneris* is spherical. As the larva matures, the animal elongates until it is ribbon-shaped. The four comb rows fluoresce blue-green to blue under mechanical stimulation. The venus girdle "stands" in the water by placing its pharynx plane vertical. It passively drifts in the current or actively swims. During certain seasons, along some coasts, large aggregations are not unusual.

T: 10°–26°C, L: 150 cm, AV: 4, D: 4

Beroe sp.
Beroe's comb jelly

Hab.: Pacific, California.

Sex.: Hermaphrodites.

M.: Unknown.

F.: C; comb jellies.

S.: Beroe's comb jelly is planktonic. With the aid of a long, broad pharynx and strong musculature, it engulfes whole conspecifics. The central stomach is largely vestigial and has shifted towards the oral pole. Juveniles are vitreous and do not have the ability to luminesce; adults are pink and capable of intense bioluminescence. It is extremely difficult to identify species within the genus *Beroe*.

T: 18°–26°C, L: 12 cm, AV: 4, D: 4

Beroe cucumis, Arctic.

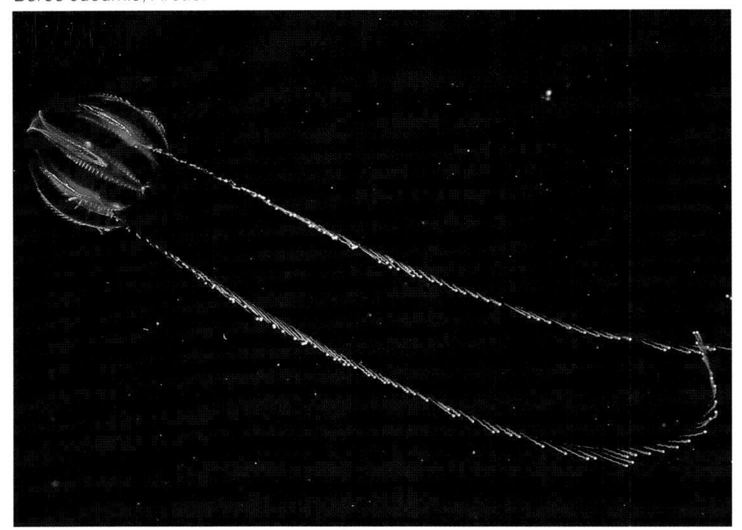

The gooseberry jelly *Mertensia ovum*, Arctic.

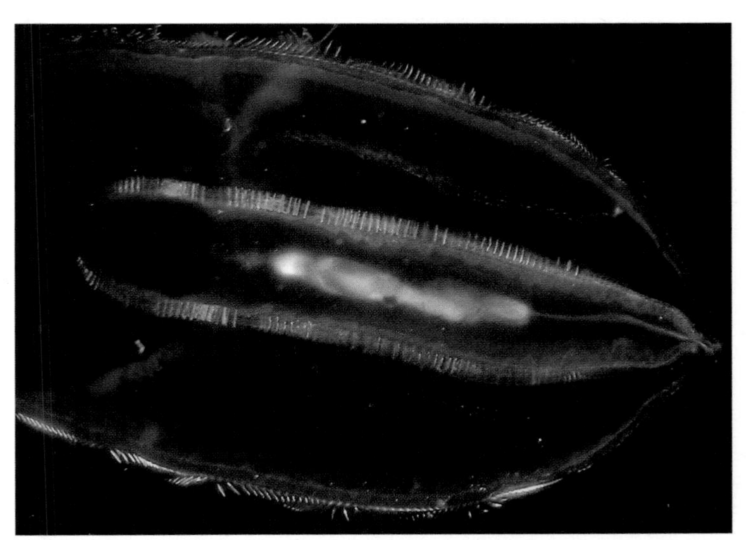

Mertensia ovum with krill in its gastric cavity.

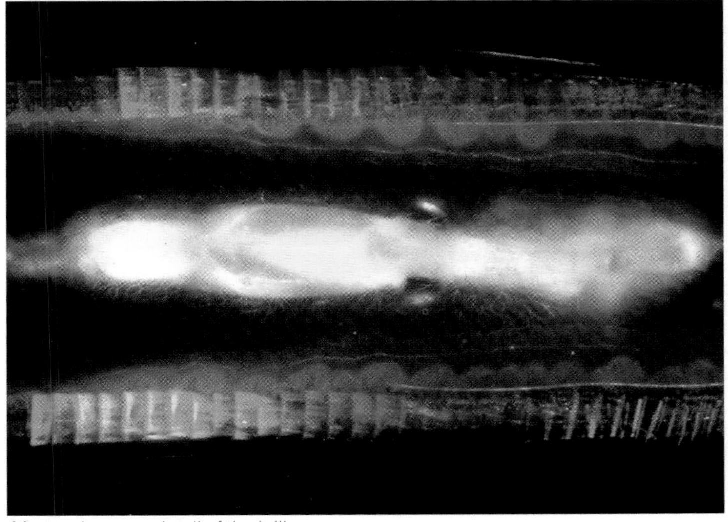

Mertensia ovum, detail of the krill.

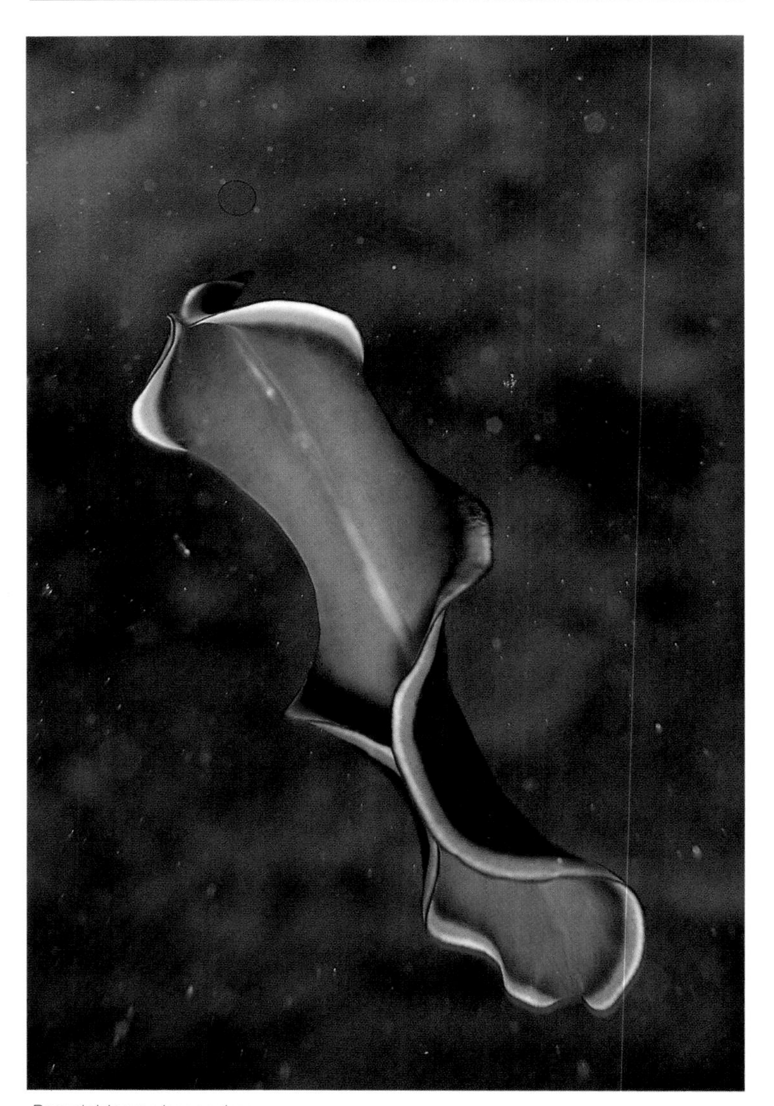

Pseudobiceros hancockanus

Order Polycladida

Polycladida flatworms from the order Polycladida are exclusively found in marine environments, most at 0–100 m depth and some beyond. Species diversity for polyclad flatworms peaks in tropical and subtropical seas. Because of their cryptic habits—they usually live under stones or in cracks and crevices of the littoral zone—polyclads are little known. Rarely are these organisms seen crawling on stones, sand, corals, or seaweed meadows or swimming freely about in their habitat. At first glance, these animals are easily confused with nudibranchs. However, upon close scrutiny, the lack of anal gills and a stout, muscular foot are noted. Polyclads have a broad, dorsoventrally compressed body. They are unsegmented, elongate-ovoid, and frequently have a lobed to undulate body margin.

Tropical and subtropical species are startlingly colorful. In contrast, coldwater species are cryptically colored, frequently transparent.

Size is another characteristic that is highly divergent within this order. Some species are minuscule, only growing to a few millimeters in length, while others grow to the substantial length of 15 cm.

Polyclads have an incredible capacity of regeneration. Injuries heal quickly, and body parts can give rise to entire organisms.

Identifying polyclads is exceedingly difficult, and there are few authorities in the world. Although tropical species can often be identified on the basis of color, other classifying criteria are less obvious, commonly necessitating a microscope. The position and arrangement of the eyes and tentacles and the location of the genital opening and pharynx are examples of the latter. Because many species cannot be identified by photos, some had to be excluded from this text and a few could only be identified to genus.

Anatomy

The entire body surface is covered by fine cilia, and its texture varies from smooth to warty or bristly. Locomotion is the main function of the cilia. Tentacles, when present, are paired, simple, and lobed. The eyes play an important role in sensory perception of light. In polyclads there may be just two eyes, a cluster of eyes, or several rows along the body's axis.

Reproduction

All polyclads are hermaphrodites; therefore, every animal has both female (ovary) and male (testis) gonads. Both the male and female system has independent ducts. Unlike more advanced members of this class, polyclads do not have yolk glands. The eggs themselves contain the yolk. Upon copulation, the sperm migrate to the ovaries and internal fertilization results. Encapsulated, fertilized eggs are adhered to the substrate singly or in spiral strings. Development may be direct or indirect. If direct, a miniature polyclad hatches from the egg. Other species have indirect development, and their eggs hatch into planktonic Müller's larvae. After a few days, the Müller's larva adopts a benthic existence and metamorphoses into a flatworm.

Nutrition

These accomplished predators can even consume organisms larger than themselves. Their diet includes unicellular organisms, sponges, worms, sea squirts, bivalves, carrion of small invertebrates and, occasionally, microalgae. The pharynx is a complex, protrusive organ that participates in feeding. Food is introduced into the mouth, passes through the pharynx, and enters the stomach. Digestion takes place in the diverticular intestine. In view of the absence of an anus, indigestible food is expelled through the pharynx. Many species are feeding specialists and live as commensals on specific animals or colonies.

Movement

Locomotion is via a combination of cilia, muscular contractions, and mucus excretions. The longitudinal and ring musculature of their body makes it possible for the animals to stretch, contract, and curl up. The undulating movements of the lateral edges of the body are particularly evident during locomotion. Many small planktonic species are proficient swimmers.

Polyclads in the Aquarium

Successfully maintaining polyclads in captive environments is still, for the most part, cloaked in mystery. Normally these animals are incidental introductions; "live rock" and corals generally have a few hidden amongst their crannies. Polyclads are not collected and introduced into the trade specifically.

Since turbellarians are largely reclusive animals, they hold little interest for marine aquarists. Further, the attractive tropical polyclads from the genus *Pseudoceros* have dietary requirements that are not easily satisfied in aquaria. Death is ensured if they cannot be offered sponges, sea squirts, and other invertebrates. Caution! It has been recognized that *Pseudoceros zebra*, the zebra polyclad, grazes on soft corals of the genus *Sinularia*. Without leaving any visible marks, *P. zebra* eats away the mucus coating of the coral. Whether or not this activity is detrimental to the coral is unknown. Further study is merited.

Red planarians can become a plague in marine aquaria. These dwarf pests enter the aquarium on live rock, and under metal halide illumination, they thrive and multiply explosively. Although all their nutritional needs are met by symbiotic zooxanthellae, their mucus secretions are so copious that everything in the aquarium is soon covered and asphyxiated. If allowed to reproduce unchecked, the population of planarians will begin to cover the aquarium's occupants, thereby blocking light and killing organisms dependent on zooxanthellae. Concurat L, a product from veterinary medicine, is an effective agent against planarians, and fortunately it is relatively harmless to aquarium inhabitants. While thoroughly cleaning the aquarium and all its fixtures helps combat these interlopers, it is by no means an effective, long-lasting measure; just consider their seemingly supernatural rate of reproduction.

Taxonomy

Phylum: PLATHELMINTHES	Flatworms
Class: TURBELLARIA	Turbellarians
Order: POLYCLADIDA	Polyclads
Suborder: Cotylea	

Family: Pericelididae
Genus: *Pericelis*

Family: Pseudocerotidae
Genera: *Acanthozoon*
Prostheceraeus
Pseudoceros
Pseudobiceros

Acanthozoon sp., Maldives.

Pericelis sp.
Leopard polyclad

Hab.: Caribbean, Florida, Bahamas, Indo-Pacific.

Sex.: Hermaphrodites; therefore, there are no differences.

M.: Unknown.

F.: C; they graze on sea squirts.

S.: Most specimens are found beneath stones and coral fragments, where they are well camouflaged. The large, broadly elliptical, dark brown body is covered by a multitude of big, round to oval spots which vary from yellow-brown to black-brown. Two short sensorial antennae are located on the head.

T: 20°–28°C, L: 6 cm, AV: 4, D: 4

Order: POLYCLADIDA
Fam.: Pseudocerotidae

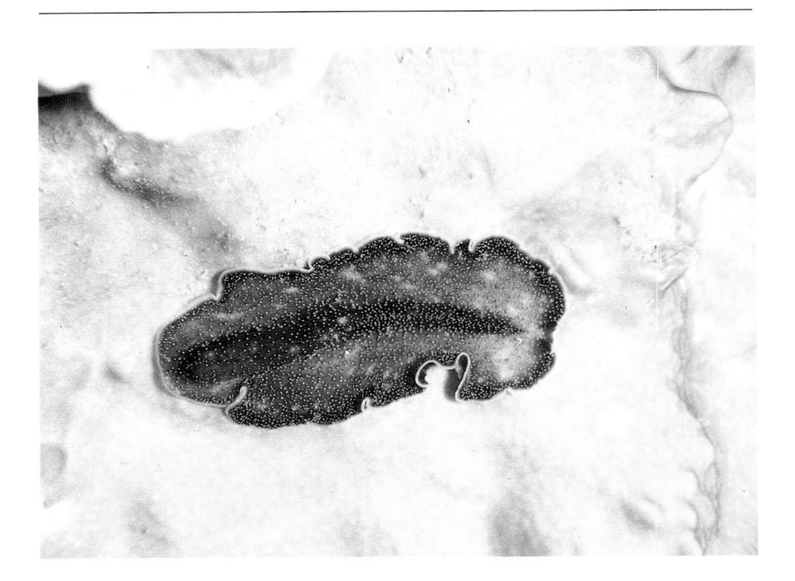

Acanthozoon sp.
Dotted polyclad

Hab.: Indian Ocean, Maldives.

Sex.: Hermaphrodites, i.e., there are no sexual differences.

M.: Unknown.

F.: C; various colonial invertebrates.

S.: Superiorly, the elongate, oval body of *Acanthozoon* sp. is dark brown to black with numerous golden dots, light spots, and bands. The inferior side is ash gray to black with a white stripe along its margin. Both ends of the body are rounded, and the edge is always undulant. The dotted polyclad has one pair of antennae.

T: 20°–28°C, L: 4 cm , AV: 4, D: 4

Indian Ocean, Baa Atoll, Maldives.

Prostheceraeus giesbrechtii LANG, 1884
Band planaria

Hab.: Mediterranean.

Sex.: None; all animals are hermaphrodites.

M.: Unknown.

F.: C; this species grazes on sea squirts.

S.: *P. giesbrechtii* is flat and ribbon-shaped with a sinuous fringe. The anterior body has well-developed tentacles and numerous eyes; the posterior end of the body is pointed. Coloration is variable. Superiorly, the body is off white with a bisecting orange to lemon yellow stripe and pink, reddish, and dark brown to black stripes between the center stripe and the lateral edges of the body. The inferior side of the body is white. Generally this species occurs among plants, colonial animals, or *Ciona*, a genus of sea squirts.

T: 8°–22°C, L: 3 cm, AV: 4, D: 4

Pseudoceros sp.
Geode polyclad

Hab.: Tropical Indo-Pacific, Maldives.

Sex.: None; all animals are hermaphrodites.

M.: Unknown.

F.: C; bryozoans.

S.: This polyclad has an elongated body, a pair of tentacles on its anterior end, and intense blue, slightly undulant edges. A broad orange-red stripe with a bisecting yellow line runs down the center of the body. Its markings are startlingly beautiful.

T: 20°–26°C, L: 4 cm, AV: 4, D: 4

Pseudoceros ferrugineus HYMAN, 1959
Red-purple polyclad

Hab.: Tropical Indo-Pacific, Sri Lanka, Indonesia, the Philippines.

Sex.: None, since all animals are hermaphrodites.

M.: Unknown.

F.: C; small invertebrates.

S.: The head is clearly distinguishable by the paired antennae. This very slender polyclad has a golden, undulant fringe and golden spots on its red-purple body. A broad, dark red band separates the body from the undulant fringe.

T: 20°–28°C, L: 8 cm, AV: 4, D: 4

Pseudoceros fuscopunctatus PRUDHOE, 1977
Dark-spotted polyclad

Hab.: Indo-Pacific, Indonesia, Australia.

Sex.: None, since all animals are hermaphrodites.

M.: Unknown.

F.: C; various invertebrates.

S.: The body is dorsoventrally compressed and elongated to oval with a semicircular head and tail. Along the body's fringe, there are short dark lines, spots, and dots. The superior side of the body is pearl white with a multitude of irregular brown and golden brown spots. *P. fuscopunctatus* has well-developed tentacles and rows of eyes. There are clusters of numerous cephalic eyes behind the tentacles. The mouth is centrally located on the inferior side of the body.

T: 20°–28°C, **L:** 2–4 cm, **AV:** 4, **D:** 4

Pseudoceros sp.
Gold-dot polyclad

Hab.: Mediterranean, Red Sea, Indo-Pacific.

Sex.: None; all animals are hermaphrodites.

M.: Unknown.

F.: C; various sea squirt colonies.

S.: Longer than it is broad, *Pseudoceros* sp. has a light, undulant fringe, clusters of cephalic eyes, and tentacles. The superior side of the body is riddled with a multitude of golden yellow papillae and tubercles. Coloration is extremely variable and may be habitat dependant.

T: 20°–28°C, L: 4 cm, AV: 4, D: 4

Pseudoceros sp. (sp. nov.)
Gold-spot polyclad

Hab.: Indian Ocean, Thailand.

Sex.: None, since all animals are hermaphrodites.

M.: Unknown.

F.: C; colonial invertebrates.

S.: This as yet undescribed species has a broad golden yellow fringe and pronounced irregular golden markings around its circumference and along its median. During the day, it hides beneath stones and corals, only emerging at night to creep along corals to reach its feeding grounds.

T: 24°–28°C, L: 4 cm, AV: 4, D: 4

Order: POLYCLADIDA
Fam.: Pseudocerotidae

Pseudoceros sp. (sp. nov.)
Yellow-line polyclad

Hab.: Indo-Pacific, Philippines.

Sex.: None, since all animals are hermaphrodites.

M.: Unknown.

F.: C; unknown.

S.: Pictured are two polyclads grazing upon polyps of a soft coral. They may even be feeding on the soft coral itself. The oval, black body has an impressive design of fine yellow lines that originate along the median line and repeatedly divide towards the edge of the animal. This species has not been described.

T: 24°–28°C, L: 3 cm, AV: 4, D: 4

Pseudoceros sp. (sp. nov.)
Yellow-striped polyclad

Hab.: Red Sea.

Sex.: None, since all animals are hermaphrodites.

M.: Unknown.

F.: C; unknown.

S.: Markings on this ovoid black polyclad include a medial yellow line, an orange band circumventing the body, and yellow lines that originate at the outer edge of the animal and radiate inward. These striations are of variable length and terminate prior to the medial yellow line. Two invaginations on the anterior edge of the body form two orange horns. The pictured animal is a new, as yet undescribed, species.

T: 20°–26°C, L: 3 cm, AV: 4, D: 4

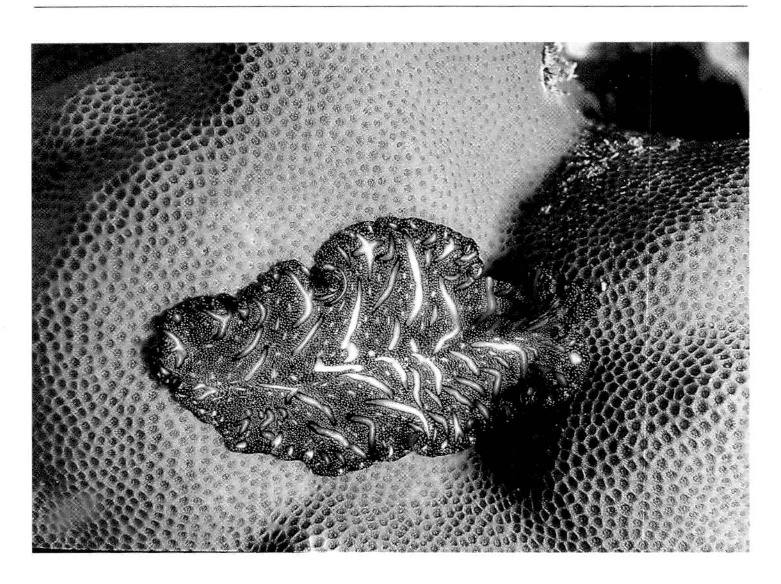

Pseudobiceros bedfordi LAIDLAW, 1903
Bedford's polyclad

Hab.: Tropical Indo-Pacific, Maldives, Australia, Indonesia.

Sex.: None, since polyclads are hermaphrodites.

M.: Unknown.

F.: C; sea squirts, amphipods, small invertebrates.

S.: Because of its conspicuous, distinctive design, this polyclad is facilely identified. Though widely distributed, design and coloration vary little. *P. bedfordi* is not a feeding specialist; prey is coated in mucus then swallowed whole. It is found in diversified habitats. Short distances can be transversed by swimming.

T: 20°–28°C, L: 5 cm, AV: 4, D: 4

Pseudobiceros gloriosus NEWMANN & CANNON, 1994
Red-framed polyclad

Hab.: Red Sea, Indo-Pacific, western Pacific, Caribbean.

Sex.: None, since all polyclads are hermaphrodites.

M.: Unknown.

F.: C; sea squirts and carrion.

S.: *P. hancockanus* is intense blue to black with white and orange peripheral bands and a purple fringe. The two short cephalic antennae are easily overlooked because they are the same color as the body. Inferiorly, this species is purple with a medial line. Usually it lives on corals, among coral fragments, or under stones. Large distances are transversed by swimming, whereby the edges of the body undulate to propel the polyclad.

T: 20°–28°C, L: 6 cm, AV: 4, D: 4

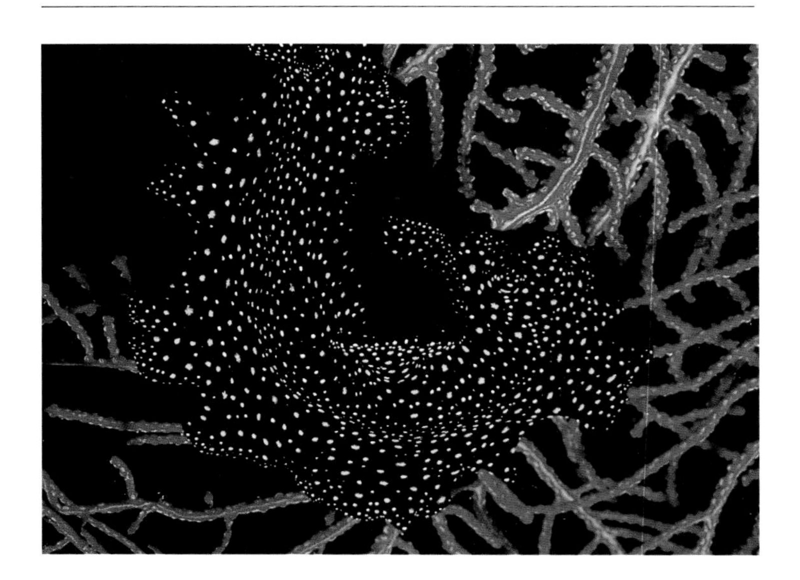

Pseudobiceros sp. (sp. nov.)
Spotted polyclad

Hab.: Pacific coast of the United States.

Sex.: None, since all animals are hermaphrodites.

M.: Unknown.

F.: C; unknown.

S.: The body's margin has broad lobes which are particularly well developed along the head region. The surface of the elongated black body is sown with white, yellow, and red spots of variable size which are concentrated along the center of the body. The pictured animal has not been described; it is grazing on a gorgonian, possibly its feeding substrate.

T: 18°–24°C, L: 8 cm, AV: 4, D: 4

Reteporellina sp.

CLASS BRYOZOA

Bryozoans have been classified into their own phylum, Tentaculata. Found in freshwater and marine environments, the majority of the more than 5000 species are cosmopolitan. They have a broad depth distribution, occurring from shallow water to depths of more than 6000 m.
With few exceptions, bryozoans are colonial, sessile organisms that colonize stones and rock walls, bivalve and gastropod shells, ships, pilings, and other harbor installations. Seaweed and sea-grass beds, sand, and mud habitats are conquered. Some are even capable of chemically boring into crustacean shells. A scant few are capable of limited movement.
Despite the microscopic size of individual animals (zooids)—0.3–1 mm—the colonies of some species grow to a respectable size.

Anatomy

Every bryozoan colony is a consortium of thousands of tiny individual animals. Often the zooids are modified to better fulfill certain functions such as defense and reproduction, i.e., colonies are polymorphic.
Using their ciliated tentacle crown, zooids generate a current to carry food particles to their oral opening. Ingested food is digested in the stomach and the U-shaped intestine.
Zooids can be divided into two body sections, the polypide and the cystid. The former refers to the viscera, muscles, and other content of the body wall, while the latter term is in reference to the chitinous and calcareous exoskeleton of the posterior body and the body wall. The circular or semicircular crown of tentacles is expanded and retracted by strong muscle fibers and hydrostatic pressure. Commonly an operculum, or lid, closes off the opening as the tentacles are retracted, sealing the soft body parts within the exoskeleton. There are autozooids, the so-called normal zooids, whose main role is to capture, digest, and distribute food for the colony, and heterozooids. Heterozooid is a collective

term to describe all the modified zooids. Typically, heterozooids have vestigial or modified tentacles. Avicularia, a defense heterozooid, has a beaklike pair of pinchers which grab and deter small organisms from settling on or overgrowing the colony. Another defense zooid, the vibraculum, has a long mobile bristle instead of a pair of pinchers. Its function is to remove sedimentation from the colony's surface. Kenozooids fasten the colony to the substrate, whereas gonozooids are specialized reproductive zooids. Gonozooids produce gametes and ripen the eggs in a special external chamber called the ovicell. Each modified animal in the colony has a clearly defined function. Since the specialized zooids are unable to capture food, the autozooids fulfill their nutritional needs. Nutrients are passed to heterozooids through pores in the body walls.
The consistency of the colony is species specific. The proportion of calcium versus chitin components in the exoskeleton determines whether the colony is gelatinous, flexible, or extremely brittle.
Because many colonies (zooaria) within this phylum resemble delicate mosses, they have been referred to as moss animals. But bryozoans have a wide array of growth forms: namely, richly branched, reticulate, encrusting, bushy, foliaceous, antlerlike, and coral-like. Bryozoans are commonly confused with hydroid colonies and delicate corals.

Reproduction

A predominant number of marine bryozoan colonies are hermaphrodites. Some even practice brood care. The free-swimming ciliated larvae, cyphonautes, distribute the species. After a brief planktonic phase, a new bryozoan colony forms as the larva settles and begins to bud. Freshwater bryozoans can reproduce through statoblasts, cysts that are produced when the maternal colony is stressed, e.g., low temperatures at the onset of winter.
While most animals have a life expectancy of a few weeks, others live up to two years. Species that live on algae and seaweeds typically have a lifespan of one year. Winter and bad weather drastically curtail growth.

Bryozoans in the Aquarium

Bryozoans are rare guests in aquaria. The lifespan of zooids rarely exceeds a few months, a year at the outside. Additionally, the calcareous skeleton of most species is very brittle and quickly becomes fouled with algae. Only very robust species, like *Myriapora truncata,* are suitable for aquaria. Their exoskeleton is often confused with a stony coral.

Taxonomy

Phylum: Tentaculata — Tentacled Animals
Class: Bryozoa — Bryozoans

Subclass: Stenolaemata

Order: Stenostomata

Suborder: Cancellata
Family: Horneridae
Genus: *Hornera*

Subclass: Gymnolaemata

Order: Cheilostomata

Suborder: Anasca
Family: Membraniporidae
Genus: *Membranipora*
Family: Calpensiidae
Genus: *Calpensia*

Suborder: Ascophora
Family: Hippoporinidae
Genera: *Hippodiplosia*
Pentapora
Family: Smittinidae
Genus: *Porella*
Family: Schizoporellidae
Genus: *Schizobrachiella*
Family: Reteporidae
Genera: *Reteporella*
Reteporellina
Family: Sertellidae
Genus: *Sertella*
Family: Myriaporidae
Genus: *Myriapora*

Hornera frondiculata LAMOUROUX, 1821
Fern bryozoan

Hab.: Mediterranean.

Sex.: Hermaphrodites.

M.: Unknown.

F.: O; plankton, detritus.

S.: These fascicular colonies have intertwined, fan-shaped, dichotomous branches. Zooids are arranged in longitudinal rows along the branches.Fren byozoans grow on hard substrates, only becoming numerous at depths below 20 m.

T: 12°–18°C, Ø: 5 cm, AV: 4, D: 4

Order: CHEILOSTOMATA
Fam.: Membraniporidae

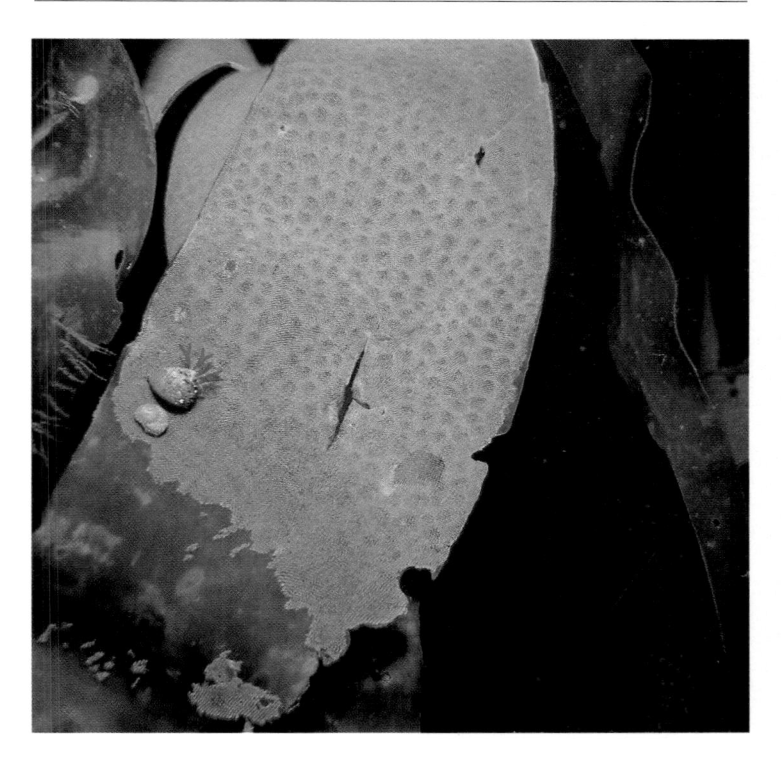

Membranipora membranacea (LINNAEUS, 1767)
Sea-mat

Hab.: Atlantic, North Sea, western Baltic Sea.

Sex.: Hermaphrodites.

M.: Unknown.

F.: O; plankton, detritus.

S.: Large colonies are composed of millions of rectangular zooids which are separated by calcareous walls and aligned in parallel rows. Colonies are membranaceous, encrusting, or matlike with a coarsely granulate surface. Their gray-white growth covers hard substrates such as stones, bivalve and gastropod shells, and seaweed. Almost circular when young, older specimens have irregular borders. Sea-mats are prey for the nudibranch *Polycera quadrilineata*.

T: 10°–18°C, Ø: 60 cm, AV: 4, D: 4

Order: Cheilostomata
Fam.: Calpensiidae

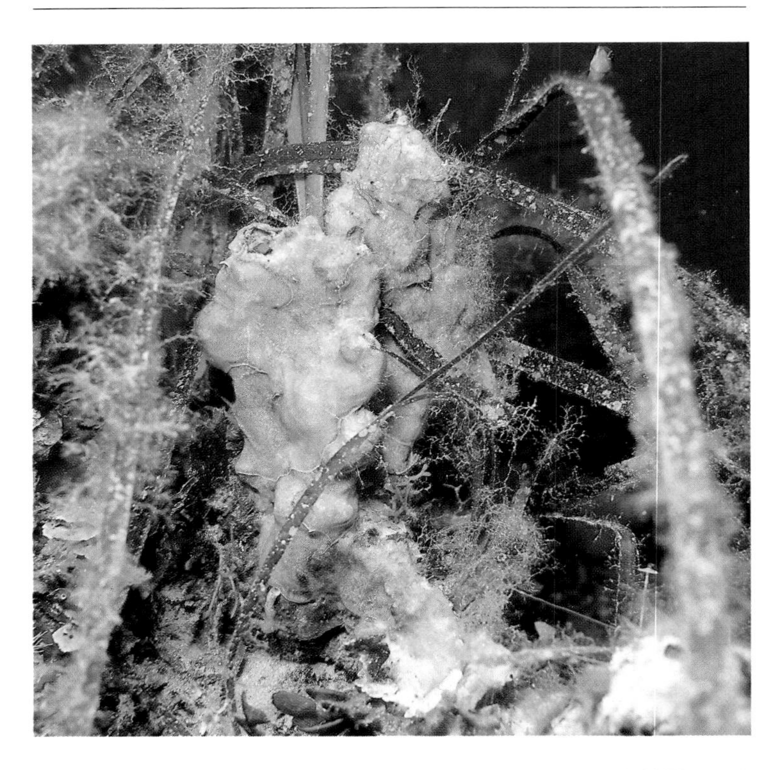

Calpensia nobilis (ESPER, 1796)
Crusty bryozoan

Hab.: Mediterranean.

Sex.: Hermaphrodites.

M.: Unknown.

F.: O; plankton, detritus.

S.: Virtually all types of substrates are overgrown by these encrusting colonies. They are frequent commensals on seagrasses, covering the leaves and roots of *Posidonia* with their durable growth.

T: 10°–18°C, L: 6 cm, AV: 4, D: 4

Hippodiplosia foliacea (ELLIS & SOLANDER, 1761)
Antler bryozoan

Hab.: Mediterranean.

Sex.: Hermaphrodites.

M.: Unknown.

F.: O; plankton, detritus.

S.: *H. foliacea* forms fascicular colonies. Each of the numerous branches is flat, foliaceous to palmately branched, and brittle because of its highly calcified exoskeleton. The flat branches are in fact two sheets of zooids arranged back to back. *H. foliacea* is almost exclusively found on hard substrates; massive stands may be encountered below 25 m. Dead sections of the colony are quickly fouled by algae and other organisms.

T: 10°–18°C, L: 20 cm, AV: 4, D: 4

Pentapora foliacea (ELLIS & SOLANDER, 1761)
Sea rose

Hab.: Atlantic, Mediterranean.

Sex.: Hermaphrodites.

M.: Unknown.

F.: O; plankton, detritus.

S.: On favorable sites, this conspicuous colony may form mats. Generally *P. foliacea* occurs on hard substrates along current-swept regions of the rocky littoral. The meandering foliaceous to spiral septa are formed by two sheets of zooids back to back. Growth rate and longevity are unmentioned in the literature. With a little latitude, one can say that a colony resembles a rose.

T: 10°–18°C, Ø: 15 cm, AV: 4, D: 4

Pentapora fascialis (PALLAS, 1766)
Band bryozoan

Hab.: Atlantic, Mediterranean.

Sex.: Hermaphrodites.

M.: See *Sertella beaniana*.

F.: O; plankton, detritus.

S.: Colonies are fascicular with foliaceous, flat, dichotomous branches. Massive numbers of these colonies live on secondary hard bottoms at great depths. The base of the colonies is usually dead and covered by algae. Live sections can be distinguished by their salmon color.

T: 10°–18°C, **L:** 20 cm, **AV:** 4, **D:** 4

Porella (= Smittina) cervicornis (PALLAS, 1766)
Staghorn bryozoan

Hab.: Mediterranean.

Sex.: Hermaphrodites.

M.: Unknown.

F.: O; plankton, detritus.

S.: Erect, filigree, and fascicular, *P. cervicornis* resembles antlers. Its surface is rough. Branches have dichotomous tips and an ovoid cross section. This species mostly settles on hard substrates and secondary hard substrates in calm water at depths greater than 20 m, but it occasionally occurs in shaded shallow areas of the rocky littoral zone.

T: 10°–18°C, L: 15 cm, AV: 4, D: 4

Schizobrachiella sanguinea (NORMAN, 1868)
Blood-red encrusting bryozoan

Hab.: Mediterranean.

Sex.: Hermaphrodites.

M.: Unknown.

F.: O; plankton, detritus.

S.: There are a variety of different growth forms, but the preponderance of colonies are flat and encrusting. In shallow water, they cover practically all smooth surfaces, e.g., harbor installations, buoys, shipwrecks, stones, and cave entrances. Blood-red encrusting bryozoans form lumpy colonies around algae, fan corals, and even the thorns of sea urchins in deep water. Carotenoid inclusions in the skeletal matrix make some colonies almost blood red, hence the name.

T: 10°–18°C, L: 8 cm, AV: 4, D: 4

Reteporella graeffei KIRCHENPAUER, 1864
Yellow fan bryozoan

Hab.: Indo-Pacific, Australia.

Sex.: Hermaphrodites, hence there are no sexual differences.

M.: Unknown.

F.: O; plankton, detritus.

S.: Colonies are in the form of small, intertwined fans that are filigree and richly branched. Shaded sites along coral reefs and vertical walls, within caves, and beneath overhangs are their preferred habitat. The intense yellow coloration draws attention, even at great depths.

T: 22°–26°C, Ø: 8 cm, AV: 4, D: 4

Reteporellina sp. (*evelinae*?)
White fan bryozoan

Hab.: Caribbean, Bermuda, Bahamas, Colombia.

Sex.: Hermaphrodites.

M.: Unknown.

F.: O; plankton, detritus.

S.: *Reteporellina* sp. forms richly branched, filigree, fan-shaped colonies. The delicate branches are lined with zooids. These very brittle colonies live on sponges and dead gorgonians, beneath corals, in caves, and under ledges in calm, deep environments; only at depths greater than 20 m do they become common.

T: 20°–24°C, Ø: 6 cm, AV: 4, D: 4

Order: Cheilostomata
Fam.: Sertellidae

Sertella beaniana (KING, 1846)
Net coral

Hab.: Atlantic, Mediterranean.

Sex.: Hermaphrodites.

M.: Ne corals are difficult to maintain. Collection and transport generally result in breakage, and captive animals have an abbreviated lifespan. Algae quickly foul their exterior.

F.: O; plankton, detritus.

S.: Colonies are very filigree and brittle. Arranged with little intervening space, the folia are finely reticulate, foliaceous, curled, and funnel-shaped. The zooids are exclusively located on the inner surface of the folia. Dead sections are immediately covered by algae. Dimly lit sites beneath overhanging ledges, within caves, and on coral substrates in deep water are where net corals generally occur. Occasionally they are observed on dead corals or on the roots of *Posidonia*.

T: 10°–18°C, Ø: 12 cm, AV: 4, D: 4

Myriapora truncata (PALLAS, 1766)
False coral

Hab.: Mediterranean.

Sex.: Hermaphrodites.

M.: Although these colonies are ccasionally available in pet stores, aquarium maintenance—especially in regard to dietary requirements—is extremely difficult. Algae quickly foul the colonies. Skeletons are used as decorations.

F.: O; plankton, detritus.

S.: The false coral is frequently confused with *Corallium rubrum*, especially since coloration and habitat of the two species often overlap. The branches of this stalwart colony are smooth and round in cross section. Branch tips are usually forked and blunt. From shallow waters to great depths, *M. truncata* inhabits dimly lit areas: for example, rocky caves, cave entrances, and beneath overhanging ledges.

T: 10°–18°C, L: 12 cm, AV: 3, D: 4

Phylum Mollusca, the Molluscs

Of the approximately 128,000 species of molluscs, the bulk inhabit diverse ecological niches of the sea, from sediments of the deep sea to the pounding surf zone and open waters. An astonishing number of forms have been adopted to fill the sea's ecological niches. Terrestrial and freshwater environments have molluscan representatives as well.
Molluscs have three principle body segments—the head, foot, and visceral mass. All are enveloped within the mantle.

The head is situated at the anterior end of the body; it contains the mouth, important sensory organs, and the brain. Aside from filter-feeding molluscs, every member of this phylum has an organ exclusive to molluscs, the radula. The radula or rasping tongue is the tool by which the organisms graze upon algae. It is located in the anterior intestine.
The complexity and development of the visual sensory organs, the eyes, differs from species to species. They range from simple sensory pits that just detect presence or absence of light to organs with a complex lens, as found in cephalopods. Tentacles are tactile organs capable of reacting to chemical stimuli. They, too, are located on the head. In Opisthobranchia, the tentacles are modified into rhinophores.

Locomotion is achieved via a ventral, muscular foot. Adaptations have led to a certain degree of disparity between the classes in regard to the foot. Bivalves often have a small, finger-shaped foot, snails have a large and flat foot, and cephalopods have modified their foot into capture arms. The foot is primarily composed of musculature, nerves, and fluid-filled interstices.

Most organs are located in the visceral mass. The intestinal tract and its addenda, the genital tract, and the heart can all be found in this segment. The intestine is the recipient of digestive enzymes and the site of digestion and nutrient absorption. Though several glands secrete into the intestine, the secretions of the central intestinal gland are the most vital.

Murex pecten, Indonesia (text on p. 615).

Only cephalopods have advanced their circulatory system slightly above that of their mulluscan brothers to the level of primitive capillaries. The rest have an open circulatory system.

The mantle surrounds the dorsal body and the mantle cavity. The gills are located in the latter. Shell formation is the main function of the mantle.

Radula—the Molluscan Oral Tool

All molluscs, with the exception of bivalves, have a rasping tongue, or radula. Made of cartilage and muscle, the radula is a mobile organ with a membrane and rows of teeth covering its surface. Over the breadth of the phylum, the radula has adapted to various feeding niches, enabling molluscs to invade and successfully populate a wide range of habitats.
The radula is kept in the radula sac, an evagination of the pharynx. It is the posterior section of the radula sac that secretes a membrane of protein and chitin, the anchor point for the teeth. New teeth are transparent, while older teeth are dark brown due to inclusions.
Like sharks, molluscs can shift new rows of teeth forward to replace old, worn down or broken teeth.
The odontophore is the cartilaginous base of the radula. It consists of two symmetrical plates interconnected with musculature. Muscles that fasten to the radula membrane pull the radula back and forth over the odontophore. Due to the mechanics of the system, the teeth along the radula's edge are pushed erect. The radula is pressed onto the substrate by the odontophore, where it rasps the surface as it is pushed and pulled.
The radula can adapt to serve the needs of both carnivores and herbivores; for example, limpets have an especially powerful, long radula that allows them to scrape blue-green algae, their preferred diet, from rocks (see photo). Predatory Muricidae have a radula that slices large chunks of flesh from their prey (see photo). In cone snails (p. 623), the radula has but one single harpoon-shaped poisonous tooth.

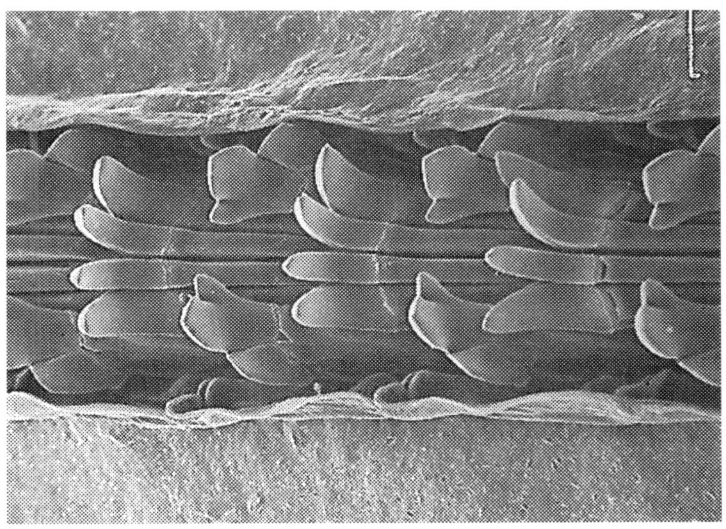

Radula of an algivorous limpet, *Patella* sp.

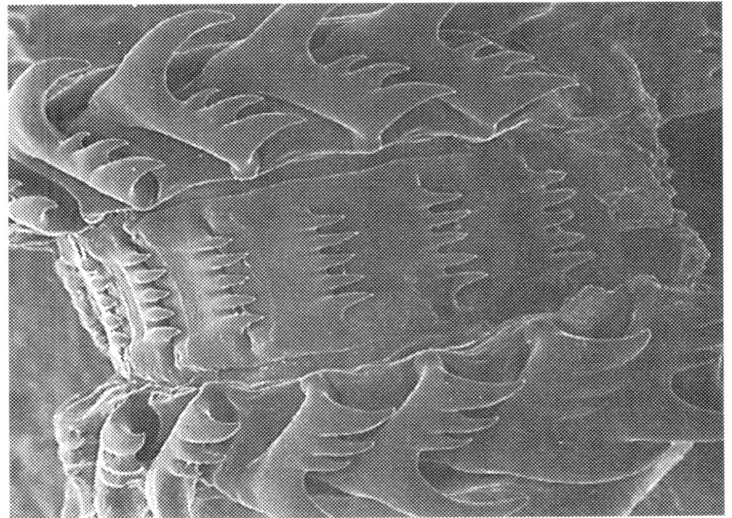

Radula of a predatory Muricidae, *Hexaplex trunculus*.

Taxonomy

Phylum: MOLLUSCA — Molluscs
  Subphylum: ACULIFERA

Class: CAUDOFOVEATA

Class: SOLENOGASTRES

Class: POLYPLACOPHORA — Chitons

Order: ISCHNOCHITONIDA
  Family: Chitonidae
    Genera: *Acanthopleura*
      Callochiton
      Chiton
  Family: Ischnochitonidae
    Genus: *Stenochiton*

Order: ACANTHOCHITONIDA
  Family: Acanthochitonidae
    Genera: *Acanthochitonia*
      Cryptoplax

continued p. 556

Caudofoveata Solenogastres
From: SALVINI-PLAVEN: Schild- und Furchenfüßer, Brehm-Bücherei, 1971.

SUBPHYLUM ACULIFERA

All representatives of the subphylum Aculifera are bilaterally symmetrical. Three classes of this subphylum are at home in marine environments—Caudofoveata, Solenogastres, and Polyplacophora. Cephalic sensory organs such as eyes and tentacles are lacking.

Class Caudofoveata

Caudofoveata are small, worm-shaped molluscs; their entire body is covered by a scaly integument. The anterior end is equipped with a digging plate, whereas the posterior end has a mantle cavity and a pair of gills. The vast majority of the more than 60 species of this class are found on deep (≥100 m), muddy substrates in which they burrow. Caudofoveatans are identified on the basis of their radula and their dermal scales. The radula is often reduced to a few transverse rows of teeth.
In their natural habitat, caudofoveatans feed on mud-dwelling microbes. Little is known about behavior and reproductive biology other than that they are dioecious and gametes are released into the open water. There do not seem to be secondary sexual characteristics. So far, the planktonic larvae have not been closely studied.
Aquarium maintenance is extremely difficult, since the mud substrate quickly fouls the water.

Class Solenogastres

A worm-shaped body and scaly integument are diagnostic of this class as well. Instead of a foot, Solenogastres have a ciliated longitudinal groove on their ventral side. With the aid of the ventral groove these animals glide over the substrate. Most species are just a few millimeters long, but a few grow to a length of 5 cm or more. Little is known concerning their habitat or biology. So far, about 200 species have been described, but there are probably significantly more. Some species live on soft bottoms, while others are epizoic (living on animals), creeping on gorgonians and hydrozoans and probably feeding on the polyps.
To the best of our knowledge, all, including nonepizoic species, feed on cnidaria. By using their suction "pharynx," Solenogastres seem to be capable of ingesting nematocysts without prompting a discharge, similar to select nudibranchs. Some species are feeding specialists that only feed on a certain species.
In contrast to Caudofoveata, Solenogastres are hermaphroditic.

Eggs are laid in spawning cords or small balls. Few practice brood care. They can be introduced into aquaria containing gorgonians. If their host dies, they too will die.

Class Polyplacophora, Chitons

Typically, chitons have eight overlapping calcareous dorsal plates, that are arranged like shingles on a roof. Because the plates are articulated, chitons can roll into a ball when they become separated from the substrate, thereby protecting their body from predation. The periphery of the plates is surrounded by a girdle, which is either naked or armed with embedded calcareous scales or needles. The ventral broad, creeping foot allows chitons to cling strongly to the substrate. Between the foot and the mantle lies the pallial groove. Depending on the species, there are between 6 and 88 pairs of bipectinate gills within the groove. The ciliated epithelium covering the surface of the gills creates a strong water current.
The sensory organs of this group are especially noteworthy. In lieu of cephalic eyes, chitons have esthetes. These specialized photic organs densely cover the plates (up to 70 per mm^2). The paired subradula organ, responsible for taste, is located in the buccal cavity.
Most chitons live in shallow coastal waters, although a small number occur in deep water environments, and tropical species tend to inhabit tidal zones. Thanks to their dorsoventrally compressed body and strong foot, chitons can hold to the substrate, withstanding even the strongest surge. Animals that inhabit the surge zone are generally positively phototactic, i.e., they seek light-bathed regions. Others hide beneath stones during the day, only emerging at night to feed (negatively phototactic).
Chitons rasp algae with their strong radula. Harmless creatures such as unicellular organisms, isopods, or mites commonly inhabit the pallial groove. Infrequent residents include scaleworms and peacrabs.
Much of the reproductive biology of chitons remains shrouded in mystery; however, it is known that almost all chitons are dioecious, between 500 and 1,500 eggs are released into the open water and fertilized there, and brood care is an uncommon attribute.

Acanthopleura haddoni, Red Sea.

Acanthopleura haddoni WINCKWORTH, 1927
Haddon's chiton

Hab.: Very abundant in the tidal zone of the Red Sea. It is generally found on rocks, dead coral stalks, and harbor walls.

Sex.: There are no external distinguishing characteristics.

Soc.B./Assoc.: Small groups may occasionally be encountered, but for the most part, *A. haddoni* lives singly.

M.: Rock aquarium. At times it sits on the aquarium pane. LANGE & KAISER (1991) recommend a surf tank. Using a pump and a reservoir tank, a surf zone can be simulated. Water is pumped from the aquarium into the reservoir. When the water in the reservoir reaches a predetermined level, it empties back into the aquarium, creating a rich spray.

Light: Sunlight zone.

B./Rep.: This species passes through a free-swimming larval stage.

F.: H; Haddon's chiton feeds on filamentous green algae that grow on the upper side of stones and coral blocks.

S.: In contrast to other chitons, this species is positively phototactic, i.e., it seeks sunny sites.

T: 22°–28°C, **L**: 7 cm, **TL**: from 100 cm, **WM**: s, **WR**: t, **AV**: 3, **D**: 3

Acanthopleura granulata, Caribbean.

Order: Ischnochitonida
Fam.: Chitotonidae

Callochiton septemvalis, Mediterranean.

Callochiton septemvalis (MONTAGU, 1803)
Coral chiton

Hab.: With the exception of Iceland, *C. septemvalis* inhabits all European coasts. It occurs on primary and secondary hard substrates from shallow water to a depth of 100 m.

Sex.: There are no external sexual differences.

Soc.B./Assoc.: *C. septemvalis* normally occurs on various encrusting red algae, e.g., *Lithophyllum*, *Corallina*, or *Peysonellia*, or the underside of stones.

M.: *C. septemvalis* is largely introduced into aquaria accidentally with stones or sediment. During the day it retires into rock crevices or beneath stones.

Light: Dim light.

B./Rep.: Nothing is known about the reproductive biology of this species.

F.: H; since the coral chiton is almost always encountered on particular red algae (see photo and Soc.B./Assoc.), it can be assumed that these plants make up at least part of its diet. It is unknown whether or not green, blue-green, or other types of red algae are consumed.

S.: Coloration ranges from orange to dark red or a combination of red-brown and green. Some specimens have light spots or stripes. Due to their variable coloration, distinct forms have been described. This species was previously described as *Chiton laevis*, then *C. achatinus*. Animals from the Atlantic are occasionally considered a subspecies (*Callochiton septemvalis euplaeae*) or an autonomous species.

T: 12°–20°C, L: 2 cm, TL: from 70 cm, WM: m, WF [illegible], A: 3, D: 3

Chiton affinis, aquarium photo.

Chiton affinis ISSEL, 1896
Red Sea chiton

Hab.: Entire coastal region of the Red Sea. There it can be encountered on dead corals along reef platforms as well as within crevices and under stones and rocks.

Sex.: There are no external differentiating characteristics.

Soc.B./Assoc.: With the notable exception of predatory starfishes, this solitary animal can be associated with many aquarium inhabitants.

M.: Rock or reef aquaria provide sufficient hiding places. To transport, remove the animal from the water and wrap it in a damp rag. Be careful not to damage its pallial groove when prying it from the substrate.

Light: Dim light.

B./Rep.: Propagation and development are equivalent to that of *C. olivaceus,* a Mediterranean species.

F.: H; algae are scraped from the upper side of stones with the strong radula. While *C. affinis* does not need to be specifically fed, successful care hinges on a mature aquarium with an abundance of green algae.

S.: As is the case with all members of the genus *Chiton*, the cuticle of the girdle contains calcareous spicules.

T: 20°–28°C, **L**: 4 cm, **TL**: from 70 cm, **WM**: m, **WR**: b, **AV**: 2, **D**: 1–2

Order: ISCHNOCHITONIDA
Fam.: Chitotonidae

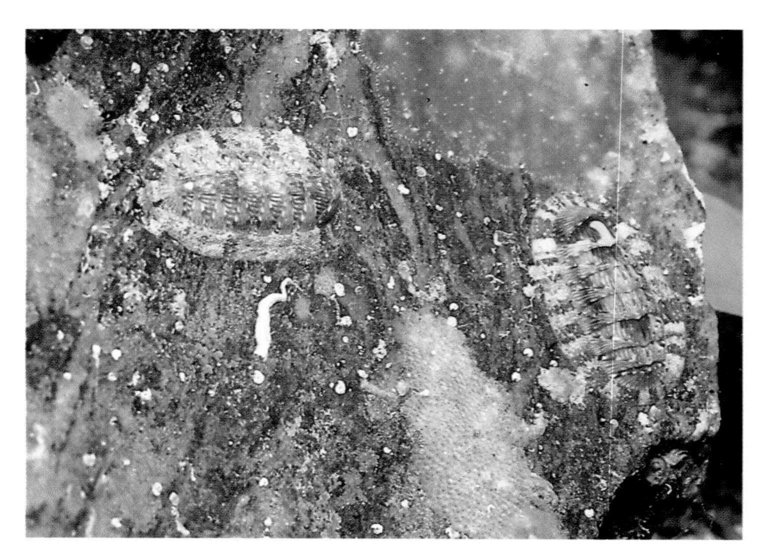

Chiton olivaceus, Mediterranean.

Chiton olivaceus SPENGLER, 1797
Olive chiton

Hab.: Common throughout the Mediterranean. From the tidal zone to shallow depths, *C. olivaceus* primarily resides on the underside of stones, within bivalve and gastropod shells, or in rock crevices.

Sex.: Separate sexes (dioecious).

Soc.B./Assoc.: Virtually all invertebrates and fishes are appropriate tankmates.

M.: House in a rock aquarium with a multitude of crevices and stones suitable for *C. olivaceus* to crawl among and under, respectively. Only aquaria that have algae-covered stones will suffice.

Light: Dim light.

B./Rep.: The gametes are released into the open water, where they are fertilized. The eggs are pelagic until they hatch into planktonic larvae about a day later. After about 10 days, the larvae metamorphose into benthic, 0.5 mm juveniles.

F.: H; the long, strong radula is used to rasp algae from stones.

S.: Negatively phototactic, i.e., it avoids light, this species only abandons its hiding place at night to search for food. All daylight hours are passed in dim secluded areas. This behavior can be used to distinguish the olive chiton from the otherwise similar *Middendorffia capraerum* which can be seen in tidal pools along rocky coasts.

T: 15°–24°C, **L:** 4 cm, **TL:** from 70 cm, **WM:** m, **WR:** b, **AV:** 2, **D:** 1–2

Chiton olivaceus

Chiton torrianus, aquarium photo.

Chiton torrianus HEDLEY & HULL, 1910
Rock chiton

Hab.: West coast of the United States. *C. torrianus* is found firmly adhered to substrates below the intertidal zone. During the day it retires into rock crevices and beneath stones.

Sex.: Dioecious. There are no external differentiating characteristics.

Soc.B./Assoc.: Various invertebrates and fishes make suitable tankmates for this solitary abiding animal.

M.: Hiding places where it can retire during the day are a necessity.

Light: Dim light.

B./Rep.: Eggs and larvae are planktonic. Aquarium breeding has not been successful, largely because of complications involving the planktonic phases.

F.: H; low-growing green algae are rasped from stones as it creeps along.

S.: As with all chitons, caution has to be practiced when collecting; injuries frequently occur when an unwieldy hand attempts to break the animal's strong grip on the substrate. Use a push of the thumb or, better yet, transport the chiton with its substrate. Injured girdles rarely heal in the aquarium, and this type of injury commonly proves lethal over the short term.

T: 22°–28°C, **L:** 4 cm, **TL:** from 70 cm, **WM:** m, **WR:** b, **AV:** 3, **D:** 1–2

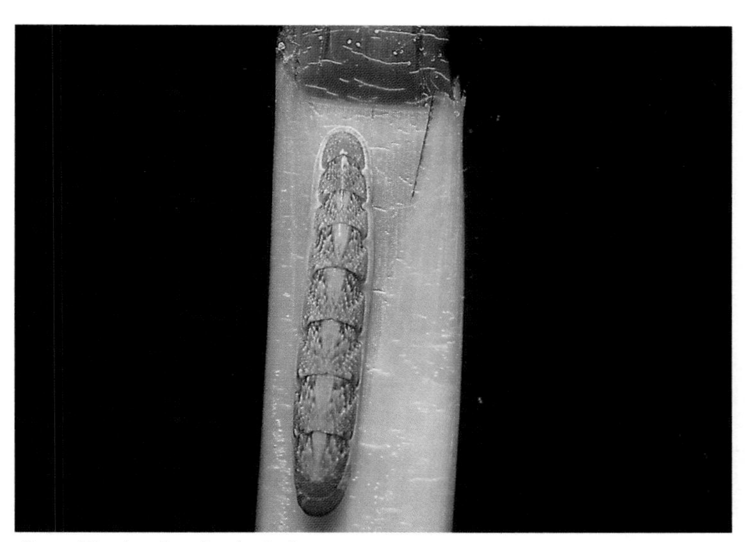

Stenochiton longicymba, Australia.

Stenochiton longicymba (BLAINVILLE, 1825)
Australian seagrass chiton

Hab.: Common along southwestern Australia and northern Tasmania. Normally, the Australian seagrass chiton is found on leaves of the Australian seagrass *Posidonia australis*, but it also occurs on rocks and bivalve shells in seagrass meadows.

Sex.: Dioecious. There are no external differentiating characteristics.

Soc.B./Assoc.: For the most part, *S. longicymba* lives alone, but sometimes aggregations are encountered. It has few enemies.

M.: Since seagrass has never been successfully cultivated in an aquarium and *S. longicymba* probably derives its nourishment from *Posidonia*, it is hardly surprising that aquarium care has not been successful.

Light: Moderate light.

B./Rep.: Little is known about the reproduction and development of this species.

F.: H; supposedly, this species lives exclusively from the Australian seagrass *Posidonia australis*.

S.: All *Stenochiton* spp. live on seagrasses. They are particularly well-adapted to their epiphytic existence, i.e., their body is long and slender and has a narrow girdle. *S. pilsbryanus* is also found on Australian seagrass; however, it is smaller and paler than *S. longicymba*.

T: 18°–22°C, L: 4.5 cm

Order: Acanthochitonida
Fam.: Acanthochitonidae

Acanthochitonia crinita, aquarium photo.

Acanthochitonia crinita (PENNANT, 1777)
Creased chiton

Hab.: Mediterranean and the Atlantic from Scandinavia to Cape Verde. It is encountered at shallow depths on the underside of stones.

Sex.: With the exception of a single species (*Lepidochitona raymondi*), all chitons are dioecious. There are no external sexual differences.

Soc.B./Assoc.: This chiton can be housed with any fish or invertebrate.

M.: *A. crinita* is very undemanding. It is occasionally introduced into aquaria on stones. Hiding places in the form of rock crevices are very important.

Light: Dim light.

B./Rep.: Eggs and sperm are released into the open water, where they are fertilized. The resulting free-swimming larvae later metamorphose into benthic juveniles. Details concerning development are sketchy.

F.: H; the long radula scrapes algae from stones.

S.: *Acanthochitonia* spp. are easily recognized by the 14 clusters of spines. Coloration within this species can vary, and sometimes calcareous red algae may foul the plates (photo). The knobs on the shell of *A. fascicularis,* a similar species, are finer.

T: 15°–20°C, L: 2 cm, TL: from 70 cm, WM: m, WR: b, AV: 2, D: 1

Cryptoplax sp., Japan.

Cryptoplax sp.

Hab.: East coast of Japan. During the day, *Cryptoplax* sp. retires beneath stones or into rock crevices.

Sex.: It is not possible to distinguish the sexes from external characteristics.

Soc.B./Assoc.: *Cryptoplax* sp. is almost always found with sponges, its prey. Fishes as well as invertebrates make suitable tankmates.

M.: There are no particular demands placed on water quality, but its dietary requirements are difficult to fulfill (see below).

Light: Moderate light. This chiton is negatively phototactic—it only emerges from its hiding place at night.

B./Rep.: Little is known about the reproduction of *Cryptoplax*. Like other chitons, species within this genus pass through a planktonic larval phase.

F.: C; to the best of our knowledge, all *Cryptoplax* spp. are feeding specialists that only feed on sponges. It is assumed that this species will not accept substitute foods in captivity.

S.: The girdle, which surrounds the 8 shell plates, is particularly broad in this genus. This family includes a genus in which the plates are totally covered by the girdle. Another member, the Californian *Cryptochiton stelleri*, is the largest chiton with a length of 40 cm. It is an appreciated laboratory animal for physiological research.

T: 17°–24°C, **L:** 5 cm, **TL:** from 100 cm, **WM:** m, **WR:** b, **AV:** 4, **D:** 4

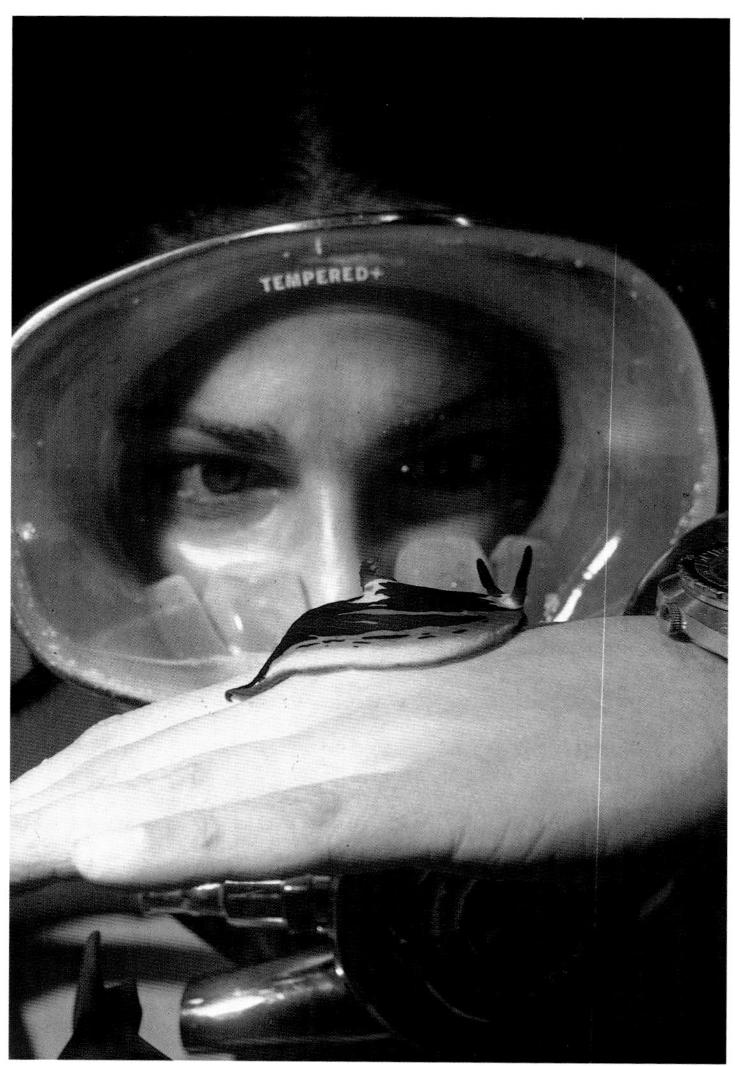

Nembrotha sp., see page 670.

CLASS MONOPLACOPHORA

Until 1952, this class was thought to be extinct. The only proof of its existence was its fossil record. During the "Galathea Expedition" along the western coast of Central America, the first live species, *Neopilina galatheae*, was found at a depth of 3,590 m. Over the next few years, five additional *Neopilina* species were discovered, all on mud substrates at depths of 1,600 to 6,500 m. Their present known distribution is limited to latitudes between 30° S and 30° N. Today, very few species fill the ranks of the Monoplacophora. Representatives of this group, like others that were once thought extinct, are called relicts.

Anatomy

As implied by the name, monoplacophorans have one 3–4 cm shell. The animals are bilaterally symmetrical with 8 pairs of muscles connecting the foot and shell. Because some of the organs are arranged in rows, it has been hypothesized that monoplacophorans evolved from Annelida, but few scientists are convinced of this relationship.

Nutrition

The diet consists of minute organisms, such as diatoms, rotifers, and foraminiferans they root from the mud.
Statocysts, tentacles, and the subradular organ (detects chemical stimuli) are the extent of their sensory organs. They do not have eyes.

Reproduction

Although *Neopilina* are dioecious, there are no external sexual differences. The 0.3 × 0.15 mm eggs are unattached in the ovary. Since copulatory organs seem to be missing, it is assumed that eggs and sperm are released in the open water where fertilization occurs. There is no information concerning more advanced developmental stages; however, a veligerlike larval stage is speculated upon.

These deep sea inhabitants are unlikely to be suitable aquarium animals.

Taxonomy

Phylum: MOLLUSCA	Molluscs
Subphylum: CONCHIFERA	Conchs
Class: MONOPLACOPHORA	Monoplacophorans
Class: GASTROPODA	Gastropods, snails
Subclass: PROSOBRANCHIA	Prosobranchs
Order: ARCHAEOGASTROPODA	Primitive gastropods
Family: Haliotidae	Abalones
Genus: *Haliotis*	
Family: Fisurellidae	Keyhole limpets
Genus: *Megathura*	
Family: Patellidae	True limpets
Genus: *Patella*	
Family: Trochidae	Top shells
Genera: *Clanculus*	
Monodonta	
Tectus	
Family: Turbinidae	Turban shells
Genus: *Turbo*	
Family: Neritidae	Nerites
Genus: *Nerita*	
Order: MESOGASTROPODA	Intermediate gastropods
Family: Littorinidae	Winkles
Genus: *Littorina*	
Family: Vermetidae	Worm shells
Genera: *Dendropoma*	
Sperbulorbis	
Vermetus	
Family: Janthinidae	Purple sea snails
Genus: *Janthina*	
Family: Aporrhaidae	Pelican's foot shells
Genus: *Aporrhais*	
Family: Strombidae	True conchs
Genera: *Lambis*	
Strombus	

Family: Cypraeidae	Cowries
Genera: *Cypraea*	
Monetaria	
Family: Ovulidae	Shuttle shells
Genera: *Calpurnus*	
Cyphoma	
Ovula	
Pseudosimnia	
Volva	
Family: Cassidae	Helmet shells
Genus: *Cassis*	
Family: Cymatiidae	Trumpet shells
Genus: *Charonia*	
Family: Tonnidae	Tuns
Genus: *Tonna*	
Order: Neogastropoda	Advanced gastropods
Family: Muricidae	Murexes, drills
Genera: *Chicoreus*	
Hexaplex	
Murex	
Family: Nassariidae	Nassa mud snails
Genus: *Nassarius*	
Family: Fasciolariidae	Tulip shells
Genus: *Fasciolaria*	
Family: Olividae	Olive shells
Genus: *Oliva*	
Family: Vasidae	Chanks
Genus: *Turbinella*	
Family: Harpidae	Harps
Genus: *Harpa*	
Family: Volutidae	Volutes
Genus: *Cymbiola*	
Family: Conidae	Cones
Genus: *Conus*	
Family: Terebridae	Auger shells
Genus: *Terebra*	

Taxonomy

Subclass: OPISTHOBRANCHIA	Opisthobranchs
Order: CEPHALASPIDEA	Headshield slugs
Family: Aglajidae	Aglajon
Genera: *Chelidonura*	
Navanax	
Order: ANASPIDEA	Sea hares
Family: Aplysiidae	True sea hares
Genus: *Aplysia*	
Family: Dolabriferidae	
Genus: *Phyllaplysia*	
Order: SACOGLOSSA	Sea slugs
Family: Elysiidae	
Genera: *Elysia*	
Tridachia	
Order: NOTASPIDEA	Sidegill slugs
Family: Pleurobranchiidae	
Genus: *Pleurobranchus*	
Order: NUDIBRANCHIA	Nudibranchs
Family: Chromodorididae	
Genera: *Chromodoris*	
Hypselodoris	
Risbecia	
Family: Dendrodorididae	
Genera: *Dendrodoris*	
Doriopsilla	
Family: Polyceridae	
Genera: *Gymnodoris*	
Nembrotha	
Polycera	
Roboastra	
Tambja	
Family: Asteronotidae	
Genus: *Halgerda*	
Family: Kentrodorididae	

Genera: *Jorunna*
Kentrodoris
Family: Hexabranchidae
Genus: *Hexabranchus*
Family: Discodorididae
Genus: *Peltodoris*
Family: Platydorididae
Genus: *Platydoris*
Family: Phyllidiidae
Genera: *Phyllidia*
Phyllidiopsis
Reyfria
Family: Tethyidae
Genus: *Tethys*
Family: Dendronotidae
Genus: *Dendronotus*
Family: Flabellinidae
Genera: *Flabellina*
Family: Glaucidae
Genera: *Glaucus*
Phidiana
Pteraeolidia

Class: Scaphopoda — Tusk shells

Continued on p. 745, Vol. 3

Torsion in Gastropods

Torsion is the twisting of the visceral sac by 90° to 180° over the course of development in the class Gastropoda (snails). The organs that are originally on the left-hand side of the body are shifted to the right side, and posterior organs move forward. This explains why the genital organs of many snails are positioned anteriorly. The spiral windings of the shell are also a consequence of torsion. Almost all snail shells are coiled to the right, but there are a few left-handed snails. Dextral species rarely coil left and sinistral species are equally careful about trespasses. Approximately 1:1,000,000 *Amphidromus* (Pulmonata) are left-handed.

Class: Gastropoda

About the class Gastropoda, snails

With more than 40,000 species, this is by far the largest class of molluscs. Of the various subclasses of gastropods, shelled prosobranchs and shell-less opisthobranchs hold the most interest to marine aquarists, divers, and snorklers. Pulmonata are almost exclusively terrestrial or freshwater inhabitants. While opisthobranchs (pp. 632–635) have highly appealing forms and colors, they are unsuitable for aquarium maintenance.

Anatomy

The following chapter covers the Prosobranchia, worldwide inhabitants found in such diverse biotopes as coastal waters and the deep sea. Some species are capable of swimming, namely *Janthina*. With few exceptions, e.g., *Patella*, the calcareous shell is spirally wound. Virtually all species can be classified by their shell. The number of whorls, the height/width ratio, presence or absence of an umbilicus, a siphonal channel, or a lip, i.e., the thick rim of the shell's aperture as well as coloration and design are all important criteria for identification. Many species have an operculum, a horny or calcareous disc attached to the dorsal side of the foot that effectively closes the aperture as a measure of defense.

Nutrition

Feeding strategies and diets are sundry. There are algivores, predators, scavengers, planktivores, and parasites. In contrast to Opisthobranchia, Prosobranchia have few feeding specialists. With the exception of a few parasitic forms, all have a relatively well-developed radula.

Reproduction

Members of the subclass Prosobranchia are dioecious. Some males have a penis on the right side of the body. After copulation, the eggs are deposited in a nest. The subsequent planktonic larval phase has a duration of a few hours to a few weeks. Development passes through a trochophore then a veliger larval stage. Veliger larvae have a two- or four-lobed sail (velum), and mature veliger larvae even have a foot and an embryonic shell. When the larvae settle into a benthic lifestyle, the sail is discarded.

Snails in the Aquarium

Aquarium maintenance of Opisthobranchia is described on pages 633–635. The following pages deal wholly with Prosobranchia.

Nutrition

Since diets and feeding methods are so diverse, generalities must be restricted. A healthy layer of green algae is a necessity for algivores. Predatory species often attack their tankmates—bivalves, snails, sea squirts, various echinoderms, and even fishes resting on the bottom will become prey. Virtually all are scavengers; an occasional piece of shrimp, mussel, or fish meat is appreciated. Via a pipette, deliver plankton substitutes or mussel milk to the capture net of planktivores such as worm shells. Some littoral forms tend to climb out of the water and fall over the tank's edge. A good cover or an inwardly projecting horizontal rim is required for such species. For proper shell development, snails require calcium. Because other members of an invertebrate or reef tank (e.g., corals, crustacea, and tubeworms) compete for the calcium, calcium hydroxide must be regularly added to the water. Replacing evaporated water with a saturated solution of calcium hydroxide is advised. (Add more than is dissolved. A sediment forms on the bottom, and the milky supernatant is poured in the aquarium.)

Reproduction

Species that have a very short larval phase can be bred in the aquarium. Additional details can be found in the individual species descriptions.

A group of *Murex brandaris* spawning in the Mediterranean (see p. 614).

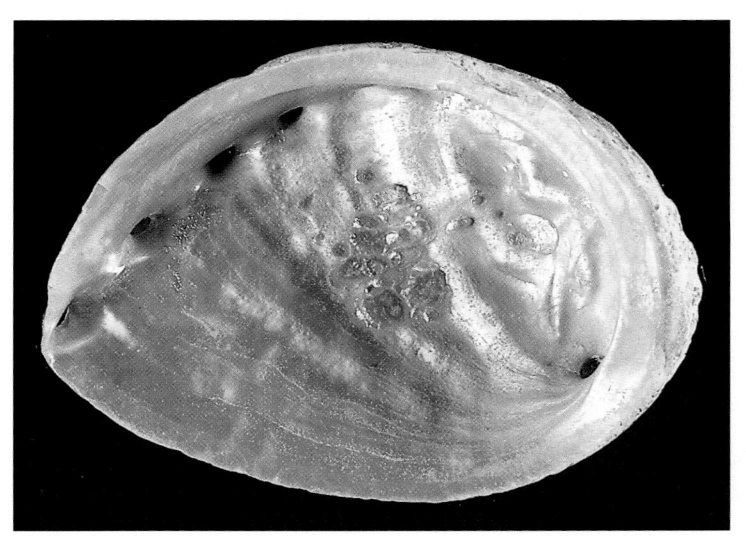

Haliotis (Notohaliotis) gigantea, Japan.

Haliotis (Notohaliotis) gigantea GMELIN, 1791
Japanese abalone, giant abalone

Hab.: Pacific coast of Japan. *H. gigantea* occurs on hard substrates in shallow water (few meters).

Sex.: There are separate sexes (dioecious), but no external differentiating characteristics.

Soc.B./Assoc.: Only during the spawning season are pairs found. Most invertebrates and all fishes make acceptable tankmates.

M.: A hard, algae-covered substrate is a necessity.

Light: Moderate light.

B./Rep.: This species is being cultivated in Japanese breeding stations, suggesting that aquarium rearing should not be difficult. PEÑA (Aquaculture 52:1986, 35–41) has described how to induce spawning.

F.: H; occasionally feeds on carrion. Large specimens should be offered blanched lettuce leaves and one food tablet every two weeks.

S.: Japanese abalones are considered a delicacy. Some marketed animals are captive bred and raised, but the balance are still collected from the wild. *H. gigantea*'s foot is divided into a left and right side. During locomotion, both halves are displaced in relation to the other, enabling the snail to crawl. When threatened, the foot firmly grips the substrate. A crowbar seems to be the only way to pry the animal loose, but this frequently proves injurious.

T: 18°–24°C, L: 15 cm, TL: from 120 cm, WM: m, WR: m, AV: 3, D: 1–2

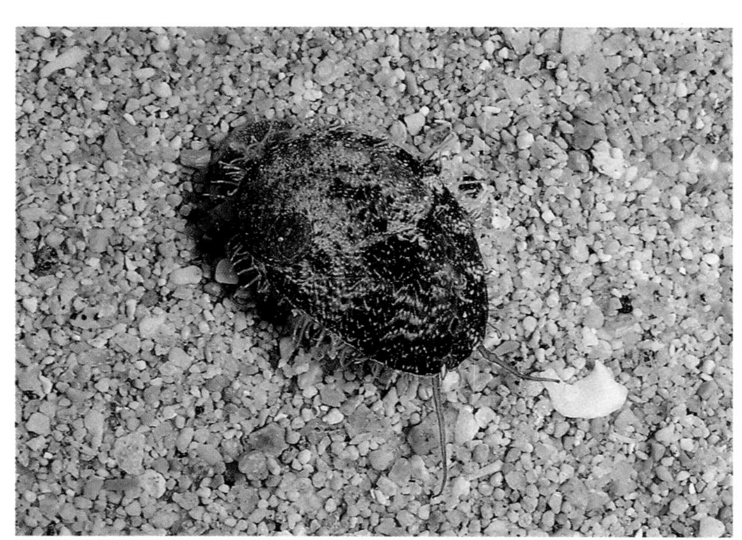

Haliotis multiperforata, aquarium photo.

Haliotis multiperforata REEVE, 1846
Multiperforated abalone

Hab.: Coastal regions of the Red Sea. During the day *H. multiperforata* sequesters itself beneath stones or dead coral blocks.

Sex.: Dioecious, but there are no external differentiating characteristics.

Soc.B./Assoc.: It associates well with many invertebrates and fishes.

M.: During the day, holes and crevices among rocks or corals are sought. Proper shell development demands regular calcium hydroxide additions.

Light: Dim light.

B./Rep.: SCHOMISCH (Das Aquarium, 4: 1993, 35–37) witnessed a successful spawning sequence in an aquarium; during the night the specimens, a different tropical species, rose to about 80% of their body length and released the gametes into the current. Rearing was successful.

F.: H; it feeds almost exclusively on various green algae. Growth corresponds to the quantity of algae in the tank. Occasionally, parboiled spinach and lettuce as well as tablet foods that have a relatively high proportion of vegetable constituents are accepted.

S.: Due to the shell's mother-of-pearl interior, it is coveted by collectors.

T: 20°–28°C, L: 4 cm, TL: from 100 cm, WM: m, WR: b–m, AV: 2, D: 1–2

Haliotis tuberculata f. *lamellosa*, Mediterranean.

Haliotis tuberculata f. *lamellosa* LAMARCK, 1822
Common ormer, ear-shell

Hab.: Throughout the Mediterranean on hard substrates. During the day it strongly adheres to the underside of stones.

Sex.: There are no external sexual differences.

Soc.B./Assoc.: This abalone is a solitary creature. It gets along well with fishes and invertebrates.

M.: Provide stones blanketed in green algae; these will be appreciatively grazed. The common ormer frequently sits on the aquarium glass.

Light: Dim to moderate light zones. This species is nocturnal.

B./Rep.: *Haliotis* is one of the very few Prosobranchia genera that breeds in the aquarium. Due to the short larval phase, it can also be bred in a filtered tank. Juveniles grow relatively quickly.

F.: H; green algae are grazed from the upper side of stones during the night. Lettuce and spinach leaves (blanched) and an occasional food tablet should be offered.

S.: The Atlantic *Haliotis tuberculata* LINNAEUS, 1758 attains a shell length of more than 10 cm. An especially relished food, it is protected in France to prevent overharvesting.

T: 16°–22°C, L: 5 cm, TL: from 40 cm, WM: m, WR: b, m, AV: 1, D: 1

Keyhole limpet, Californian coast.

A keyhole limpet defending itself against a starfish.

The mantle extends.

The starfish retreats.

Megathura crenulata, Baja California.

Megathura crenulata (SOWERBY, 1825)
Giant keyhole limpet

Hab.: Along the southern Pacific coast of North America and the Gulf of California. *M. crenulata* lives in the upper sublittoral zone, where it retires beneath stones and in rock crevices during the day.

Sex.: There are no external differentiating characteristics.

Soc.B./Assoc.: This solitary animal can be kept with other organisms. However, it is easy prey for predacious starfishes and snails.

M.: Rock aquarium. The hiding places, where it sequesters itself during the day, are abandoned at night.

Light: Dim light.

B./Rep.: Relatively little is known about reproduction—namely that eggs are laid and that the emerging planktonic larvae have winglike lobes.

F.: H; nocturnal algivore. Occasionally food tablets or dry foods will be consumed from the bottom.

S.: Most species within this family have an opening at the apical end of the shell which serves as the excurrent and the anal opening. The inner rim of the hole is callused. Keyhole limpets do not have an operculum to close their shells, and unlike true limpets, the entirety of their soft body will not fit under the shell.

T: 18°–22°C, L: 10 cm, TL: from 120 cm, WM: m, WR: m, AV: 2, D: 1–2

Patella coerulea, Mediterranean.

Patella coerulea LINNAEUS, 1758
Blue limpet

Hab.: Mediterranean, primarily in the western section. Blue limpets inhabit the intertidal zone and the upper water strata of rocky coasts.

Sex.: There are no external differentiating characteristics.

Soc.B./Assoc.: These solitary animals are very common in some areas. They are good associates for fishes and, with the exception of predatory starfishes, invertebrates.

M.: According to LANGE & KAISER, 1991, a surf aquarium—a relatively elaborate system—best suits its needs. Survival is frequently brief in aquaria lacking strong water movement, but animals able to adapt to aquarium conditions often have a longevity in excess of 10 years.

Light: Sunlight zone.

B./Rep.: Some males transform into females over the course of growth. Planktonic larvae hatch from oligolecithal eggs (eggs with a small amount of yolk).

F.: H; exceeding their body in length, the powerful radula is used to rasp green algae and endolithic (embedded in the surface of stones) blue-green algae.

S.: The shell is bowl-shaped and lacks an operculum. *P. coerulea* establishes a "home." Orienting itself by mucus tracts, the blue limpet returns to the same place during low tide.

T: 15°–24°C, L: 5 cm, TL: from 100 cm, WM: s, WR: t, AV: 1, D: 1–2

Patella rustica, Mediterranean.

Patella rustica LINNAEUS, 1758
Rustic limpet

Hab.: Common throughout the Mediterranean and from Mauritania to Biarritz in the Atlantic. The intertidal zones of rocky coasts are its preferred habitat.

Sex.: There are no external differentiating characteristics.

Soc.B./Assoc.: *P. rustica* is frequently found in great densities (photo). It is a good tankmate for invertebrates and fishes.

M.: An ideal inhabitant of tidal or surf aquaria (LANGE & KAISER, 1991), since the animal demands alternating submerse and emerse conditions.

Light: Sunlight zone.

B./Rep.: Trochophora larvae hatch from the eggs.

F.: H; the coiled radula is several times the length of the body. It is used to rasp blue-green algae from rocky surfaces.

S.: *P. rustica* is easily distinguished from other *Patella* by the dark striations on the inferior side of the shell and, on the upper half of the shell, by ribs that dissolve into nodules. Because of its impressive adhesion to the substrate, it is largely impervious to predators. The shell is adapted to the uneven surfaces of rocky substrates. During high tide it searches for food, facilitating capture.

T: 15°–24°C, L: 5 cm, TL: from 100 cm, WM: s, WR: t, AV: 1, D: 2–3

Clanculus pharaonius, aquarium photo.

Clanculus pharaonius (LINNAEUS, 1758)
Strawberry shell

Hab.: Red Sea, eastern Africa, and the Indian Ocean. It occurs along the lower littoral zone and among coral rubble.

Sex.: There are no external differentiating characteristics.

Soc.B./Assoc.: *C. pharaonius* can be kept with other tropical organisms.

M.: Other than rocks the animal can crawl beneath, there are no special requirements.

Light: Moderate light zone.

B./Rep.: Fertilized eggs are deposited in a capsule. After a brief veliger stage, the young metamorphose into benthic juveniles. Captive breeding has not been reported.

F.: H; nocturnal algivore. Its intended home must have a cover of green algae.

S.: Because of its red color, *C. pharaonius* has been called the strawberry shell. The underside of the shell has a so-called false umbilicus, a cavelike indention over the umbilicus that is closed with a calcareous callus.

T: 21°–28°C, L: 2 cm, TL: from 70 cm, WM: m, WR: m, AV: 2, D: 1–2

Monodonta turbinata, Mediterranean.

Monodonta turbinata (BORN, 1780)
Turban shell

Hab.: Mediterranean. Shallow water inhabitants of rocky coastlines.

Sex.: There are no external differentiating characteristics.

Soc.B./Assoc.: Although they are very common, social behavior remains a mystery. Excellent tankmates for fishes and invertebrates.

M.: Turban shells climb on glass panes and rock edifications; hence a cover is required to prevent them from tumbling out of the tank. According to LANGE & KAISER a surf tank meets their needs. Rather than submersed, transport the animals in a damp cloth.

Light: Sunlight zone.

B./Rep.: The larval phase is short. *M. turbinata* can be bred in mature aquaria, provided that filtration is not too strong.

F.: H; algae are grazed from stones and aquarium panes.

S.: Identification is confounded because of intraspecific variability and the fact that *M. articulata* and *M. mutabilis*, two similar species, also occur within *M. turbinata*'s distribution range.

T: 16°–24°C, L: 3 cm, TL: from 30 cm, WM: m–s, WR: t, AV: 1, D: 1

Tectus dentatus, Red Sea.

Tectus dentatus (FORSSKÅL, 1775)
Dentate top shell

Hab.: Common in the Red Sea, the Indian Ocean, and the Indo-Pacific, where they are encountered along the upper zones of coral reefs near the water surface.

Sex.: There are no external differentiating characteristics.

Soc.B./Assoc.: These solitary animals are easily associated with virtually all fishes and invertebrates. Sensitive sessile animals (e.g., tubeworms) are occasionally disquieted by *T. dentatus*.

M.: House in a rock aquarium that has sufficient open spaces for crawling.

Light: Sunlight zone.

B./Rep.: The encapsulated eggs hatch into free-swimming larvae. It seems that this species has not been captive bred.

F.: H; algae are grazed from the substrate at night. An abundance of green algae is a necessity for successful maintenance, but blanched lettuce and tablet foods are suitable fare in a pinch.

S.: This top shell is commonly unrecognizable due to epizoites such as calcareous red algae, bryozoans, and other invertebrates. The shell is conical with blunt spikes and a buttonlike tip. With its foot, *T. dentatus* strongly adheres to the substrate, thereby maintaining its hold in the surf. Great quantities are collected and eaten.

T: 20°–28°C, **L**: 8 cm, **TL**: from 120 cm, **WM**: m, **WR**: t, **AV**: 2, **D**: 1–2

Tectus (Rochia) niloticus, Malaysia.

Tectus (Rochia) niloticus (LINNAEUS, 1767)
Common turban shell

Hab.: Throughout the entire Indo-Pacific region in shallow waters of coral reefs.

Sex.: Dioecious. There are no external differentiating characteristics.

Soc.B./Assoc.: *T. niloticus* can be kept with many fishes and invertebrates. Sensitive sessile animals (e.g., tubeworms and soft corals) are occasionally disturbed by their movements.

M.: Rock edifications should be far enough below the water surface that the tip of the shell cannot break out of the water.

Light: Sunlight zone.

B./Rep.: Fertilized eggs are encapsulated and deposited on hard substrates. Before settling to the substrate, the larvae pass through a short planktonic phase.

F.: H; great quantities of green algae are grazed. In the aquarium, large individuals must be fed supplements of blanched lettuce and tablets.

S.: The black and white striped shells are often so overgrown with algae that their coloration is indistinguishable. Pet stores usually carry young specimens that are about 5 cm tall. Large shells are commonly sold as souvenirs. With the exception that it only grows to about 5 cm, the red and white striped *T. conus* is similar.

T: 20°–28°C, L: 15 cm, TL: from 120 cm, WM: m, WR: t, AV: 1–2, D: 1–2

Turbo petholatus, Great Barrier Reef.

Turbo petholatus LINNAEUS, 1758
Cat's eye

Hab.: Widely distributed. From the Red Sea to eastern Africa and Polynesia, it resides in holes and crevices along outer reefs at shallow to moderate depths.

Sex.: Dioecious. There are no external differentiating characteristics.

Soc.B./Assoc.: This solitary animal makes an affable tank inhabitant.

M.: Rock aquarium. *T. pethotatus* effectively controls filamentous algae.

Light: Moderate light zone.

B./Rep.: The planktonic phase is relatively short. Aquarium breeding and rearing have not been accomplished to date, but STARK (Das Aquarium 3: 1992, 25–27) coincidentally bred a very similar species in the aquarium.

F.: H; algae. Occasionally offer tablet foods.

S.: Because of the dark mark at the center of the calcareous operculum, it has been named "cat's eye." Shells are notoriously smooth and shiny. Whorls have dark spots and light bands, but the coloration and flamelike markings are extremely variable.

T: 20°–28°C, **L:** 8 cm, **TL:** from 120 cm, **WM:** m, **WR:** m, **AV:** 3, **D:** 2

Nerita albicilla, Great Barrier Reef.

Nerita albicilla LINNAEUS, 1758
White-spotted nerite

Hab.: From eastern Africa to the Philippines and the Great Barrier Reef, *N. albicilla* inhabits intertidal zones of rocky coasts.

Sex.: There are no external differentiating characteristics.

Soc.B./Assoc.: This solitary animal may be encountered in small groups due to the sheer numbers of them. In captivity it can be associated with most fishes and all invertebrates, with the exception of predacious starfish.

M.: Algae-covered rocks for climbing and grazing and a good cover are required; the latter will prevent *N. albicilla* from falling out of the tank as it scales the glass. LANGE & KAISER maintain that this species can be cared for in a surf tank.

Light: Sunlight zone.

B./Rep.: Little is know about its reproductive biology.

F.: H; this nocturnal algivore spends the day in rock crevices. The habitat must have healthy algae growth.

S.: Shells of *N. albicilla* come in an array of colors—white, dark brown, black, or white and black striped. Since the calcareous operculum seals tightly, they resist desiccation. All neritidae only have one gill and one kidney (a pair of each is the usual arrangement).

T: 20°–28°C, L: 2.5 cm, TL: from 70 cm, WM: m, WR: t, AV: 2, D: 1–2

Littorina littorea, North Sea.

Littorina littorea (LINNAEUS, 1758)
Common periwinkle

Hab.: North and Baltic Seas and the northern Atlantic to Portugal. It has been introduced into North America. It lives on rocks and seaweeds close to shore.

Sex.: There are no external differentiating characteristics.

Soc.B./Assoc.: *L. littorea* often occurs in great numbers. Thanks to its thick shell, it has few enemies.

M.: This species is only suitable for coldwater aquaria. The snail readily climbs above the water, so make sure the aquarium has a cover to preclude it from falling outside. To transport, remove it from the water and keep it moist. *L. littorea* can be kept thus for several days.

Light: Sunlight zone.

B./Rep.: Development is direct without a free-swimming larval stage, making captive breeding relatively easy.

F.: H; algae and, supposedly, barnacles. Supplemental foods are unnecessary if the tank has plenty of green algae.

S.: To compensate for its reduced gills, the common periwinkle has a highly vascular mantle cavity, an attribute that enables this gastropod to endure desiccation for prolonged periods of time.

T: 10°–18°C (!), L: 3 cm, TL: from 70 cm, WM: m, WR: t, AV: 2, D: 1

Dendropoma maxima, Maldives.

Dendropoma maxima (SOWERBY, 1825)
Large worm shell

Hab.: Red Sea, Indian Ocean, and the Indo-Pacific. This species adheres to rocks or coral blocks of shallow reefs. The shell becomes partially overgrown by corals.

Sex.: There are no external differentiating characteristics.

Soc.B./Assoc.: Dense aggregations can sometimes be found.

M.: It is imperative that *D. maxima* is transported with its substrate. Place the specimen in the upper or middle region of the tank where it is exposed to a good current.

Light: Sunlight zone.

B./Rep.: The free-swimming larva adheres to a hard substrate then begins to form its shell.

F.: C; worm shells are planktivores. A suspended mucus net is secreted, then it and the captured plankton are retrieved and devoured. Afterwards, a new net is built. Feed moderate quantities of crumbled dry diets in the aquarium.

S.: A septum separates the forward, inhabited portion of its shell from the empty posterior section. Reportedly, the large worm shell has a lifespan of several years in an aquarium.

T: 22°–28°C, L: 15 cm, TL: from 120 cm, WM: m–s, WR: t–m, AV: 2, D: 1–2

Serpulorbis arenarius, Mediterranean.

Serpulorbis arenarius (LINNAEUS, 1758)
Mediterranean worm shell

Hab.: Rocky coasts of the Mediterranean, the eastern Atlantic, and the Canary Islands.

Sex.: There are no external differentiating characteristics.

Soc.B./Assoc.: *S. arenarius* can be associated with invertebrates and fishes.

M.: A Mediterranean aquarium make an appropriate captive environment. With its substrate, arrange the worm shell so that the opening of the shell points upward. A moderate to strong current should be present.

Light: Moderate light zone.

B./Rep.: Development includes a free-swimming larval stage. Upon contact with a hard substrate, the larvae adhere and begin to form a shell (protoconch), which is spirally wound like a "normal" shell.

F.: O; substitute plankton and crumbled flake foods. Using a mucus net, *S. arenarius* entraps suspended plankton and detritus. The net and its contents are retrieved and eaten. A new net is then secreted.

S.: Though worm shells are sessile and often confused with calcareous tubeworms, there are discernible differences. Worm shells have a three-layered shell in which the innermost layer is shiny, whereas the shell of a tubeworm is composed of just two layers, and the inner layer is matte.

T: 18°–22°C, L: 8 cm, TL: from 70 cm, WM: m–s, WR: m, AV: 2, D: 1

Order: Mesogastropoda
Fam.: Vermetidae

Serpulorbis sp.
Naked worm shell

Hab.: Coral reefs of the Red Sea and the Indian Ocean.

Sex.: There are no external differentiating characteristics.

Soc.B./Assoc.: *Serpulorbis* sp. is an appropriate tankmate for many other coral reef animals.

M.: Keep in a rock or reef aquarium, and provide a good current.

Light: Moderate light zone.

B./Rep.: As soon as the free-swimming larva settles onto a hard substrate, it begins to secrete a shell.

F.: O; like other worm shells, this species gathers its food with a mucus net.

S.: The shell is affixed to the substrate.

T: 22°–28°C, **L:** 8 cm, **TL:** from 100 cm, **WM:** m, **WR:** m, **AV:** 2, **D:** 1–2

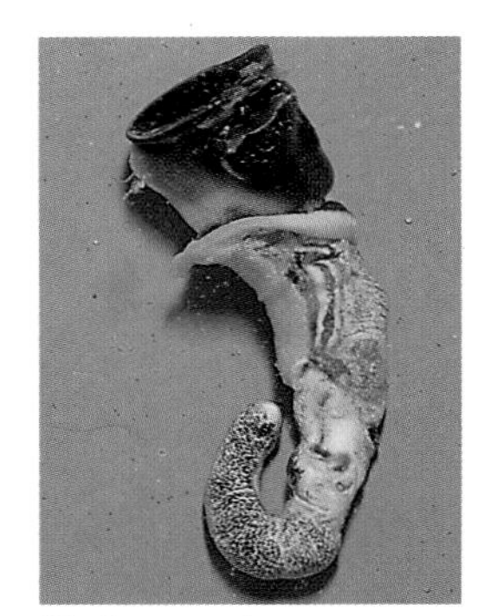
Shell removed.

Serpulorbis sp., Maldives.

Vermetus adansoni, Maldives.

Vermetus adansoni DAUDIN, 1800
Adanson's worm shell

Hab.: Entire Indian Ocean and Indo-Pacific. It is generally an epizoite of gorgonians and black corals.

Sex.: There are no external differentiating characteristics.

Soc.B./Assoc.: As seen in the photo, *V. adansoni* commonly forms colonylike aggregations. It is an appropriate tankmate for many invertebrates and fishes.

M.: Place the animals in direct current, preferably on a dead gorgonian branch.

Light: Moderate light zone.

B./Rep.: Development passes through a free-swimming veliger larval stage. Aquarium breeding is impossible.

F.: O; drifting plankton and detritus are captured in a mucus net.

S.: Tubes are of variable length, only about 5 mm in diameter, and often overgrown by hydroids. The Scandinavian scientist ADANSON (species name) was the first to recognize that worm shells are molluscs, not worms.

T: 22°–28°C, **L:** 4 cm, **TL:** from 100 cm, **WM:** m–s, **WR:** m, **AV:** 2, **D:** 1–2

Janthina janthina, aquarium photo.

Janthina janthina (LINNAEUS, 1758)
Violet snail

Hab.: Cosmopolitan inhabitants of tropical and temperate oceanic waters. Using a self-made float, these animals drift along the water surface.

Sex.: *J. janthina* is a protandric hermaphrodite (first male then female).

Soc.B./Assoc.: Benthic animals can be housed with violet snails.

M.: The current of the filter pump will carry violet snails to the aquarium pane.

Light: Sunlight zone.

B./Rep.: As the eggs hatch into veliger larvae in the mantle cavity, *J. janthina* appears to be a livebearer. Other *Janthina* species carry their eggs with them on the float, thereby practicing brood care.

F.: C; primarily Siphonophora such as *Velella* and *Physalia* (Portuguese-man-of-war). Its dietary requirements greatly complicate captive maintenance.

S.: All *Janthina* are pelagic creatures that build a float out of hardened mucus and air bubbles. If they become separated from their float, they sink.

T: 18°–28°C, L: 4 cm, TL: from 150 cm, WM: w, WR: t, AV: 3, D: 3

Aporrhais pespelicani, Mediterranean.

Aporrhais pespelicani (LINNAEUS, 1758)
Common pelican's foot

Hab.: From Iceland to Norway, Morocco, and the Mediterranean. *A. pespelicani* inhabits mud substrates at depths of 10 to 180 m.

Sex.: There are no external differentiating characteristics.

Soc.B./Assoc.: In its natural biotope, the pelican's foot shell lives in small groups. Fishes and invertebrates are suitable tankmates.

M.: A flat area of sand may be best, since the animal usually becomes caught when venturing into rock crevices. Temperatures in excess of 21°C are only tolerated for brief periods of time.

Light: Dim light zone.

B./Rep.: As with all Prosobranchia, the eggs hatch into veliger larvae. Before settling to the bottom, the veliger larvae are planktonic.

F.: O; organic compounds and leftover dry commercial diets are scavenged from the bottom.

S.: The shape of the shell is exceedingly variable. Resting the shell on the substrate, the snail creeps forward and jerks the shell along behind. People along the coast of the Adriatic Sea frequently eat this mollusc.

T: 15°–20° C, L: 4 cm, TL: from 60 cm, WM: w, WR: b, AV: 2, D: 2

Lambis chiragra, Malaysia.

Lambis chiragra (LINNAEUS, 1758)
Chiragra spider conch

Hab.: Southwest Pacific to western India. *L. chiragra* inhabits coral reefs at depths of 5 to 30 m.

Sex.: Females attain a length of 25 cm, substantially larger than the male's maximum length of 18 cm. The aperture of the shell is light red, and both knobs on the shoulder of the body whorl exceed the size of the others. In contrast, males have a dark red aperture, and all knobs are of equal size.

Soc.B./Assoc.: Other benthic organisms may be disturbed by *L. chiragra*.

M.: Only small, young individuals should be introduced into the aquarium.

Light: Moderate light zone.

B./Rep.: Larvae are free-swimming. Captive breeding is very difficult.

F.: O; primarily green algae, though commercial dry foods and tablets are also accepted.

S.: Male were erroneously described as an additional species, *L. rugosa*.

T: 22°–28°C, L: 25 cm, TL: from 100 cm, WM: m, WR: b, AV: 2, D: 2

Lambis lambis, Maldives.

Lambis lambis (LINNAEUS, 1758)
Common spider conch

Hab.: Red Sea to the western Pacific. Unlike other conchs, this species is not found in the Atlantic. It inhabits sand and hard substrates of coral reefs.

Sex.: Sex can be determined by the form of the shell. The smaller males have two single knobs on the shoulder of the body whorl and spines that extend straight out from the aperture. Females have a longitudinal hump on the shoulder of the body whorl and upward curved spines.

Soc.B./Assoc.: *L. lambis* can be housed with fishes. Caution—sessile invertebrates may become prey.

M.: Provide plenty of space. Since it is an agile climber, a cover is recommended when there are tall rock edifications.

Light: Moderate light zone.

B./Rep.: As with all Prosobranchia, development includes free-swimming veliger larvae. Unfortunately, the larvae do not mature in aquaria.

F.: O; principally algae, but carrion on occasion. Bivalve meat, especially that of mussels, and tablet foods are readily accepted.

S.: *Lambis* have well-developed eyes.

T: 22°–28°C, **L:** 15 cm, **TL:** from 100 cm, **WM:** m, **WR:** b, **AV:** 2, **D:** 2

Lambis truncata sebae, Thailand.

Lambis truncata sebae (KIENER, 1843)
Large conch

Hab.: Distribution is limited to the southwestern Pacific. At depths of 10 to 30 m, *L. truncata truncata* (HUMPHREY, 1786) inhabits reefs of the Indian Ocean.

Sex.: The sexual differences are not as pronounced as those of *L. lambis*.

Soc.B./Assoc.: All conchs can be housed with fishes. However, some sessile invertebrates will be considered food and eaten, and others may be damaged or loosened from the substrate by the conchs' clawlike extensions.

M.: As for *L. lambis*. Only juveniles are suggested for aquarium maintenance. Older specimens are not long-lived aquarium inhabitants.

Light: Moderate light zone.

B./Rep.: The large conch not been successfully bred in an aquarium.

F.: O; as for *L. lambis*.

S.: The stout apex of the shell is characteristic for this species, and it is particularly noteworthy in the subspecies *L. t. truncata*.

T: 22°–28°C, L: to 40 cm, TL: from 120 cm, WM: m, WR: b, AV: 3, D: 2

Strombus costatus, Caribbean.

Strombus costatus (GMELIN, 1791)
Small conch

Hab.: Caribbean. This species is found in seagrass lawns and on sand and mud substrates from shallow waters to a depth of 50 m.

Sex.: There are no external sexual differences.

Soc.B./Assoc.: Fishes make suitable tankmates, whereas sessile invertebrates are a recipe for disaster.

M.: A sand or crushed shell substrate is best. Unstable edifications are often destroyed.

Light: Moderate light zone.

B./Rep.: *S. costatus* has a planktonic veliger larval stage.

F.: O; algivores, but they consume carrion on occasion. Every once in awhile offer tablets and pieces of fish or mussel meat.

S.: The shell is thick and heavy and, in older specimens, often totally overgrown by algae and invertebrates. Like *S. gigas*, this species' operculum can be used as a defensive weapon. In the Mediterranean, there is a single *Strombus*; *S. decorus raybaudii* inhabits sandy substrates off the coast of Turkey. The significantly smaller *S. pipus* and *S. gibberulus*, which are better suited for aquaria, are imported from the Indo-Pacific region.

T: 22°–28°C, L: 20–30 cm, TL: from 200 cm, WM: m, WR: b, AV: 3, D: 1

Strombus gigas, Caribbean.

Strombus gigas LINNAEUS, 1758
Queen conch

Hab.: Southern Florida, Bahamas, Caribbean. *S. gigas* lives on soft bottoms from a depth of a few meters.

Sex.: There are no external sexual differences.

Soc.B./Assoc.: Large specimens should only be kept in spacious aquaria. Benthic fishes are contraindicated. Although the queen conch will not attack invertebrates, its movements are disturbing.

M.: A large aquarium with an even bottom is needed. Limit maintenance to juveniles.

Light: Moderate to sunlight zone.

B./Rep.: *S. gigas* is now bred and raised in the Caribbean. The larvae advance to a size of 1 mm within one month (HEYMAN et al.: Aquaculture 77:1989, 277–285).

F.: O; pieces of fish and mussel flesh are readily accepted.

S.: Because of its ability to use the sharp lanceolate operculum as a defense mechanism, it has earned the name "fencing shell" in Germany. Raw or cooked, the meat of the queen conch is quite toothsome; hence it is avidly collected throughout its habitat.

T: 22°–28°C, L: 20–30 cm, TL: from 200 cm, WM: m, WR: b, AV: 3, D: 1

Strombus gigas, eye.

A pile of *Strombus* shells, Isla Providencia, Colombia.

Strombus mutabilis, Maldives.

Strombus sp., Red Sea.

Family Cypraeidae

Sometimes called porcelain shells because of their luster, the colorful cowries are the most prized mollusc. The shine, which looks as if the shells have been coated in lacquer, stems from the fact that the animals completely cover their shell with their mantle, thereby preventing algae or similar fouling agents from attaching to the shell. The mantle has many lobed appendages that offer perfect camouflage, and it is only retracted in the event of danger. The narrow aperture of the shell, unlike almost all Prosobranchia, lacks an operculum, but it is surrounded by toothlike projections.

By the way, these shells have given porcelain its name, not the other way around. In Italy they are called "porcellana" (small pig). When the first porcelain was brought to Italy from China in the middle ages, the material was named after the shell due to its delicacy and sheen. Some species of cowries were the basis of monetary systems in parts of Asia and Africa; the genus name *Monetaria* serves as a reminder of those times. In Cyprus, the shell of the panther cowry (*Lyncina pantherina*) was dedicated to the goddess of love, Aphrodite. Other species were worn as amulets against sterility and sexual diseases.

Though they are found in all warm seas, the sheer number of cowries and the degree of species diversity is greatest in the Indo-Pacific. Most species remain hidden beneath stones and within caves during the day, but others sit on sponges, stones, or soft corals.

The diet for the vast majority of the almost 200 cowries remains unknown. Smaller species are either feeding specialists (soft corals, sponges) or algivores; larger specimens are commonly predators.

For the aquarium, juveniles or smaller species are recommended. With the exception of the feeding specialists, they are easily maintained. They largely assume the role of scavengers in the aquarium, consuming algae and leftovers. As such they do not need to be specifically fed.

Some authors subdivide the genus *Cypraea* into several genera.

Cypraea spedica, aquarium photo.

Cypraea carneola, aquarium photo.

Cypraea carneola (LINNAEUS, 1758)
Fox cowry

Hab.: Widely distributed on coral reefs of the Red Sea to Polynesia and Hawaii.

Sex.: There are no external sexual differences.

Soc.B./Assoc.: Since the animal is nocturnal and sensitive to disruptions, it should not be maintained with nocturnal fishes, crabs, or other large crustacea.

M.: *C. carneola* needs a reef or rock aquarium with hiding places. Daylight hours are spent in seclusion.

Light: Dim light.

B./Rep.: Due to the relatively long planktonic larval phase (veliger larva), breeding in the aquarium is extremely difficult.

F.: O; algivores, but carrion and sometimes sponges are also accepted. Small specimens do not need to be fed specifically in a mature aquarium. Occasionally offer bivalve or squid meat to large specimens.

S.: Unknown.

T: 20°–28°C, **L:** 7 cm, **TL:** from 120 cm, **WM:** m, **WR:** m, **AV:** 2, **D:** 2

Cypraea ocellata, aquarium photo.

Cypraea ocellata (LINNAEUS, 1758)
Ocellate cowry

Hab.: Gulf of Oman to western India and Sri Lanka. *C. ocellata* is found along coral reefs.

Sex.: Dioecious. However, there are no external sexual differences.

Soc.B./Assoc.: Although *C. ocellata* has a bad reputation in front of sea anemones, it is tolerant of a notable few, e.g., *Hetractis* species (see Marine Atlas, Vol. 1). Do not keep with nocturnal fishes. In the aquarium, it sometimes feeds on sponges—encrusting *Halichondria* and *Haliclona* spp. are particular favorites—and soft corals.

M.: House in a rock aquarium and provide sufficient hiding places. The ocellate cowry sequesters itself away during the day.

Light: Dim light.

B./Rep.: Like all Prosobranchia, *C. ocellatus* has a free-swimming veliger larval stage. Unfortunately, these larvae do not prosper in an aquarium.

F.: C, H; because juveniles graze the substrate, primarily feeding on diatoms, they do not need to be specifically fed. Larger specimens should occasionally be given small pieces of mussel and fish flesh after the lights have been turned off.

S.: The species *nomen* is derived from the eyelike spots on the shell.

T: 21°–26°C, L: 3 cm, TL: from 70 cm, WM: m, WR: b, AV: 3, D: 2

Cypraea guttata, aquarium photo.

Cypraea venusta with spawn.

Cypraea sp., Philippines.

Cypraea vitellus, aquarium photo.

Cypraea talpa (LINNAEUS, 1758)
Map cowry

Hab.: Coral reefs of the Indo-Pacific, from eastern Africa to the Philippines.

Sex.: There are no external distinguishing features.

Soc.B./Assoc.: If possible, avoid housing with other nocturnal animals.

M.: Rock or reef aquaria with hiding places are appropriate.

Light: Dim light.

B./Rep.: As with other cowries, the eggs hatch into free-swimming larvae.

F.: O; basically herbivores, but animal-based foods are also accepted. Stones covered by green algae are grazed during the night. Tablet and flake foods are eaten from the bottom.

S.: *C. talpa* typically has three light cross-bands on its dark shell.

T: 22°–28°C, L: 8 cm, TL: from 120 cm, WM: m, WR: b, AV: 2, D: 2

Cypraea mappa, Malaysia.

Cypraea talpa, Red Sea.

Cypraea tigris, Maldives.

Cypraea tigris LINNAEUS, 1758
Tiger cowry

Hab.: The entire Indo-Pacific from the Red Sea to the Great Barrier Reef. Shells from Hawaii are the largest.

Sex.: There are no external distinguishing features.

Soc.B./Assoc.: Most invertebrates, such as tubeworms, sea urchins, and small starfish, make good tankmates. Do not house with *Condylactis*. Large specimens have been known to nibble on sea anemones, sponges, and soft corals. The extended mantle edge is often vulnerable to fishes.

M.: Easily maintained in rock aquaria that offer sufficient cover.

Light: Moderate light zone.

B./Rep.: All attempts have failed.

F.: C; predators and scavengers. Pieces of fish and mussel flesh and TetraTips are suggested. Small tiger cowries do not need to be specifically fed.

S.: Because of the demand for its shell, this once common cowry has become rare in some areas. It is one of the largest cowries. The species exists in different color morphs, generally contingent on geographic location.

T: 22°–28°C, **L**: 5–10 cm, **TL**: from 60 cm, **WM**: w, **WR**: b, **AV**: 2, **D**: 1

Cypraea histrio, aquarium photo.

Studies along the coast of Tanzania have proven beyond a shadow of a doubt that species commercially harvested for their shells are in great decline (NEWTON et al., Biol. Conserv. 63 (1993) 241–245). Up to 18 times the number of specimens were found on the remote Mafia Island in comparison to the coasts of Dar es Salaam and Zanzibar Island, two locals where shells are intensely collected. *Cypraea* species, e.g., *C. tigris*, *C. histrio*, and *C. lynx*, are particularly affected. Apparently the demand of live snails for the marine aquarium hobby barely influences population levels. Tourist demand shells by the hundreds, placing more pressure on the population than can be borne.

Cypraea mappa

Cypraea sp., aquarium photo.

Monetaria annulus, aquarium photo.

Monetaria annulus (LINNAEUS, 1758)
Ringed money cowry

Hab.: Common throughout the Indo-Pacific. The ringed money cowry inhabits rocks of the intertidal zone.

Sex.: As with all porcelain shells, there are no external distinguishing features.

Soc.B./Assoc.: Money cowries, *M. annulus* and *M. moneta*, can be readily associated with all invertebrates and fishes.

M.: Rocky areas and an abundance of green algae are its only requirements.

Light: Sunlight zone.

B./Rep.: About 500 eggs are adhered to protected reef sites near the low tide mark. Apparently this species has not been bred in an aquarium.

F.: H. Using its radula, *M. annulus* rasps algae from rocks.

S.: Small porcelain snails are often inadvertently introduced with live rock. This species—easily identified by the yellow to greenish ring—and its congener, *M. moneta*, can be kept under the same conditions. *M. moneta* has several color variants.

T: 22°–28°C, L: 2.5 cm, TL: from 60 cm, WM: m–s, WR: m, AV: 1, D: 1

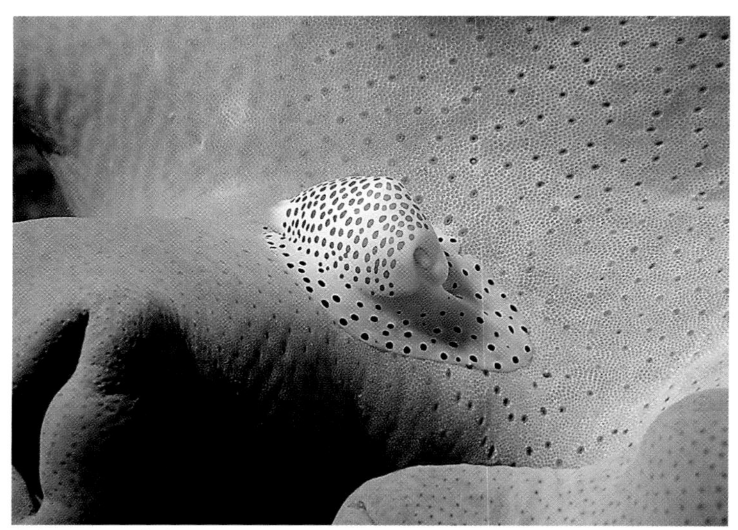

Calpurnus sp., Indonesia.

Calpurnus sp.
Small egg shell

Hab.: This unidentified species hails from the coral reefs of the Indo-Pacific.

Sex.: There are no external distinguishing features.

Soc.B./Assoc.: *Calpurnus* sp. is a solitary abiding, nocturnal animal that feeds on soft corals (photo).

M.: Successful care hinges upon the presence of soft corals.

Light: Dim light.

B./Rep.: Reproduction and development are unknown.

F.: C; feeding specialist. *Calpurnus* sp. feeds exclusively on polyps of various soft corals; substitute foodstuffs are not accepted.

S.: Remaining hidden during the day, *Calpurnus* sp. emerges at night to search for food. Its congener, *C. verrucosus*, has a wide shell with pink ends and lives in the Red Sea. It, too, is found on soft corals, primarily *Sacrophyton*.

T: 22°–28°C, **L:** 3 cm, **TL:** from 120 cm, **WM:** m, **WR:** m, **AV:** 3, **D:** 3

Cyphoma gibbosum (LINNAEUS, 1758)
Flamingo tongue

Hab.: Southeast North America and throughout the Caribbean. It is frequently found at shallow depths on gorgonian fans.

Sex.: There are no external distinguishing features.

Soc.B./Assoc.: *C. gibbosum* is solely found in conjunction with *Gorgonia flabellum* and *G. ventalina*. It is not uncommon to see several flamingo tongues on one fan.

M.: Maintain with gorgonians.

Light: Sunlight zone.

B./Rep.: The reproductive and developmental biology of the flamingo tongue remains unknown.

F.: C; feeding specialist. Its diet is limited to *Gorgonia flabellum* and *G. ventalina*.

S.: Unknown.

T: 22°–28°C, **L:** 3 cm, **TL:** from 100 cm, **WM:** s, **WR:** m, **AV:** 3, **D:** 4

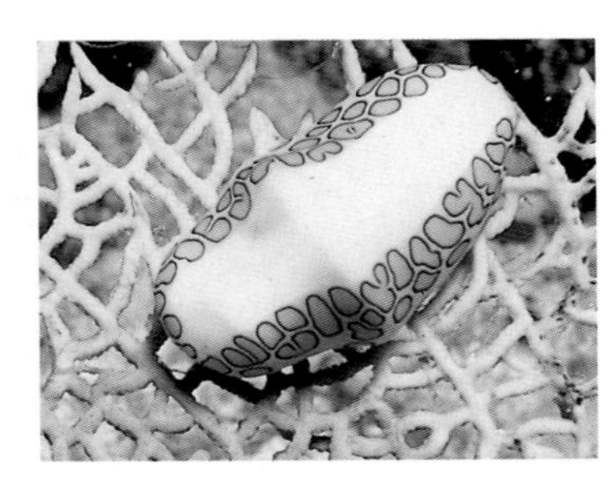

Cyphoma sp., Caribbean.

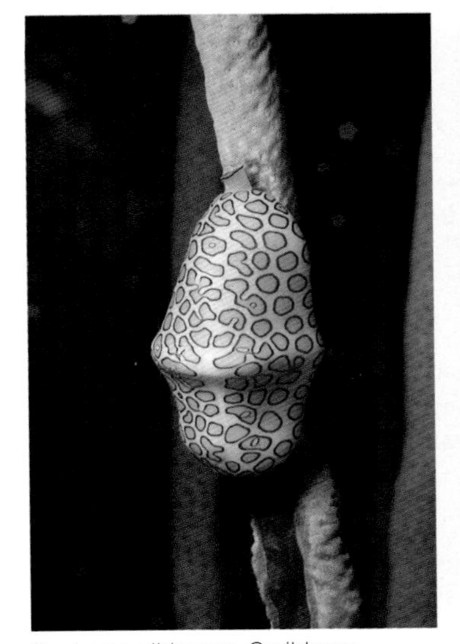

Cyphoma gibbosum, Caribbean.

Cyphoma signatum

Ovula ovum, Kenya.

Ovula ovum (LINNAEUS, 1758)

Egg cowry, common egg cowry

Hab.: Coral reefs of the Indo-Pacific. From a depth of a few meters, *O. ovum* lives on hard substrates.

Sex.: There are no external distinguishing features.

Soc.B./Assoc.: A solitary living organism. Beware! Large predatory snails and starfish are capable of overcoming this species.

M.: Keep *O. ovum* in a rock or reef aquarium.

Light: Moderate light.

B./Rep.: Details are unknown.

F.: C; various soft corals. Substitute fare is not accepted.

S.: The shell is white exteriorly and orange-brown interiorly.

T: 22°–28°C, **L:** 8 cm, **TL:** from 120 cm, **WM:** m, **WR:** b, **AV:** 2–3, **D:** 3

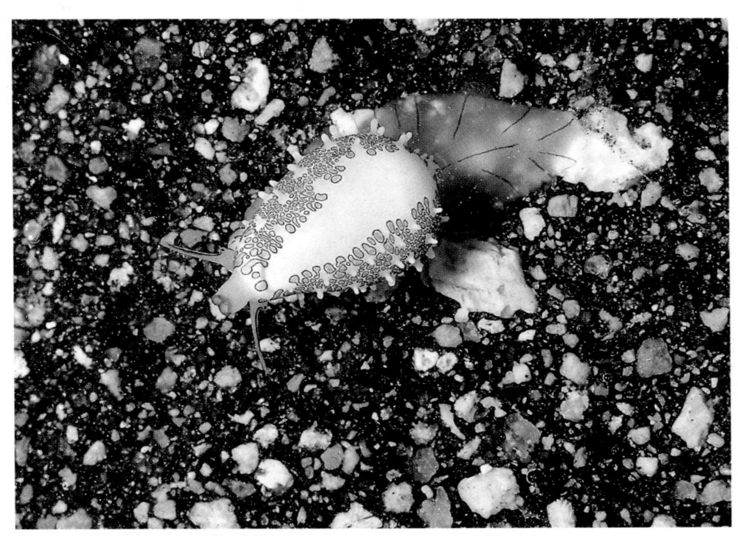

Pseudosimnia carnea, Red Sea

Pseudosimnia carnea (POIRET, 1789)
Dwarf red ovula

Hab.: Mediterranean, the Azores, Madeira, and the Canary Islands. *P. carnea* is distributed on hard substrates at depths of 20 to 120 m.

Sex.: There are no external distinguishing features.

Soc.B./Assoc.: It lives in association with *Corallium rubrum* and, occasionally, gorgonians such as *Lophogorgia sarmentosa*.

M.: *P. carnea* must be housed in a Mediterranean tank containing live *Corallium* or gorgonians.

Light: Dim light.

B./Rep.: Most details on reproduction and development are unknown, except that *P. carnea* does have a free-swimming veliger larval phase.

F.: C; feeding specialist. The diet consists exclusively of octocorals, particularly the polyps of *Corallium*. Substitute fare is refused.

S.: Coloration varies according to host—red on *Corallium* and orange, yellow, or white on gorgonians. In southern Spain, *P. carnea* is commonly encountered by divers.

T: 17°–22°C, L: 1.5 cm, TL: from 100 cm, WM: m, WR: m, AV: 3, D: 3

Volva volva (LINNAEUS, 1758)
Shuttle volva

Hab.: From eastern Africa to the Philippines and Taiwan. It is usually found on gorgonians and soft corals (Alcyonacea).

Sex.: There are no external distinguishing features.

Soc.B./Assoc.: Congenial tankmates for fishes and invertebrates.

M.: Live gorgonians or soft corals must be provided.

Light: Moderate light.

B./Rep.: Development includes a free-swimming larval stage.

F.: C; *V. volva* feeds on its host.

S.: The mantle often adopts the color of the host. The shell's aperture does not have labial teeth.

T: 22°–28°C, L: 10 cm, TL: from 120 cm, WM: m, WR: m, AV: 2, D: 3

Phaenacovolva birostris

Volva volva, Maldives.

Cassis cornuta, Malaysia

Cassis cornuta (LINNAEUS, 1758)
Horned helmet

Hab.: Widely distributed from the Red Sea through eastern Africa to Polynesia, Japan, and Australia. Sandy bottoms at moderate depths are its preferred habitat.

Sex.: In comparison to the male, the female's shell is larger with numerous small shoulder knobs.

Soc.B./Assoc.: Because of its predacious nature, tankmates should be limited to large fishes.

M.: Sandy substrate, a strong filter, and a protein skimmer are conditional for successful care.

Light: Moderate light.

B./Rep.: The diminutive eggs are stacked into towering piles. From the eggs emerge veliger larvae. A relatively long planktonic phase follows.

F.: C; predators. The diet is principally composed of sea urchins. Prey are immobilized by a neurotoxinlike compound. Because of the sulfuric acid laden saliva, not even calcareous armor impedes *C. cornuta*. Mussel or squid meat is generally accepted.

S.: This is one of the few species that has external sexual differences. The horny operculum is long and slender. Although there are several related species, none attains *C. cornuta*'s impressive size. Large shells are often sold in souvenir shops.

T: 22°–28°C, L: 30 cm, TL: from 150 cm, WM: m, WR: b, AV: 2, D: 2–3

Order: Mesogastropoda
Fam.: Cymatiidae

Charonia lampas (LINNAEUS, 1758)
Atlantic trumpet

Hab.: The Azores, Madeira, Canary Islands, Atlantic coast of Spain and France, and the Mediterranean, principally the western section. *C. lampas* is found form 10 to 700 m, chiefly on hard substrates.

Sex.: There are no external distinguishing features.

Soc.B./Assoc.: Maintain in a manner in keeping with its solitary nature. Echinoderms (sea urchins, starfish) are unsuitable associates for large specimens.

M.: Other than the need for ample free space, there are no particular demands placed upon the aquarium's decor. A protein skimmer is required to counteract this animal's high rate of metabolism.

Light: Moderate light.

B./Rep.: Little is known about reproduction and development of this species.

F.: C; the red starfish, *Echinaster sepositus*, is the main component of its diet. In the aquarium bivalve and squid meat are accepted.

S.: Shells are treasured collector's objects. In large specimens, older whorls are often overgrown with calcareous red algae, eroded, or even broken away. Parts of the shell may be green from algal growth.

T: 15°–20°C, L: 25 cm, TL: from 120 cm, WM: m, WR: b, AV: 3, D: 2–3

Charonia tritonis, Mediterranean.

Charonia tritonis (LAMARCK, 1816)
Trumpet triton

Hab.: Indo-Pacific as well as the tropical and subtropical Atlantic and eastern Mediterranean.

Sex.: Dioecious, but it is impossible to distinguish the sexes based on external differences.

Soc.B./Assoc.: Keep this species singly. Sensitive invertebrates are disturbed by its wanderings, and echinoderms are preyed upon.

M.: To fulfill its need to wander about, leave plenty of open space.

Light: Moderate light.

B./Rep.: Eggs hatch into veliger larvae.

F.: C; this predator feeds on starfishes and sea urchins. Because of the sulfuric acid component, *C. tritonis*'s saliva is able to dissolve calcareous carapaces. Mussel and squid meat are accepted in captivity.

S.: By punching a hole in the upper region of the shell, ancient fishermen created a signal horn. This is one of the very few enemies of the crown-of-thorn starfish, *Acanthaster planci*. The shell of *C. tritonis* is usually in better condition than its congener, *C. lampas*.

T: 18°–25°C, L: to 40 cm, TL: from 120 cm, WM: m, WR: b, AV: 3, D: 2–3

Tonna galea, Mediterranean.

Tonna galea (LINNAEUS, 1758)
Giant tun

Hab.: Mediterranean and Canary Islands, chiefly in the Adriatic Sea and along the coast of Greece. *T. galea* inhabits hard and soft substrates at 20–80 m depth.

Sex.: There are no external differences.

Soc.B./Assoc.: This solitary animal cannot be maintained with echinoderms or other molluscs.

M.: A sufficiently flat sandy or rocky bottom, a good filter, and a protein skimmer are required.

Light: Dim light.

B./Rep.: Details on reproduction and development are unknown.

F.: C; *T. galea* is a voracious predator that primarily feeds on echinoderms (starfishes, sea urchins), but bivalves and snails are also components of the diet. Using a measure of sulfuric acid, the saliva dissolves the calcareous shells of its prey. Small prey is swallowed whole, whereas larger organisms are shredded by the radula. Substitute foods are not accepted.

S.: The shell has thin walls and a wide aperture, but no operculum. It is commonly used as an object d' décor. In Sicily this snail is considered edible.

T: 18°–20°C, **L:** 20 cm, **TL:** from 150 cm, **WM:** m, **WR:** b, **AV:** 3, **D:** 3

Chicoreus ramosus, Great Barrier Reef.

Chicoreus ramosus (LINNAEUS, 1758)
Branched murex

Hab.: Throughout the Indo-Pacific and along the Great Barrier Reef. It resides on rock, sand, and mud bottoms at shallow depths.

Sex.: There are no distinguishing characteristics on the shell. Occasionally the male's large penis can be seen to the right of the head.

Soc.B./Assoc.: Large specimens are best kept singly or with fishes that are not bottom oriented.

M.: Due to its penchant for crawling about, the aquarium should have a flat substrate.

Light: Moderate to sunlight zone.

B./Rep.: The female lays her eggs in thick, brittle capsules. The capsules are then adhered to each other, forming a nest.

F.: C; predator. Using its radula, gastropod and bivalve shells are compromised. In the aquarium it accepts fish, shrimp, squid, and bivalve meat.

S.: With its shiny pink lip and white interior, this is the largest and most familiar murex. There are a number of smaller, darker *Chicoreus* species. *C. ramosus* is a commonly sold souvenir.

T: 22°–28°C, **L**: 18 cm, **TL**: from 150 cm, **WM**: m, **WR**: b, **AV**: 2–3, **D**: 1–2

Chicoreus steeriae, Malaysia.

Chicoreus steeriae (REEVE, 1845)
Steer's murex

Hab.: Coral reefs of the Indo-Pacific.

Sex.: There are separate sexes, but it is impossible to distinguish them based on external differences.

Soc.B./Assoc.: Because of its predacious nature, this murex is best kept singly or with fishes that are not bottom-oriented.

M.: House in a rock aquarium that has a flat bottom. Small specimens live for several years.

Light: Moderate light.

B./Rep.: Eggs hatch into veliger larvae. It is during this planktonic larval stage that an "embryonic shell" is formed. The ciliated velum is cast off when the larva settles onto the substrate.

F.: C; using its radula like a drill, *C. steeriae* penetrates mollusc shells. Thawed frozen foods are the best substitute fare.

S.: The long inhalant siphon—responsible for respiration and prey detection—is protected by the inferiorly elongated siphonal canal.

T: 22°–28°C, **L:** 10 cm, **TL:** from 120 cm, **WM:** m, **WR:** b, **AV:** 2–3, **D:** 1–2

Hexaplex trunculus, Mediterranean.

Hexaplex trunculus (LINNAEUS, 1758)
Trunk murex

Hab.: Mediterranean. Casual inhabitants of the Canary Islands and the Portuguese and Moroccan coasts.

Sex.: Dioecious.

Soc.B./Assoc.: This solitary abiding animal should not be associate with echinoderms, molluscs, or sessile animals.

M.: *H. trunculus* can be easily kept in any Mediterranean rock aquarium.

Light: It only emerges at night in brightly lit aquaria.

B./Rep.: In nature, up to several hundred individuals congregate and create a spawning ball as part of the spawning sequence, but in captive environments, individual females form small egg balls. Newly hatched young can be fed TetraTips.

F.: C; a scavenger for the most part, although slow-moving or sessile animals may be attacked. Starfishes are considered delicacies. Occasional pieces of fish or mussel flesh are accepted, as are tablet foods.

S.: In ancient times, *H. trunculus* was collected for its purple pigment. One thousand animals were needed to obtain one gram of the precious substance. Today *H. trunculus* is more valued for its edible flesh.

T: 15°–25°C, L: 5–8 cm, TL: from 50 cm, WM: m, WR: b, AV: 3, D: 1

Murex brandaris, Mediterranean.

Murex brandaris LINNAEUS, 1758
Spring murex

Hab.: Soft substrates of the Mediterranean and the eastern Atlantic coast.

Sex.: There are no external sexual differences.

Soc.B./Assoc.: The spring murex feeds on sessile invertebrates, echinoderms, and other molluscs. However, it is a very compatible tankmate for fishes.

M.: Maintenance is facile as long as the temperature remains 22°C or less most of the time. Although it lives on soft substrates in nature, rocks and glass panes of the aquarium are scaled.

Light: Moderate light.

B./Rep.: Numerous animals assemble to spawn, forming a large spawning ball (see p. 562). G. MÄGERLEIN (TI) reported that newly hatched *Murex* refused beef heart and fish flesh, but eagerly consumed TetraTips.

F.: C; a predator and a scavenger. Occasionally offer pieces of bivalve or fish flesh and tablet foods.

S.: Like *Hexaplex trunculus*, this species also renders a purple pigment. Individuals from the northern Adriatic Sea are virtually always host to *Calliactis parasitica*, a commensal sea anemone which provides defense against predacious octopi.

T: 15°–22°C, L: 8 cm, TL: from 50 cm, WM: m, WR: b, AV: 2, D: 1

Murex sp., Indonesia.

Murex sp.
Comb murex

Hab.: Indonesia. Like all long-spined *Murex* species, this one lives on deep, soft substrates of fine sand or mud.

Sex.: The male's large penis is visible on the right side of the body beneath the cephalic tentacles.

Soc.B./Assoc.: While this species should not be associated with bivalves or sessile invertebrates (e.g., tubeworms), it is a good tankmate for fishes.

M.: A substrate of fine sand is required.

Light: Moderate light.

B./Rep.: The eggs, which are deposited in balls, hatch into planktonic larvae.

F.: C; using its radula, *Murex* sp. drills a hole through the shell of its prey—bivalves and snails. An anesthetic produced in the salivary gland (purpurin) is then injected into the perforation. In the aquarium, thawed frozen fare (mussel, squid, shrimp, or fish flesh) are acceptable substitutes.

S.: Individuals from protected regions have especially long spines. Each row of spines corresponds to the position of an enlargement on the outer lip of the shell's aperture. The extremely elongated "stalk" is the siphonal canal, an appendage that supports the inhalant siphon. The congener *M. pecten* has particularly beautiful spines (p. 539).

T: 22°–28°C, **L:** 10 cm, **TL:** from 120 cm, **WM:** m, **WR:** b, **AV:** 3, **D:** 1–2

Nassarius variegatus (A. ADAMS, 1852)
Variegated nassa

Hab.: Red Sea and the western Indo-Pacific. It inhabits sandy substrates.

Sex.: There are no sexual differences.

Soc.B./Assoc.: This species can easily be associated with fishes and invertebrates.

M.: A rock or reef aquarium with a sandy substrate is appropriate.

Light: Moderate light.

B./Rep.: *N. vareigatus* has not been successfully bred in an aquarium.

F.: C; carrion. Every once in a while offer pieces of frozen foods.

S.: This species has a well-developed sense of smell. At a distance of up to 30 m, carrion can be detected using its extremely elongated siphon (photo)!

T: 20°–28°C, L: 3 cm, TL: from 100 cm, WM: m, WR: b, AV: 2, D: 1

Nassarius coronatus, Maldives.

Nassarius variegatus, Red Sea.

Fasciolaria tulipa, Colombia.

Fasciolaria tulipa (LINNAEUS, 1758)
Tulip snail, Caribbean whelk, true tulip

Hab.: Only in the Caribbean. To a depth of 10 m, *F. tulipa* is found in regions with a high proportion of organic components.

Sex.: There are no external sexual differences.

Soc.B./Assoc.: Fellow molluscs are consumed with predilection; hence, they should be excluded from the list of possible tankmates.

M.: Provide sufficient open space. A protein skimmer is recommended.

Light: Moderate light.

B./Rep.: Veliger larvae, which hatch from the eggs, sometime pass through a free-swimming planktonic phase.

F.: C; *F. tulipa* primarily preys upon gastropods; even equal-sized individuals are not exempt from being attacked. Various frozen foods are accepted in the aquarium.

S.: The dark orange foot is used to plow the substrate. Its periostracum (the shell's external organic layer) is typically very thin and may even be absent in older specimens. The shell's markings are variable

T: 21°–27°C, L: 18 cm, TL: from 150 cm, WM: m, WR: b, AV: 3, D: 1–2

Oliva spicata, Baja California.

Oliva spicata RÖDING, 1798
Californian olive

Hab.: West coast of southern North America, particularly the Gulf of California. It lives in or slightly below the intertidal zone.

Sex.: There are no external differences.

Soc.B./Assoc.: Do not keep with invertebrates or fishes that live on or within sandy substrates.

M.: Provide a sand substrate suitable for burrowing; *O. spicata* prefers to pass the day hidden within a sandy medium.

Light: Moderate light.

B./Rep.: Clusters of lentil-shaped, egg-bearing capsules are adhered to a hard substrate. Each capsule contains about 8 eggs. After hatching, the young are immediately benthic, i.e., development does not pass through a free-swimming larval stage like most other molluscs.

F.: C; carrion and organisms found on or within the sand. In the aquarium, fish and mussel flesh are accepted.

S.: Like *Cypraea*, *O. spicata* has a shiny enameled layer that is secreted by the shell's enveloping mantle lobes. During the day, the animal is hidden in the sand; it hunts for prey at night.

T: 20°–25°C, **L:** 5 cm, **TL:** from 100 cm, **WM:** m, **WR:** b, **AV:** 3, **D:** 1

Oliva sp., Colombia.

Oliva sp.
Olive

Hab.: Shallow waters of the Caribbean. *Oliva* sp. occurs on or burrowed within sandy substrates.

Sex.: Olives do not have external sexual differences.

Soc.B./Assoc.: House in community aquaria containing anemones, sea urchins, brittle stars, crustacea, and fishes.

M.: Provide a layer of sand sufficiently deep to allow the olive to completely bury itself.

Light: Moderate light.

B./Rep.: As for the vast majority of Prosobranchia, eggs of *Olivia* sp. hatch into planktonic larvae. Details, however, are unknown.

F.: C; small molluscs and bristleworms encountered in the substrate are consumed. Aquarium specimens can be fed bivalve meat.

S.: The general shape of the shell is reminiscent of an olive, and the spire is flat with a pointed apex. *Oliva* spp. are found on all the world's coral reefs.

T: 21°–28°C, **L:** 4 cm, **TL:** from 100 cm, **WM:** m, **WR:** b, **AV:** 3, **D:** 1

Turbinella angulatus, Caribbean.

Turbinella angulatus SOLANDER, 1786
Angular chank

Hab.: Bahia Concha, the Caribbean, and the coast of Colombia. This species inhabits soft substrates.

Sex.: The shell bears no distinguishing features.

Soc.B./Assoc.: Large specimens are only suited for show aquaria. Invertebrate tankmates are not recommended, since *T. angulatus*'s creeping is generally disruptive.

M.: A large tank and a sandy substrate are suggested. The latter permits the snail's strong foot to plow.

Light: Moderate light.

B./Rep.: The female deposits hard egg capsules. Free-swimming larvae emerge.

F.: O; algivores. Pieces of fish and mussel flesh are also accepted.

S.: The shell is coated with a thick, dark brown periostracum; this layer scales off dead animals. Northern Colombians use the shell as a wind instrument and—finely ground—as an additive to coca leaves.

T: 21°–27°C, L: 20 cm, TL: from 150 cm, WM: m, WR: b, AV: 3, D: 1–2

Harpa amouretta, Red Sea.

Harpa amouretta RÖDING, 1798
Small harp

Hab.: Red Sea to the Indian Ocean and Hawaii. It is encountered in various habitats from shallow to deep waters.

Sex.: There are no external sexual differences.

Soc.B./Assoc.: *H. amouretta* can readily be associated with fishes as well as many invertebrates, but it frequently preys upon small crustacea.

M.: The substrate must contain a proportion of sand to allow *H. amouretta* to burrow.

Light: Moderate light.

B./Rep.: Development includes a free-swimming veliger larval stage.

F.: C; primarily small crustacea captured from the sand. In the aquarium, it accepts pieces of carrion: mussel, fish, and shrimp meat.

S.: *Harpa* shells have an unusually long, broad foot. When extended, it may exceed the shell's dimensions by 100%. In a process called autotomy, the posterior end of the foot may be sacrificed, thereby startling a pursuing predator and allowing *H. amouretta* to escape. This self-mutilation is employed by crabs, daddy longleg spiders, and lizards. There are several similar species that generally grow larger than *H. amouretta*.

T: 22°–28°C, **L**: 5 cm, **TL**: from 100 cm, **WM**: m, **WR**: b, **AV**: 3, **D**: 1–2

Cymbiola vespertilio, aquarium photo.

Cymbiola vespertilio (LINNAEUS, 1758)
Vespertine volute

Hab.: The Philippines and northern Australia. *C. vespertilio* is generally found along coral reefs at moderate depths.

Sex.: There are no external sexual differences.

Soc.B./Assoc.: Large specimens should either be maintained alone or with free-swimming fishes.

M.: A rock aquarium equipped with a good filter and protein skimmer is recommended. Since it is relatively fast, the aquarium must be large enough to accommodate its active lifestyle.

Light: Moderate light.

B./Rep.: Most volutes—this species included—do not have a free-swimming larval phase. Distribution is therefore restricted to advances *C. vespertilio* makes by crawling. Aquarium breeding and rearing should not be overly difficult.

F.: C; predators. Feed frozen foods in the aquarium (shrimp, squid, mussels).

S.: This species varies greatly in regard to shape and color. Due to its limited distribution (see above), local races are common. The shell has a blunt apex, three knotty whorls, and pointed tubercles on the shoulder. The light aperture has four columellar folds.

T: 22°–28°C, L: 10 cm, TL: from 120 cm, WM: m, WR: b–m, AV: 2, D: 1–2

Genus *Conus*, cone shells

Conus is one of the largest genera of the gastropods. The radula sports one highly modified harpoon-shaped tooth. Reserve teeth are to be found within the radula sac. The tooth is barbed (one or more), hollow, and attached to a poison gland. The neurotoxin is extremely potent—more so in some species than in others. It is dangerous, sometimes even lethal, to humans. The most dangerous species are: *C. aulicus*, *C. geographus*, *C. marmoreus*, *C. striatus*, *C. textile*, and *C. tulipa*.
The radula tooth is thrust from the proboscis and fired into the prey. It is only used once; if the trajectory is off and the shot misses its mark, the tooth is lost. It is replaced by a tooth from the radula sac.
Most species have specialized diets. Prey include gastropods, cephalopods, bristleworms, crustacea, and fishes. Some cone shells stalk sleeping fishes at night. After the fish has been detected and approached, the long proboscis is extended and the radula tooth is fired into the fish.
Extreme caution is advised when handling these organisms. Never touch them with your bare hands! Use a net!
Locomotion is unusual. The shell is pulled along with arrhythmic motions. Other Neogastropoda such as *Aporrhai* (p. 583) and *Strombus* (p. 588) employ similar methods.
Cone shells are often valuable collector items. The rare *Conus gloriamaris*, of which only a few specimens have been found, is worth several thousand dollars.

Conus arenatus

Conus captianeus

Conus geographus, Okinawa.

Conus geographus LINNAEUS, 1758
Geography cone

Hab.: Throughout the Indo-West Pacific. The geography cone lives on coral rubble.

Sex.: Dioecious, yet sexes cannot be distinguished from external characteristics.

Soc.B./Assoc.: Few invertebrates and fishes are appropriate tankmates. Fishes that sleep on the bottom at night are vulnerable and may be consumed.

M.: Hiding places are required. Caution!—this species has a poisonous radula tooth. Only handle with a net.

Light: Dim light.

B./Rep.: Planktonic larvae hatch from the eggs. This phase lasts for some time before the larvae subsequently change over to a benthic lifestyle. Aquarium breeding has not been successful.

F.: C; predators. Sleeping fishes are its chief prey.

S.: The shell is relatively thin with knots along the shoulder. This species, *C. aulicus*, *C. marmoreus*, *C. striatus*, *C. textile*, and *C. tulipa* are some of the most venomous gastropods.

T: 22°–28°C, **L:** 10 cm, **TL:** from 100 cm, **WM:** m, **WR:** b, **AV:** 2, **D:** 4 (poisonous!)

Conus litteratus, Great Barrier Reef.

Conus litteratus LINNAEUS, 1758
Literary cone

Hab.: From eastern Africa to the Great Barrier Reef and Polynesia. It occurs on sand or coral rubble at shallow depths.

Sex.: There are no external sexual differences.

Soc.B./Assoc.: Very few invertebrates and fishes make suitable tankmates. Fishes that sleep on the bottom are attacked.

M.: Because the animals burrow to some extent, a sand substrate is required. Caution!—this species has a poisonous radula tooth. Use a net when handling.

Light: Moderate to sunlight zone.

B./Rep.: Egg capsules hatch into planktonic veliger larvae, which then develop into a veliconcha; though this phase has retained its ability to swim, it primarily crawls along the bottom.

F.: C; this voracious predator presents a danger to many invertebrates and fishes.

S.: A frequent import. The surface of the shell has a dense pattern of dark brown or black rectangular spots running in circumferential bands. The spots are larger, sometimes coalescent, on the shoulder. There are a number of similar species with different designs and patterns.

T: 22°–28°C, **L:** 13 cm, **TL:** from 100 cm, **WM:** m, **WR:** b, **AV:** 1, **D:** 4 (poisonous!)

Conus marmoreus, Great Barrier Reef.

Conus marmoreus LINNAEUS, 1758
Marmorated cone

Hab.: Indo-Pacific. It lives in lagoons and on reef platforms and shallow reef slopes.

Sex.: Sex cannot be distinguished by external characteristics.

Soc.B./Assoc.: Fishes that sleep on the bottom and many invertebrates, particularly snails and bivalves, are preyed upon.

M.: Caution!—the radula is capable of injecting a potent, lethal (even to humans) toxin. Use a net when handling. *C. marmoreus* should be provided with a sand substrate and hiding places where it can retire during the day.

Light: Moderate light.

B./Rep.: *C. marmoreus* reproduces like *C. litteratus*, passing through free-swimming veliger larval and veliconchal phases. Unlike the aforementioned species, *C. omaria* does not pass through a free-swimming larval stage; therefore, aquarium rearing of *C. omaria* should be possible.

F.: C; predator. It primarily feeds on snails and bivalves.

S.: Generally this organism only emerges from secluded sites amidst coral rubble or within sand substrates at night to search for food. Roundish to triangular white spots, a low spire, and small tubercles on the shoulder are typical for *C. marmoreus*. *C. marmoreus* is the type species of the subgenus *Conus*.

T: 22°–28°C, **L:** 10 cm, **TL:** from 100 cm, **WM:** m, **WR:** b, **AV:** 1, **D:** 4 (poisonous!)

Conus textile LINNAEUS, 1758
Textile cone

Hab.: Red Sea and Indo-Pacific. Coral rubble and sand substrates of lagoons and shallow waters are *C. textile*'s preferred habitat. During the day, *C. textile* burrows within the substrate, only emerging at night to search for food.

Sex.: No external differences.

Soc.B./Assoc.: A very limited number of invertebrates and fishes are suitable tankmates.

M.: To permit *C. textile* its penchant for burrowing, the aquarium must have a coarse sand, crushed coral, or shell substrate. Caution!—this species has a poisonous radula tooth. Use a net when handling.

Light: Moderate light.

B./Rep.: Eggs are laid in flat, pouchlike capsules. They later hatch into planktonic veliger larvae.

F.: C; predators. Snails, generally species lacking an operculum, are its chief prey. The prey is "felt" with the long proboscis before the poisonous radula tooth is discharged.

S.: A relatively common import despite the fact that it is one of the most toxic species of the genus.

T: 22°–28°C, **L**: 10 cm, **TL**: from 100 cm, **WM**: m, **WR**: b, **AV**: 2, **D**: 4 (poisonous!)

Terebra guttata, aquarium photo.

Terebra guttata (RÖDING, 1798)
Slender auger

Hab.: Indo-Pacific region on sand substrates of reef platforms and the reef bottom.

Sex.: There are no external sexual differences.

Soc.B./Assoc.: This is a solitary abiding animal. Associating *T. guttata* with fishes and invertebrates in the aquarium generally does not present problems, though benthic organisms may occasionally be subject to predation.

M.: Several centimeters of medium-fine sand are needed, since this species tends to burrow as its crawls about the substrate.

Light: Moderate light.

B./Rep.: Unknown. Aquarium breeding is probably very difficult.

F.: C; predators. The upper sand layers are scoured, primarily for polychaetes. In captivity, mussel and shrimp meat and an occasional earthworm are consumed.

S.: Many Indo-Pacific auger shells have a tapered, pointed shell. The primary differentiating characteristics are design and coloration.

T: 22°–28°C, **L:** 10 cm, **TL:** from 100 cm, **WM:** m, **WR:** b, **AV:** 3, **D:** 1–2

Terebra maculata, Red Sea.

Terebra maculata (LINNAEUS, 1758)
Marlinspike, spotted auger

Hab.: Red Sea and throughout the western Pacific. It lives in sand substrates of shallow lagoons and on coral reefs to a depth of 30 m. As it crawls, a broad track is left behind.

Sex.: There are no external differences.

Soc.B./Assoc.: Since it is confined to the sandy bottom, lithophilic invertebrates can be housed in the same aquarium. Likewise, most fishes are suitable tankmates.

M.: A sandy substrate with an extension at least ten times the length of the shell should be provided.

Light: Moderate light.

B./Rep.: Planktonic veliger larvae hatch from the eggs.

F.: C; predators. Sand substrates are scoured for edibles, especially bristleworms. Various frozen foods are generally accepted.

S.: The basic coloration of the shell is a creamy white. There are bands of dark brown, flame-shaped spots. Axial sculpturing is only present on the whorls of the tip.

T: 22°–28°C, L: 11 cm, TL: from 120 cm, WM: m, WR: b, AV: 3, D: 1–2

Casmaria ponderosa digs into the substra

The escape into the sediment only takes 10 seconds.

Subclass Opisthobranchia

Because of their vivid colors, the most colorful of the phylum, opisthobranchs have the distinction of being the joy of every diver and aquarist. However, aquarium maintenance is extremely difficult due to their specialized dietary needs. Opisthobranchs are a heterogenous group composed of various orders, the most significant of which are presented on the following pages. There are approximately 13,000 opisthobranchs; the majority of the attractive, eye-catching species are nudibranchs (order Nudibranchia). The gills are usually positioned posterior, but degenerate in higher forms. Some species have retractable secondary anal gills that disappear into sheaths, e.g. *Chromedoris* spp. Normally opisthobranchs totally lack a shell, but in rare cases there may be an external shell or a thin vestigial shell overgrown by the mantle. Even when adults lack a shell, it is present in the free-swimming veliger larval stage. When the larva settles to the substrate, the shell is discarded. In addition to cephalic tentacles, the preponderance of opisthobranchs have rhinophores, densely ciliated, plicate tentacles. Their purpose is to detect chemical stimuli and currents. Like the secondary anal gills, rhinophores can be retracted into sheaths.

Virtually all Opisthobranchia make their home in the sea; a few genera occur in brackish waters, and only two have adapted to freshwater. They are principally found on rocky coasts of tropical and temperate seas, but deep sea biotopes have been invaded to a depth of 2,000 m. Opisthobranchs have adapted to diverse habitats ranging from the open water (free-swimming forms such as *Glaucus*, p. 702), to the benthos (burrowing species that utilize soft substrates) and the mesopsammon (the interstitial water among grains of sand that is inhabited by some diminutive species). Some even parasitize organisms such as jellyfish, snails, bivalves, and tubeworms. In turn, they bear commensals and parasites, e.g., unicellular organisms and various worms and crustacea. The relationship between the Spanish dancer, *Hexabranchus sanguineus,* and a *Periclimenes* shrimp (p. 682) is particularly noteworthy. The subclass Opisthobranchia is preyed upon by sea anemones, crabs, starfish, fishes, and sea birds. In response, a variety of different defense mechanisms have been devised, namely toxins and nematocysts accumulated from plants and animals they graze and prey upon.

Aquarium Maintenance of Opisthobranchia

Few aquarists are able to gaze upon the bizarre, beautiful colors of these creatures and not have an overcoming desire to place one in their home aquarium. Beware! Many of them are feeding specialists that demand a very selective diet—whether it be herbivorous or carnivorous. Unfortunately, these resplendent animals are common imports. Either unwittingly or with unconscionable knowledge, they are sold with the aside that a mature aquarium is sure to provide enough edibles to support them.

Nutrition

Since opisthobranchs are unable to adapt to a different diet, the majority of cases are destined to a slow death by starvation. But this is not to say that all species are unsuitable for aquaria. The dietary requirements of some can be met by diligent hobbyists. Almost all sea hares (Anaspidea) are included in this group. Algae (*Caulerpa*) and parboiled, unfertilized lettuce sustain these animals. Much to the dismay of aquarists that have tediously raised their aquatic garden, sea hares have an incredible appetite that is only satisfied by copious quantities of green stuff. Their appetite, unfortunately, is on a par with growth; for example, the Californian sea hare, *Aplysia californica*, grows to a length of 1 m and a weight of 7 kg! Although Sacoglossa are also herbivores, they are more difficult to maintain then sea hares, as they feed on particular plants. Dedicated hobbyists will be able to raise green algae from the genera *Halimeda* or *Udotea* to feed them. Before the animals are introduced, ensure that there is a good stand of one of the above mentioned genera. Sacoglossa secrete compounds to protect them from predators such as fishes and shrimp. Notaspidea prey on invertebrates such as cnidarians, bryozoans, and sea squirts. Fortunately the diets of Notaspidea do not seem

to be as specialized as those of other members of this subclass. Hence, substitute fare such as bivalve meat and rock anemones (*Aiptasia*)—reared in a separate tank—are generally accepted.
Exceeding the above mentioned orders in both beauty and difficulty are the nudibranchs (Nudibranchia). All are predators and many are feeding specialists that only prey on a specific species or genus. For instance, some feed on one particular species of sponge. If it cannot be offered, the animals starve. That such feeding specialists accept tablet foods is a tale touted to gullible hobbyists. But who is to say that nudibranchs will never be aquarium animals; what seems impossible today may be within our grasp tomorrow. Just a few years ago maintaining stony or soft corals was beyond our ability, but thanks to advances in technology, it is almost commonplace now. Once the dietary requirements for individual nudibranchs have been established, maintenance should not be difficult. Growing sponges—a difficult feat at best—for the benefit of nudibranchs is dubious. All that effort to sacrifice them to a nudibranch? It is certainly less complicated to nourish nudibranchs that feed on cnidarians. Some are even satisfied with *Aiptasia*, although their beautiful colors frequently fade. Conditional on the availability of plenty of the hydrozoan *Eudendrium*, the Mediterranean *Flabellina affinis* can be maintained. *Eudendrium* are relatively easily cared for with *Artemia* nauplii and mussel milk.
Cooperation is the first step in overcoming the difficulties associated with delicate feeding specialists. The collector should relay where the organism was collected (on sponges, algae, etc.) to the exporter, who passes the information to the merchant. Through this chain, the information should reach the most vital link—you, the hobbyist. There is a paucity of information in the literature. The above proposed chain of events is one of the few hopes for successful care.

Reproduction

Breeding Opisthobranchia will probably remain a dream for quite a while yet. They are hermaphrodites—each individual has both male and female gonads. Sperm is transferred during copulation. After fertilization, the eggs are deposited in a characteristic man-

ner, frequently in the vicinity of the species' prey. Commonly the spawn is the same color of the opisthobranch or its prey. The minute eggs are often less than 0.1 mm in diameter. Hatching veliger larvae pass part of their phase in the plankton. It is this developmental stage that does not survive in the aquarium. However, many aquarists have reported the sudden appearance of numerous juveniles. It is suspected that these juveniles were introduced at a point past the planktonic stage, probably as the veliger larvae settled onto the substrate.
While it is certainly possible to successfully maintain opisthobranchs in captivity, forethought and preparation are an indisputable necessity.

Order Cephalaspidea, Headshield Slugs

Headshield slugs are cosmopolitan. This group, the most primitive order within the Opistobranchia, inhabits shallow to deep biotopes in all the world's seas. Like prosobranchs, but in contrast to their more advanced cousins (e.g., nudibranchs), headshield slugs often have a shell and sometimes even an operculum. Over the course of evolution, however, the shell's importance diminished, until some species just have a reduced interior shell covered by the mantle. As a result of fused tentacles, the majority of Cephalaspidea have a characteristically broadened head. A specialized elongated sensory organ, the Hancock organ, lies between the cephalic shield and the foot. The cephalic shield is used to burrow into soft substrates, aiding in the quest for foraminiferans, bristleworms, bivalves, and snails. Some species are capable of engulfing their prey whole, only fragmenting the prey after it enters the stomach, e.g., the European *Scaphander lignarius*.
Aside from a few species that have entered tropical freshwater, all Cephalaspidea live in the sea. Most are benthic, but some are

planktonic. The free-swimming veliger larvae have a small spiral shell and an operculum. Veliger larvae are the rule, but there are a few exceptions.
Sea butterflies are also members of the order Cephalaspidea. Swarms of these organisms live among the plankton. They, together with krill, make up a large portion of the diet of whales.

Chelidonura flavolobata HELLER & THOMPSON, 1983
Light-lobed bubble shell

Hab.: Coral reefs of the Red Sea.

Sex.: Hermaphrodites.

Soc.B./Assoc.: *C. flavolobata* is generally found in groups rather than alone. Little is known about its behavior in an aquarium. Limit invertebrate tankmates to hardy species that can resist attacks by this species.

M.: Rock aquarium.

Light: Moderate light.

B./Rep.: *Chelidonura* spp. lay numerous small eggs that subsequently hatch into veliger larvae. Since these are planktonic, reproduction in an aquarium is very difficult.

F.: C; predators, though it is not known exactly what this bubble shell feeds upon.

S.: *C. flavolobata* is normally found on large brain corals, sea squirts, and soft corals. It is unknown whether or not it is a feeding specialist, but aquarium maintenance is not recommended.

T: 22°–28°C, L: 3 cm, TL: from 60 cm, WM: m, WR: b, AV: 4, D: 3–4

Chelidonura flavolobata on a star coral.

Navanax polyalphus

Order Anaspidea, sea hares

Sea hares have an elongated body, parapodia, rhinophores, and oral tentacles. The parapodia are lateral lobes on the body. With undulating movements and up and down sweeps of the parapodia, sea hares swim through the water. These animals commonly grow to massive proportions, and in extreme cases (e.g., *Aplysia californica*), they may attain a length of 1 m and weigh 7 kg. The reduced shell may be calcareous or membranous and either partially or totally covered by the mantle. Occasionally the shell may be totally absent. The rolled, erect rhinophores (organs of taste and current detection) and the oral tentacles give anaspideans a harelike appearance. This similarity is denoted by their common name.

Purely herbivorous, anaspideans consume vast amounts of red, green, and brown algae as well as seagrasses. Large pieces of plant are swallowed and then ground in the gizzard. The purple gland, which is located in the intestinal pouch, produces a dark, often purple, liquid. As a defense mechanism, a cloud of the ink is released when the animals are disturbed (see photo on the right). This ink is an effective fish repellant.

Some species form chains when spawning. Long threads of very small eggs are deposited in piles. One *Aplysia* can produce close to 500,000,000 eggs in a space of 5 months! Detailed descriptions about the behavior of the Californian sea hare, *Aplysia californica*, can be found in LEONARD & LUKOWIAK (Behaviour 98/1986, 320 ff.).

Aplysia californica
Spotted sea hare

Hab.: Indo-Pacific. *A. californica* inhabits seagrass lawns and dense algae stands.

Sex.: Like all Opisthobranchia, sea hares are hermaphrodites.

Soc.B./Assoc.: The spotted sea hare is sometimes gregarious. In aquaria, it can be associated with robust invertebrates and fishes. Its movements are disruptive to delicate animals.

M.: There are no particular demands.

Light: Moderate light.

B./Rep.: Long, spaghetti-like egg threads are deposited in captivity. Eggs and the subsequent larvae are diminutive. Since development includes a long planktonic phase, aquarium breeding is problematic.

F.: H; large quantities of algae (primarily green). *Caulerpa* stands will be summarily consumed. Clean, parboiled lettuce is an appropriate substitute food.

S.: Unknown.

T: 18°–25°C, **L:** 38 cm, **TL:** from 150 cm, **WM:** w, **WR:** b, **AV:** 3, **D:** 2

Aplysia californica releasing its "ink" from the purple gland.

Aplysia californica

Aplysia dactylomela feeding on *Caulerpa*, until 40 cm.

Phyllaplysia taylori DALL, 1900
Green sea hare

Hab.: Eastern Pacific. *P. taylori* inhabits lawns of *Zostera* (seagrass) growing in reef lagoons.

Sex.: Since all sea hares are hermaphrodites, there are no sexual differences.

Soc.B./Assoc.: This small sea hare should not be associated with predators.

M.: The animal is at ease and maintenance is unproblematic if—and only if—the aquarium has a stand of seagrass.

Light: Moderate light.

B./Rep.: The green sea hare lays its mucus-covered spawn on blades of seagrass (see photo). Since development includes a prolonged planktonic phase, aquarium breeding has been ineffective. Small juveniles may be inadvertently introduced with *Zostera*.

F.: H; in its natural biotope, *P. taylori* selectively feeds on *Zostera*. A diet of algae (*Caulerpa*) or parboiled lettuce is usually inadequate.

S.: This bright green sea hare is perfectly adapted to its herbivorous diet. Unlike *Aplysia*, *Phyllaplysia* has reduced parapodia and is therefore unable to swim in the water column.

T: 22°–28°C, **L:** 2 cm, **TL:** from 100 cm, **WM:** w, **WR:** b, **AV:** 3, **D:** 3

An unidentified sea hare from Australia.

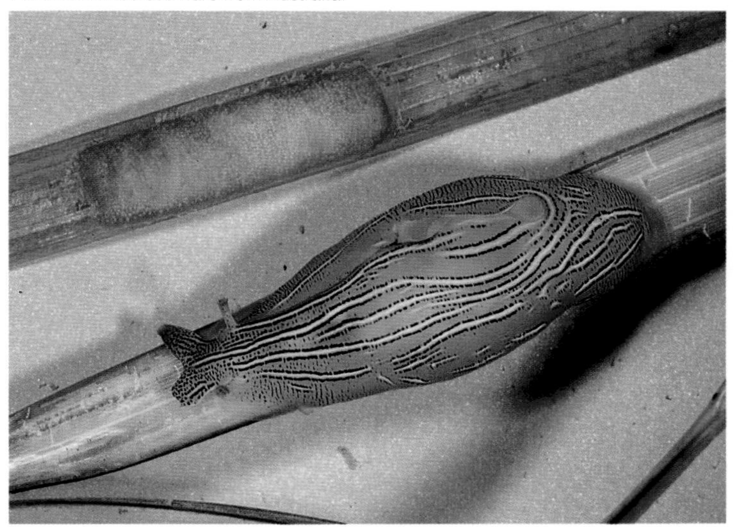

Phyllaplysia taylori with spawn on the seagrass *Zostera* sp.

ORDER SACOGLOSSA, SEA SLUGS

Commonly referred to as plant-suckers, sea slugs are highly specialized herbivores. These small, delicate animals grow to a maximum length of 4 cm, and although most lack a shell, some have a thin external or internal shell. Members of one small family, the Juliidae, even have two shells—something unique to the Bivalvia with this exception. Until the first live specimens were found in 1959, the Juliidae were considered bivalves. Sacoglossa have highly diverse body shapes—elongated and smooth to broad with a multitude of cerata (flat dorsal appendages). The cerata are not armed with nematocysts. The most predominant external characteristic that differentiates Sacoglossa from all other Opisthobranchia is the rolled rhinophores (chemical and current sensors). Oral tentacles are either absent or very small. Most species lack gills.
With few exceptions, Sacoglossa are herbivores; the majority feeds on green algae. The radula is especially modified for this pur-

Continued on p. 644

Elysia decorata HELLER & THOMPSON, 1983
Decorated sea slug

Hab.: Probably endemic to the Red Sea (only found there). It occurs in shallow water beneath coral blocks or in rock crevices.

Sex.: Since *E. decorata* is a hermaphrodite, there are no sexual differences.

Soc.B./Assoc.: This solitary animal can be kept with small fishes or invertebrates as long as sufficient hiding places are available. Its secretory defense mechanism is quite effective; it is avoided by fishes and shrimp.

M.: A rock aquarium with sufficient green algae growth is appropriate.

Light: Dim light.

B./Rep.: Veliger larvae hatch from the eggs. Because of their ciliated crown, the larvae are free-swimming. It is impossible to breed *E. decorata* in an aquarium.

F.: H; green algae. The single row of teeth rips open algal cells, and the animal then sucks the cells empty.

S.: The family Elysiidae is comprised of five genera of variable morphology. All have a smooth, elongated body and lateral lobes on the mantle. They lack a shell, dorsal appendages, and oral tentacles. The pedal tentacles are normally short, and the rhinophores are rolled.

T: 20°–27°C, **L:** max. 12 mm!, **TL:** from 60 cm, **WM:** m, **WR:** b, **AV:** 3, **D:** 2

Elysia verrucosa, Australia.

Elysia decorata, Red Sea.

Continuation of p. 642:
pose. The single row of razor sharp teeth rips algal cells, and the muscular pharynx sucks out the plant cell's fluids. After the teeth are used once, they drop off the odontophore and collect in a pouch (Sacoglossa = pouch tongues). The few nonherbivorous Sacoglossa feed on the eggs of other Opisthobranchia.
Sacoglossa are found in all tropical, temperate, and cold seas, usually in shallow coastal regions among algae and seagrasses. A meager number inhabit fresh or brackish water. Many are green like their grazing substrate, making them difficult to see. Using their broad lateral lobes (parapodia), Lobigeridae swim like sea hares (p. 638). Species can be identified and classified into clearly defined genera, but the relationships between genera and their classification within families is significantly less facile.
Although all the species presented belong to a single family, their appearance is quite divergent. Contrary to being the exception, the same can be said about the other thirty-odd genera.

Tridachia crispata (MÖRCH, 1863)
Lettuce sea slug

Hab.: Limited to the Caribbean.

Sex.: None, since *T. crispata* is a hermaphrodite.

Soc.B./Assoc.: An invertebrate tank is appropriate, although most fishes are suitable tankmates as well.

M.: House in a rock aquarium that has healthy green algae growth.

Light: Moderate light.

B./Rep.: *T. crispata* repeatedly spawns in aquaria. However, rearing the larvae has been unsuccessful to date.

F.: H; green algae grazers.

S.: Over the course of its development, *T. crispata* undergoes drastic alterations in appearance. Juveniles are similar to *Elysia*, as both sides of the body have parapodial lobes. As they mature, these lobes repeatedly fold until they have achieved the characteristic adult state pictured here.

T: 23°–28°C, **L:** 8 cm, **TL:** from 100 cm, **WM:** w, **WR:** b, **AV:** 2, **D:** 2

Tridachia crispata

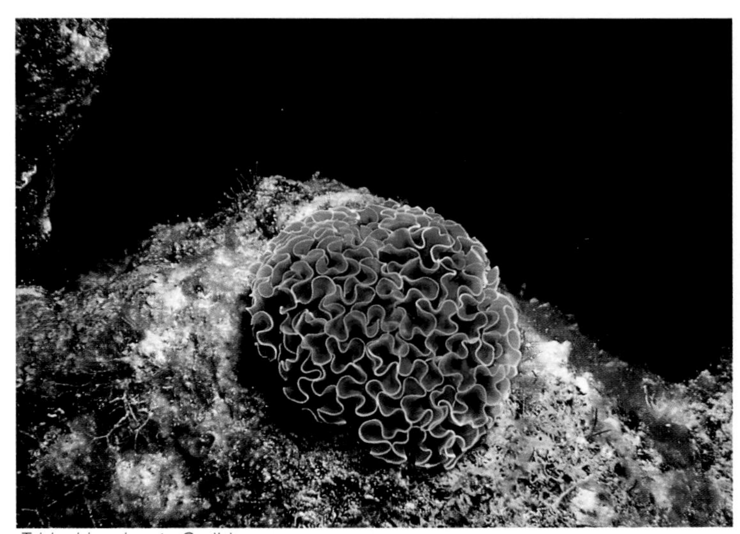
Tridachia crispata, Caribbean.

Tridachia crispata

Order: Notaspidea
Fam.: Pleurobranchiidae

Order Notaspidea, sidegill slugs

Most notaspideans have a flat, thin shell. It is located dorsally and partially overgrown by the mantle. Umbraculidae have a bowl-shaped shell covering part of the body, whereas the Pleurobranchiidae have a shell which is completely covered by the mantle or vestigial. Hardened by calcium inclusions, the mantle fringe is either unattached or fused with the head. All Notaspidea have a bipinnate gill on the right side of the body between the mantle's edge and the foot, hence the name sidegills.
Most Notaspidea inhabit shallow waters of temperate and tropical seas. Since they feed on sessile invertebrates such as sponges and sea squirts, they are "grazing" predators. Animals within this order have a broad radula and strong mandibles, tools used to tear chunks from their prey, a fin on their head, and a pair of rhinophores (chemical and current receptors). Some produce sulfuric acid which is used as a defense mechanism. Although fishes often mistake these unarmored gastropods for a tasty morsel, the error is quickly noticed and the gastropod freed.

Pleurobranchus semperi (VAYSSIERE, 1896)
Sidegill slug

Hab.: Sandy and rocky substrates of the tropical Indo-Pacific, generally at few meters depth.

Sex.: None, since the animals are hermaphrodites. The female genital opening and the penis are anterior to the gills on the right side of the body. Because of torsion, the twisting of the intestinal pouch that occurs over the course of embryonic development, the genital openings are located anteriorly.

Soc.B./Assoc.: This solitary abiding animal lives along reefs. It is unknown what, if any, animals make appropriate tankmates.

M.: Unknown.

Light: Moderate light.

B./Rep.: Because *P. demperi* passes through a planktonic veliger larval stage, reproduction in the aquarium is impossible.

F.: C; a predator. Stomach content analyses, however, are not available. It is said to accept mussel meat in the aquarium.

S.: Unknown.

T: 23°–28°C, L: 15 cm, TL: from 100 cm, WM: w, WR: b, AV: 4, D: 3

Pleurobranchus forskali from the Red Sea.

Pleurobranchus semperi

THE ORDER NUDIBRANCHIA, NUDIBRANCHS

Undoubtedly nudibranchs are the most beautiful and conspicuous of the Opisthobranchia. Depending on diet and habitat, there is a wide assortment of morphologies. All are predacious carnivores and many are feeding specialists, i.e., they only accept very specific foods such as polyps of one particular cnidaria, a certain sponge species, or fish eggs. Hence, nudibranchs are difficult to care for in an aquarium. Almost all invertebrates such as cnidaria, other molluscs, various worms, bryozoans, and fish eggs are part of their food web.
Many nudibranchs have very precisely adapted to the coloration of their prey. Others, in turn, use bright colors and designs as warning signals to indicate their unpalatability. Certain members of this group are mimics, imitating the design of poisonous species. Many species produce toxic substances; some collect unfired nematocysts from hydrozoans and use them towards their own defense.

Reproduction of Opisthobranchia (pictures on facing page)

All Opisthobranchia are hermaphrodites, and every animal has an ovotestis, an organ responsible for egg and sperm production. It, the common genital duct, and the prominent genital aperture are located on the right side of the body near the head. As evidenced by the photo, two individuals—in this case *Chromodoris amoena*—crossfertilize. After fertilization, the partners separate and creep away to spawn on a suitable substrate. This may occur the same day or up to several days following fertilization.
Many nudibranchs lay their spiral or membranous spawns only on those organisms which serve as food; typically, the eggs are the same color as their substrate. Hatching veliger larvae are planktonic, a quality that furthers species dispersal. All Nudibranchia have a shell during their veliger larval phase! After a few days or months, they settle on a suitable substrate and the shell is discarded.

Mating *Chromodoris amoena*, eastern Australia.

Chromodoris amoena laying eggs.

About the genus *Chromodoris*

Chromodoris is the largest genus of the family. While most specimens inhabit the tropical Indo-Pacific, their distribution extends to the Caribbean and Mediterranean. These stunningly colored animals, arguably the most beautiful of the family, have an elongate-oval mantle that completely encompasses the foot. Pinnate secondary anal gills are located on the extreme posterior body. There is just one pair of tentacles, the lamellate rhinophores, on the head. Rhinophores are sensory organs capable of detecting chemical stimuli and current. Only members of the subclass Opisthobranchia posses these organs. Submarginally the mantle is lined with small open glands. The secretions of these glands probably serve as a defense mechanism.

Chromodoris sp.

The Philippine species *Chromodoris annae* is easily differentiated from others.

Chromodoris annulata. This species is notoriously similar to *Risbecia* (p. 663).

Chromodoris coi is widely distributed throughout the western Pacific.

Chromodoris kuiteri from Indonesia. Dorsal markings can vary.

A few species are so similarly colored, they are hard to differentiate, while the coloration of other species is so distinctive that they can be irreputably identified solely on the basis of color. The coloration of the common *Chromodoris fidelis* (lower left) varies according to habitat. *Chromodoris geminus* (lower right), a rare find, is an inhabitant of the Red Sea and the Indian Ocean. Underwater, the mantle's colors seem to fluoresce.

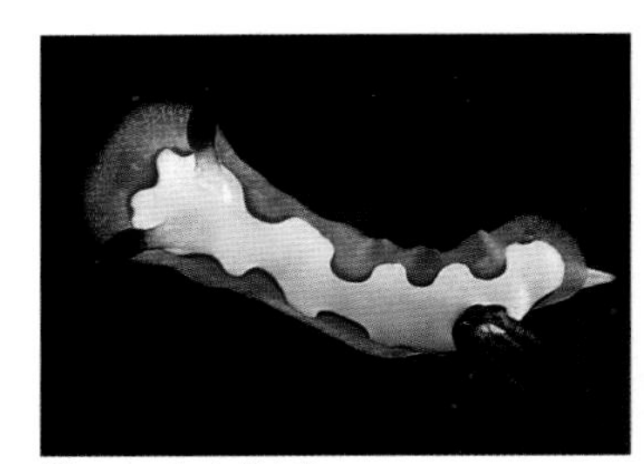

Chromodoris fidelis

Chromodoris geminus

Chromodoris kuniei, a common species from the tropical southwestern Pacific.

Chromodoris africana, 6 cm, Red Sea.

Chromodoris africana from the Red Sea.

Chromodoris quadricolor (RÜPPELL & LEUCKART, 1828), pictured on this page and the top of the facing page, is frequently encountered in the Red Sea and along the coast of eastern Africa by divers and snorkelers. The dorsal section of the mantle is blue with three black longitudinal stripes, and the outer fringe is yellow. When viewed from the side (lower right), it can be seen that the markings continue around the mantle's edge on onto the foot. Despite their obvious beauty and availability, *C. quadricolor*'s dietary requirements preclude aquarium maintenance. Starvation results.

Chromodoris quadricolor, Red Sea.

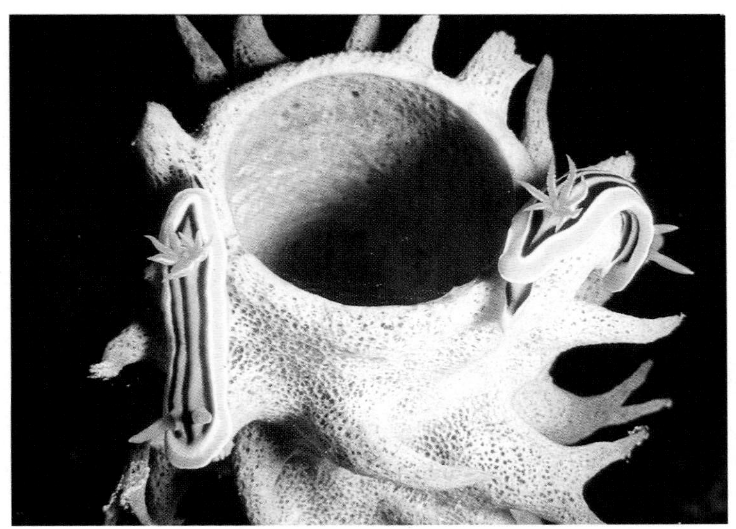

Chromodoris quadricolor on a tube sponge in the Red Sea.

Because several *Chromodoris* spp. have colors and a design similar to *C. quadricolor*, identification is not simple. Many refer to these species as the "*C. quadricolor* group." Members of the group include *C. annae* (p. 650), *C. elizabethina* (p. 652), *C. magnifica* (p. 653), and the unidentified species pictured below. Surely other, as yet undescribed, species will later join this list. Intraspecific geographic variations further confound the situation.

Chromodoris quadricolor, Philippines.

Gymnodoris aurita, 9 cm, Comores.

Chromodoris splendida, 6 cm, from the waters off New South Wales (Australia).

Chromodoris luteorosa, western Mediterranean.

Chromodoris luteorosa (v. RAPP, 1827)
Yellow-spotted nudibranch

Hab.: Only confirmed from the western Mediterranean and the Bay of Biscay. *C. luteorosa* lives on hard substrates and within seagrass meadows.

Sex.: Hermaphrodites.

Soc.B./Assoc.: A solitary abiding animal. Conspecifics are only sought during the spawning season.

M.: *C. luteorosa* is not suitable for aquarium maintenance because of its dietary requirements.

Light: Moderate light.

B./Rep.: Breeding is similar to that of its congeners. In the ovotestes, both eggs and sperm ripen, but gametes from the same animal do not combine. Sperm are exchanged during copulation, and the "foreign" sperm fertilize the eggs. Tiny eggs are deposited in a species specific spiral pattern. Before settling to the benthos, the larvae swim about in the plankton using their ciliated crowns. Aquarium rearing is not possible with today's technology.

F.: C; exclusively sponges of the genus *Spongionella*. Unfortunately, *Spongionella* spp. have never been maintained for extended periods of time in aquaria.

S.: The secondary gills and rhinophores are only retracted when touched, and each rhinophore and gill is independent of the other. Because of its coloration, this species is easily recognized.

T: 15°–22°C, **L:** 5 cm

Order: NUDIBRANCHIA
Fam.: Chromodorididae

Hypselodoris tricolor, Mediterranean.

Hypselodoris tricolor (CANTRAINE, 1835)
Tricolored nudibranch

Hab.: Western Mediterranean and the Atlantic coast of France. It is found on algae covered hard substrates and within *Posidonia* lawns at depths of 10–50 m.

Sex.: Hermaphrodites.

Soc.B./Assoc.: Solitary animals. Conspecifics are only sought during the spawning season (see photo).

M.: Unknown. Dietary requirements seem to be the chief impediment to successful maintenance (see below).

Light: Moderate light.

B./Rep.: Typical for this group, *H. tricolor* lays a 0.5 by 2.5 mm flat spiral. The diminutive eggs—0.1 mm—are laid in the Mediterranean from April to September. Newly hatched larvae are planktonic.

F.: Its diet is a complete mystery. Until more information becomes available, refrain from purchasing this species.

S.: The blue dorsal coloration can be much more intense than that of the pictured individuals. As with all similar species, the rhinophores and the dorsal secondary gills can be retracted into special sheaths. *H. tricolor's* internal anatomy, particularly the morphology of the genital organs, has been studied (SCHMEKEL and PORTMANN, 1982), but very little is known about the lifestyle and ecology.

T: 15°–22°C, L: 2.5 cm

Chromodoris willani has only been found at the New Hebrides. Due to its characteristic design, it is easily differentiated from congeners.

Genus *Hypselodoris*

The genus *Hypselodoris* is often confused with *Chromodoris,* but there are distinct characteristics that separate the two genera. In comparison to *Chromodoris*, *Hypselodoris* spp. have an elongated, tall body, a broad head, and extremely anterior rhinophores. Furthermore, *Hypselodoris* spp. have simple gills arranged in a circle around the anus and mantle glands along the mantle's edge; the latter are particularly numerous posteriorly. A few pictures of *Hypselodoris* can be found on the bottom of this page as well as on the following pages.

Hypselodoris ghiselini, western coast of North America until Clipperton Islands.

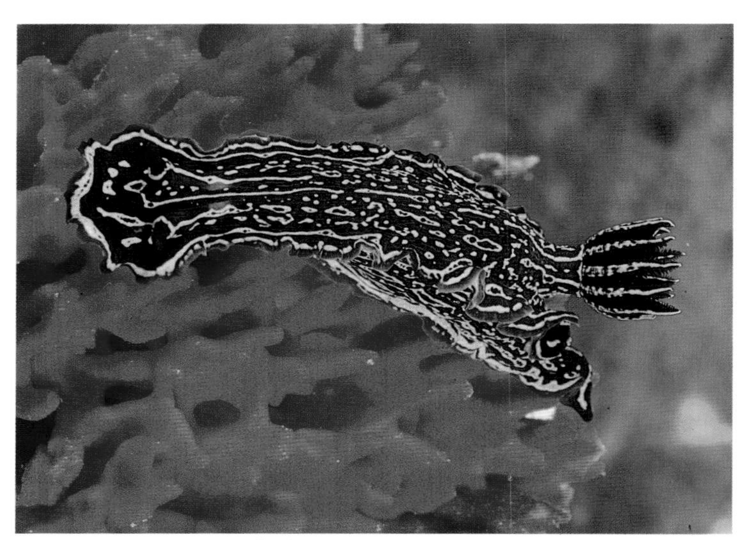

Hypselodoris edenticulata, Caribbean.

Hypselodoris edenticulata was originally described from Florida.

Hypselodoris elegans, Mediterranean.

Hypselodoris elegans (CONTRAINE, 1835)
Elegant nudibranch

Hab.: Mediterranean and neighboring eastern Atlantic, the Bay of Biscay, Canary Islands, and Gulf of Guinea, western Africa. From a depth of 5 m, *H. elegans* is found on encrusted rocks.

Sex.: Hermaphrodites.

Soc.B./Assoc.: *H. elegans* is a solitary animal which only seeks conspecifics to breed.

M.: *H. elegans* must be maintained as a transient guest because of its dietary requirements. Temperatures in excess of 22°C are only tolerated for short periods of time.

Light: Moderate light.

B./Rep.: Eggs hatch into free-swimming veliger larvae.

F.: C. Because it exclusively feeds on sponges of the genus *Ircinia*, aquarium maintenance is very difficult.

S.: While coloration varies slightly between individuals, the yellow-green design is consistent.

T: 15°–22°C, L: 12 cm, TL: from 100 cm, WM: w, WR: b, AV: 4, D: 4

Though fairly common, *Hypselodoris maridadilus* was not described until 1977.

Hypselodoris bullockii, aquarium photo.

Risbecia pulchella is quite common in the Indo-Pacific.

Risbecia pulchella is very similar to *Chromodoris annulata* (p. 651).

Genus *Risbecia* (previous page)

External characteristics of *Risbecia* fall somewhat between those of *Chromodoris* and *Hypselodoris*. Like *Chromodoris*, the body is tough and the mantle overlaps the foot. The body height, the broad head, and the fact that the posterior foot extends far beyond the mantle are all traits *Risbecia* shares with *Hypselodoris*. The gills and rhinophores are typical for this family with one additional aspect: the gills beat rhythmically above the anal papillae, even if there is no current. *Risbecia pulchella* (RÜPPELL & LEUCKART, 1828), an inhabitant of the Red Sea and the Indo-Pacific, is pictured on the previous page. There are two variants of this species—one white with large orange spots, the other purple with small spots. *Chromodoris annulata* is a very similar species from the same family (p. 651), but upon closer inspection, *Chromodoris annulata* has the typical *Chromodoris* shape and dark rings around the gills and rhinophores.

Family Dendrodorididae

Except for Phyllididae, Dendrodorididae is the only family of nudibranchs that lacks mandibles and a radula. Dendrodorididae have well-developed oral glands, a suction muscle, and a small mouth which they use to ingest encrusting sponges. Secondary gills encircle the dorsal anus. Most species inhabit tropical seas, though some are found in the Mediterranean and the Atlantic. A representative of each of the two genera of Dendrodorididae, *Dendrodoris* and *Doriopsilla*, are pictured on the facing page.
Dendrodoris is soft and slimy and usually smooth; a few of its species has soft cerata (dorsal projections). Contrary to *Dendrodoris*, *Doriopsilla* have small spiculae embedded in the mantle and a harder body.

Dendrodoris limbata, Mediterranean. The dark animal is typically colored.

Doriopsilla miniata occurs from Australia to South Africa and the eastern Mediterranean.

FAMILY POLYCERIDAE (Page 668)

Polyceridae most commonly inhabit tropical waters, but some make their home in temperate seas such as the Mediterranean. In most cases the body is elongated, the gills are on the highest point at the center of the back, and the mantle is slender and short—so short that the foot protrudes posteriorly. It is not uncommon for polycerids to have lateral or anteriorly projecting or supine appendages on the mantle's edge, either along or around the gills. The rhinophores (chemical and current receptors) are almost always foliaceous, and they can be retracted into sheaths which are often elevated. Though debatable, we believe that the following pages contain the most colorful species of the family. Many are feeding specialists, and particular bryozoans are their prey. The bottom picture on the facing page, however, proves that not all polycerids have a diet limited to bryozoans. Apparently some are not above preying on close cousins.

Gymnodoris rubropapulosa, 5 cm, western coast of Australia.

Gymnodoris ceylonica, 12 cm, occurs in the Red Sea and the Indo-Pacific.

Gymnodoris ceylonica in the process of capturing a *Glossodoris atromarginata* (aquarium photo).

Order: NUDIBRANCHIA
Fam.: Polyceridae

Genera *Nembrotha, Roboastra* (p. 672), and *Tambja* (p. 674)

These three genera can only be differentiated by examining the radula and sex organs, as external characteristics are insufficient. All are relatively large (up to 15 cm), soft, and brilliantly colored. Over substrates of crushed coral in the shallow tropical waters they inhabit, these predators concentrate on sessile animals such as bryozoans, tunicates, and cnidaria. As members of the Polyceridae, they have an extendable narrow throat that acts as a suction pump instead of a radula. Dietary requirements are not as exotic or particular when compared to other nudibranchs, facilitating aquarium maintenance. Small *Aiptasia* sp., pests in some aquaria, can be offered.

Nembrotha cristata is distributed through the western Pacific and the Philippines.

Nembrotha kubaryana, Indonesia.

Nembrotha sp. common on Philipines.
This exceptionally attractive *Nembrotha* has not been scientifically described. The photo was taken in the Red Sea. Feeds basicly tunicates.

Nembrotha lineolata from the tropical western Pacific and Indonesia.

A number of opisthobranchia swim in the open water with undulating movements of their lateral lobes (parapodia). The most familiar is probably the Spanish dancer, *Hexabranchus sanguineus* (p. 681). The pictured animal is an *Nembrotha megalocerca* from the Red Sea.

Genus *Polycera*

Most species of this genus only obtain a length of a few centimeters. Normally, they have anteriorly directed appendages on the head and similar appendages next to the gills. In contrast, *Gymnodoris* has small appendages and mantle extensions. The rhinophores of *Polycera* are lamellar and, because of the lack of sheaths, nonretractable. Prey detection and seizure are greatly enhanced by the large dimensions of the rhinophores. *Polycera* chiefly consumes bryozoans of the genus *Bugula*. Therefore, their dietary requirements impede aquarium maintenance. Some species are said to feed on hydrozoan polyps in the aquarium.

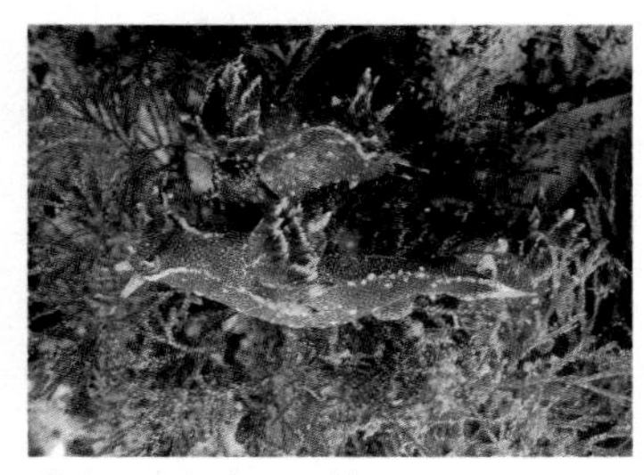

Polycera hedgepethi

Polycera capensis, 7 cm, Southafrica

Polycera capensis, Australia

Polycera capensis QUOY & GAIMARD, 1824 has been described under many synonyms (e.g., *P. nigrocrocea* BARNARD, 1927 and *P. conspicua* ALLAN, 1932). As shown in the above two photos, coloration can deviate. The body ranges from white to gray, and the black markings along the dorsum vary in intensity. The typical colors of *Polycera*—black and yellow-orange—are present. Distributed from South Africa to Australia, *P. capensis* achieves a length of 5 cm and feeds on bryozoans of the genus *Bugula*. It is found from the intertidal zone to depths of at least 30 m.

Roboastra sp.

Roboastra sp.

Tambja abdere, California.

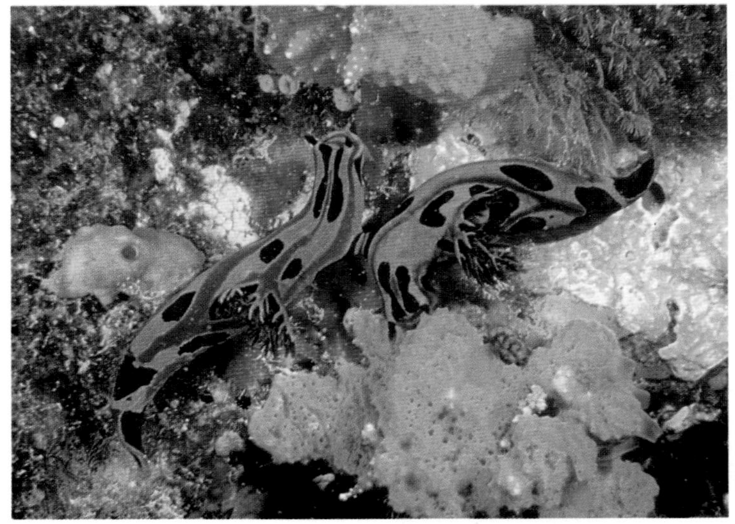

The genital organs are clearly visible on these two mating *Tambja* sp.

Tambja affinis, Red Sea.

Tambja kushimotoensis among pinnate stalks of hydrozoans.

Genus *Halgerda*

All species of the family Asteronotidae inhabit tropical waters at depths of 3–40 m. Because of their conspicuous lifestyle, they are frequently encountered by divers. *Halgerda* have a smooth, hard, sometimes slimy, hemispherical dorsum. The rest of the animal's morphological features can be likened to a mountain range, with deep canyons and coalescing crests forming a pointed range. The coloration of the rhinophores (olfactory organs) and gills is uncommon; regardless of the animal's coloration, they are white or yellow with black markings. Some species feed on sponges. *Halgerda* shares many characteristics with *Sclerodoris*, which has a similar distribution. The other genera of this family are: *Aphelodoris*, *Asteronotus*, and *Artachaea*. These animals' requirements in regard to captive maintenance are largely unknown.

Halgerda sp., Philippines.

This photo of *Halgerda* sp. was taken in the Maldives at a depth of 48 m.

Halgerda terramtuensis, a Hawaiian species, was described in 1982.

Halgerda willeyi

Halgerda willeyi, Red Sea.

FAMILY KENTRODORIDIDAE

This is a very diversified family of nudibranchs. The family name was derived from one of its genera, *Kentrodoris*, a completely atypical member of its ranks. With the exception of *Kentrodoris*, all have an oval body and sufficient extension of the mantle to cover the foot. The predominant color scheme is a white or yellow body with black to brown spots. Gills and rhinophores (receptors for chemical and current stimuli) are black. Because they have fine, somewhat protrusive spicules in their mantle, their surface is rough to the touch. All kentrodoridids feed on sponges; hence, long term maintenance is extremely difficult.

Genus *Jorunna*

With their oval body and rough mantle, *Jorunna* are much more representative of the Kentrodorididae than the genus the family was named after. They inhabit both tropical and subtropical seas. The circles of contrasting dark papillae and rings of embedded spicules are called karyophyllidia and are notorious in this genus. The animal in the top photo on the facing page has conspicuous black karyophyllidia; they are part of the animal's defense. *Jorunna funebris* (KELAART, 1858) of the Indo-West Pacific is easily confused with *J. zania* MARCUS, 1976 of South Africa.

Genus *Kentrodoris*

Kentrodoris rubescens BERGH, 1876 is the only species in this monotypic genus. Capable of achieving the respectable length of 12 cm, the very soft, slimy body is as tall as it is broad with extremely elevated secondary gills and jutting tentacle sheaths. The anterior edge of the mantle is shaped like a helmet. Throughout the Indo-Pacific, *Kentrodoris rubescens* is commonly found in shallow water of lagoons.

Jorunna funebris lives in the Indo-West Pacific.

Kentrodoris rubescens is the only species of the genus.

Order: Nudibranchia
Fam.: Hexabranchidae

The Spanish Dancer, *Hexabranchus sanguineus*

The family Hexabranchidae probably contains just one genus and one species, *Hexabranchus sanguineus* (RÜPPEL & LEUCKART, 1828). The prominent, large mantle fringe is laterally rolled when the animal crawls, but unfurled for swimming. When disturbed, *H. sanguineus* extends its mantle. The flashing bright colors and designs startle predators. The Spanish dancer's ability to swim is one of its most notable skills. Using a combination of up-and-down body movements and undulations of the mantle fringe, the organisms lifts from the substrate and swims freely. Gazing upon the Spanish dancer in "flight," one sees the resemblance between the motions of the mantle fringe and the flounce of the flamenco dancer's skirt. The rhinophores are lamellar, and the oral tentacles are large, flat lobes. Each of the partially retractable secondary gills has an individual sheath as opposed to one common sheath. The Spanish dancer is widely distributed. It is found in the Red Sea, the tropical Indo-Pacific, and even the Atlantic. While individuals from the Red Sea consistently have an intense red body, animals from the Indo-Pacific are known for their remarkable design, though they are principally red as well. The pictures on the following pages demonstrate the various colors. These color variants were previously described as an autonomous species (*H. marginatus*).
The Spanish dancer almost always serves as a host to a small, well-camouflaged shrimp, *Periclimenes imperator* (see Marine Atlas Vol. 1, p. 528). The shrimp feeds on it's host dermal mucus and substances adhered to it. However, it has been known to leave its host and feed on food it finds along the bottom or the nudibranch's excrements. If both partners are kept in an aquarium over a long period of time, the shrimp usually becomes independent of its host. Extensive research on this subject can be found in SCHUHMACHER (Mar. Biol. 22/1973, pp. 355 ff.).
The spawn of *H. sanguineus* is characteristic and frequently indicative of the presence of the Spanish dancer. The eggs are deposited in a brittle, flat spiral ribbon. Varying in color from pink to a bright dark red (bottom photo on facing page), the ribbon is anchored to the substrate by one edge.
Although it is not as choosey as some nudibranchs, little foodstuff is accepted in captivity. In its natural habitat, various invertebrates such as sponges, worms, slugs, and echinoderms are consumed.

Hexabranchus sanguineus spawning.

Many small eggs can be seen in the spawn of *Hexabranchus sanguineus*.

Hexabrachnus sanguineus is often host to the small emperor shrimp.

This photo of *Hexabranchus sanguineus* was taken in an aquarium.

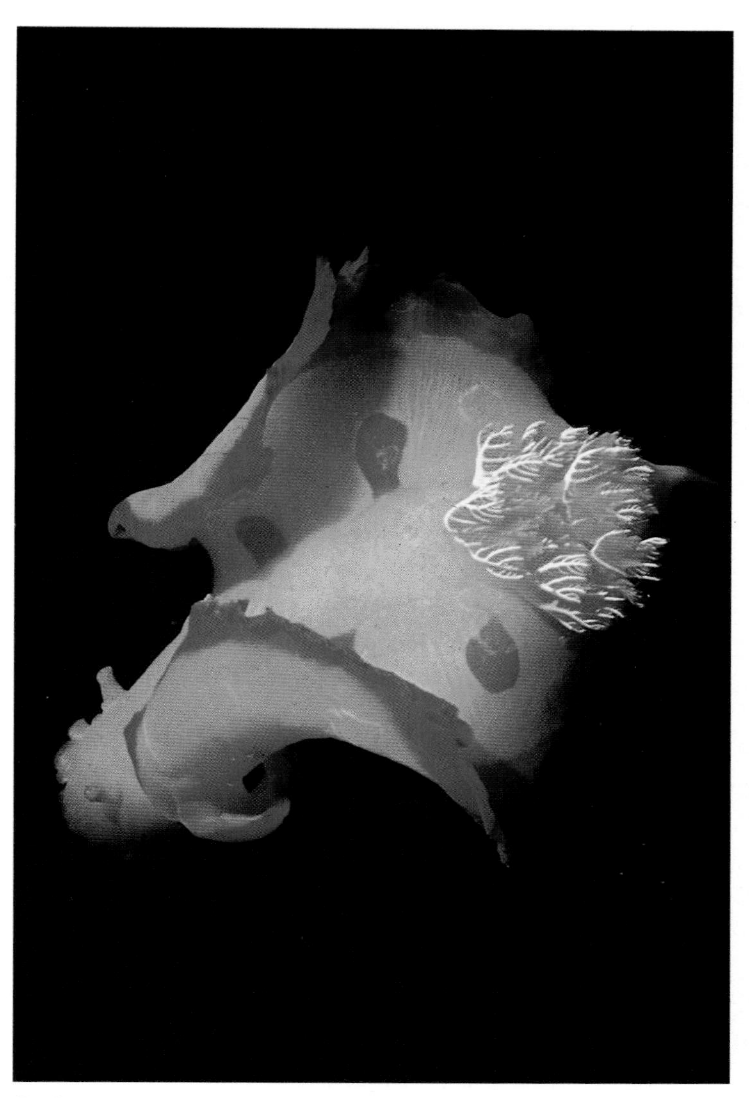

The Spanish dancer, *Hexabranchus sanguineus*, is a graceful swimmer.

Order: NUDIBRANCHIA
Fam.: Discodorididae

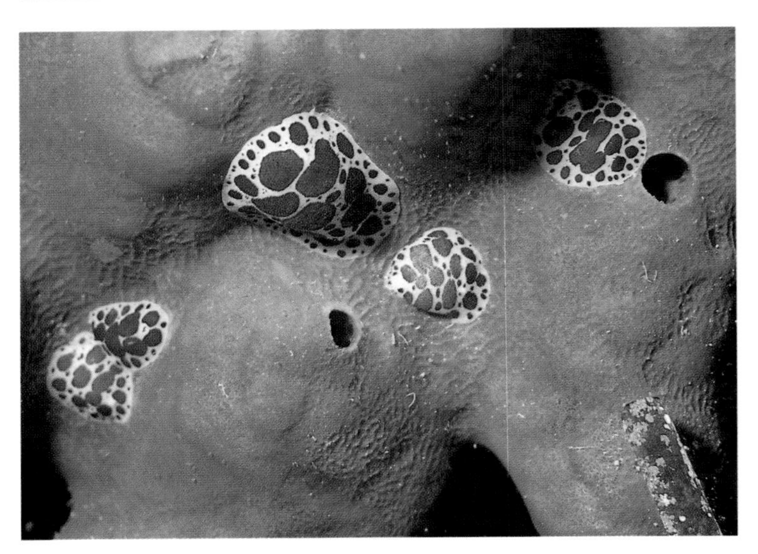

Peltodoris atromaculata, Mediterranean.

Peltodoris atromaculata BERGH, 1880
Leopard nudibranch

Hab.: Common throughout the Mediterranean on hard substrates at depths of 2 m and below. It is usually found on the sponge *Petrosia ficiformis*, which colonizes ledges and shaded vertical walls.

Sex.: None, since the animal is a hermaphrodite.

Soc.B./Assoc.: Aggregations of juveniles often sit on *Petrosia ficiformis*.

M.: Because of its dietary requirements, aquarium maintenance is virtually impossible. Long-term care requires a constant supply of fresh *Petrosia ficiformis*.

Light: Dim light.

B./Rep.: Spawns are laid in a spiral with several windings. The chief spawning season extends from May to August. The light-colored eggs have a diameter of 0.18 mm.

F.: C; exclusively *Petrosia ficiformis*, a sponge.

S.: This nudibranch is relatively hard and rough because of its cone-shaped leathery tubercles along the dorsum. Upon the slightest disturbance, the rhinophores and secondary gills are retracted into special sheaths. While young animals confine themselves to the sponge *Petrosia ficiformis*, adults often wander, probably in the search of a mate. *P. atromaculata*'s coloration makes identification irrefutable.

T: 15°–22°C, L: 5–10 cm

Platydoris argo, western Mediterranean.

Platydoris argo (LINNAEUS, 1767)
Red-brown nudibranch

Hab.: Western Mediterranean and eastern Atlantic along the coast of France. *P. argo* is generally found on hard bottoms, but it has been known to occasionally transverse expansive sand areas.

Sex.: None, since this animal is a simultaneous hermaphrodite. In other words, male and female gametes ripen at the same time (verses consecutive hermaphrodites, in which a sexual transformation takes place).

Soc.B./Assoc.: Nothing is known concerning the biology or ecology of this species.

M.: Unknown. Surely the main problem is its dietary requirements, although there is a paucity of information on the subject.

Light: Moderate to dim light.

B./Rep.: Spawning occurs during the summer. Divers often encounter pairs following each other.

F.: Detailed research is not available, but it is said to feed on algae. If this is confirmed, aquarium maintenance should be possible.

S.: Unknown.

T: 15°–22°C, L: 10 cm

Family Phyllidiidae

Members of Phyllidiidae are easily identified by their warty mantle and the placement of the gills, which is between the mantle and the foot versus the dorsum. While the three pictured genera are similar in appearance, note the position of the anus and the cephalic tentacles. Phyllidiids are black and white with orange accents, and since they rarely hide, they are commonly encountered along coral reefs. Sponges seem to be their sole source of nourishment. Some species release toxins that are lethal to small fishes and crustacea in the aquarium.

Genus *Phyllidia*

The genus *Phyllidia* is comprised of many species; most live in the Indo-Pacific, but at least two live in the Mediterranean. In this genus the anal opening is dorsal. Many species are familiar to divers, but have not been scientifically described. The photo on the bottom right demonstrates how *Phyllidia* spp. feed; the pharynx and the attached "oral glands" are extended and the sponge is consumed *in situ*.

Phyllidia arabica EHRENBERG, 1831 has a convoluted taxonomic history. It seems that specimens identified as *P. varicosa* were, in reality, *P. arabica*. The principal difference between the two species is the black line on the sole of *P. arabica*'s foot. Though frequently mentioned, *P. varicosa* has probably not been found since its original description in 1801 by LAMARCK.

Phyllidia arabica

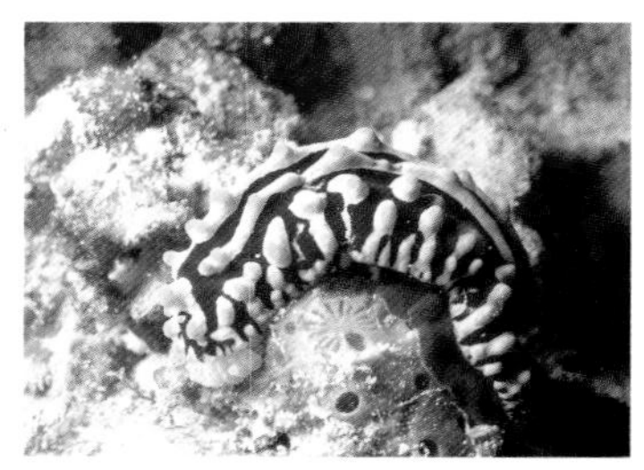

P. arabica feeding on a sponge.

Phyllidia arabica on a sponge from the Pacific.

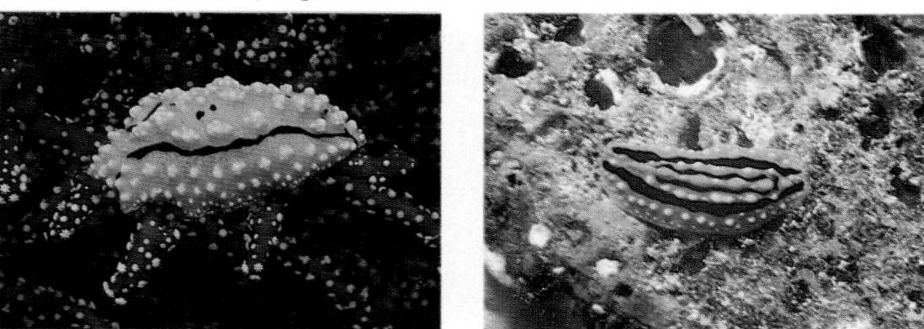

These four *Phyllidia* species from the Maldives have not been scientifically described. The genus, however, can be positively identified.

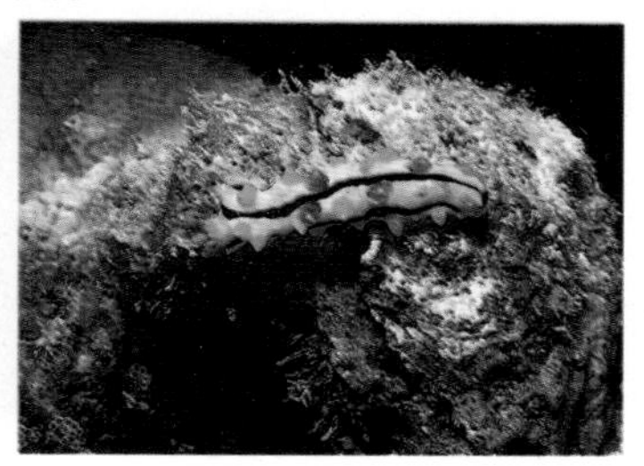

Phyllidia coelestis from the tropical Indo-Pacific.

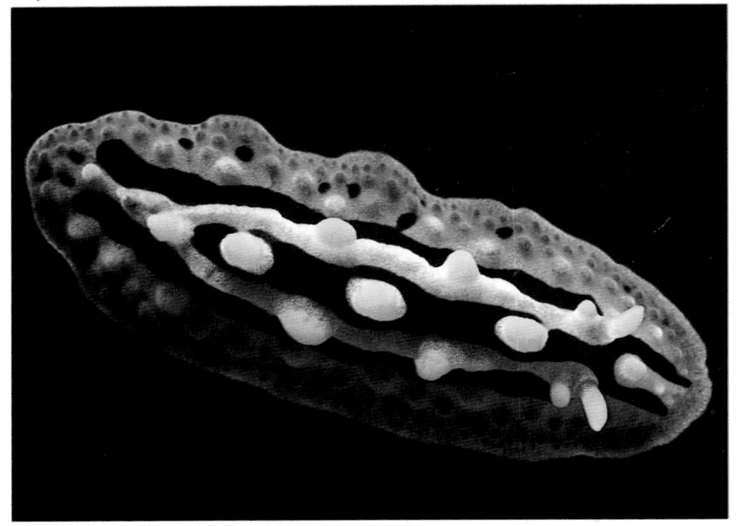

Phyllidia coelestis; the tips of the tubercles of this individual are orange.

Philidiopsis dautzenbergi from the Red Sea. Its body surface is smoother than that of its congeners. The rhinophores are not distinctly colored.

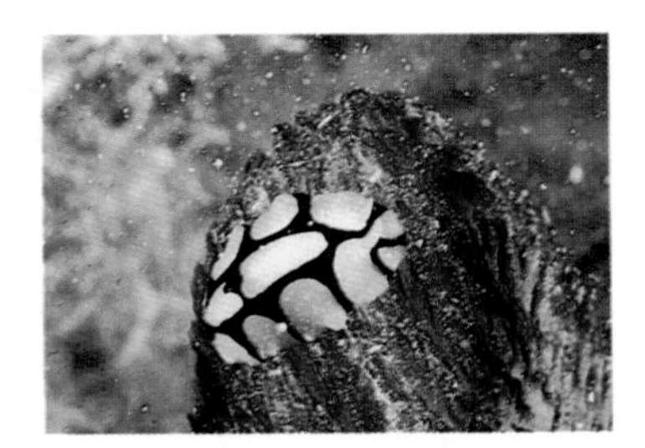

Phyllidia elegans, Philippines.

Phyllidia meandrina from the Indian Ocean. All species of this genus are unsuitable for aquarium maintenance. Some may even release toxins into the water.

Phyllidia ocellata, western Pacific.

The distribution of *Phyllidia ocellata* is limited to the Red Sea.

Phyllidia pustulosa, Red Sea.

At a length of 5 cm, *Phyllidia pustulosa* is one of the largest species of the genus. It is widely distributed throughout the Indo-Pacific and the Red Sea. Its coloration varies—some are green and black. The eastern form may be an autonomous species.

Phyllidia ocellata CUVIER, 1804 (top photo on facing page) is probably not one species, but rather a group of very similar species that all live in the Indo-Pacific region. There are differences in the design of the mantle. Detailed studies are needed to determine if the populations are of a single, highly variable species or several species. The subspecies *P. ocellata undulata* YONOW, 1986 on the bottom of the facing page is endemic to the Red Sea. Its ocelli coalesce into undulating waves that cover the entire back.

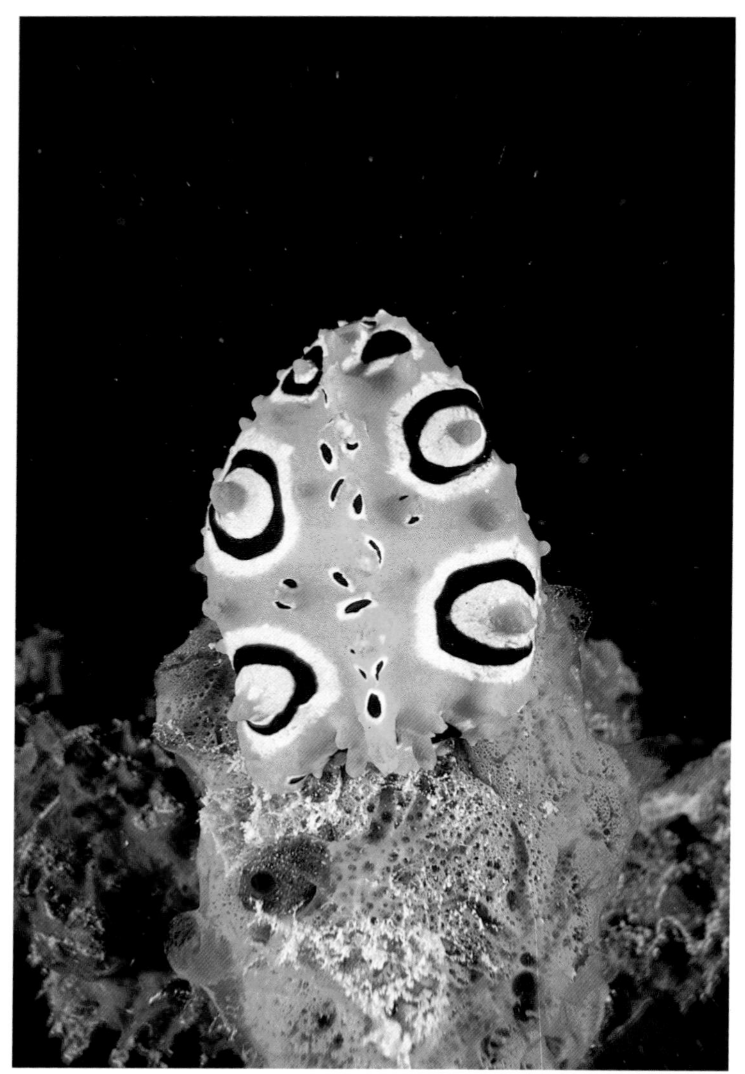

Phyllidia ocellata was not described by YONOW & DEBELIUS until 1991!

Phyllidiopsis sp. photographed in the Maldives.

Genera *Phyllidiopsis* and *Reyfria*
(this and the following page)

In contrast to the genera *Phyllidia* and *Phyllidiopsis,* the anus of *Reyfria* is positioned posteriorly between the foot and the mantle. Almost all *Reyfria* remain smaller than *Phyllidia* and *Phyllidiopsis*. Many have a repugnant odor and a coarse texture. Exterior coloration is merely an aid towards identification, not the premise. Since early identifications were based exclusively on preserved museum specimens, coloration was often unknown. Taxonomic criteria include rhinophore morphology, the number of gill lamellae, and the development of dorsal tubercles. Today's more exact methods of classification involve determining the exact location of the genital organs. Of course, this necessitates dissection. Only in recent years have color photos become an additional important clue towards classification.

Left: *Reyfria hennesi*. Described in 1990, this species inhabits the Indian Ocean.

Reyfria ruepellii lives in the Red Sea.

FAMILY TETHYIDAE

Tethyids have an extremely enlarged head that is endowed with appendages. The dorsal appendages hold the liverlike digestive diverticula. Moving like the Spanish dancer (p. 680), tethyids can swim short distances. *Tethys*, which is found at great depths over sandy substrates, reportedly has the capability to swim all the way to the water surface. Unlike other molluscs, these predators do not have a radula. Their diet consists of various benthic organisms such as mantis shrimp (*Squilla*), other molluscs, echinoderms, small fishes, and sometimes cnidaria. The Mediterranean *Tethys fimbria* has been maintained and spawned in aquaria. The small egg capsules are laid in a spawning line which is about 20 cm long and 1 cm high. Each capsule holds 15–20, 0.08 mm eggs. However, rearing the larvae to adults is impossible with today's technology because of the long planktonic phase. There are two genera—*Tethys* of the Mediterranean and Atlantic and *Melibe* of the Indo-Pacific.

This Mediterranean Tethyidae has not been scientifically described.

The Capture of Prey by *Dendrodoris iris*

Howard HALL (IKAN) took this impressive series of photographs along the Pacific coast of California. *Dendronotus* has an interesting method of acquiring its favorite prey—tube anemones. After *Dendronotus* has the scent of the anemone (photo 1), it crawls part way up the tube (picture 2), then rears up in a cobralike fashion (photo 3). Before attacking, the nudibranch repeats its touching sequence (photo 4). *Dendronotus*'s mandibles clasp the anemone—it lacks a radula—and the nudibranch allows itself to be pulled into the tube (photo 5). There the anemone is devoured.

Photo 1

Photo 2

Photo 3. Like a cobra, *Dendronotus iris* rises up in front of the tube anemone.

Photo 4. *Dendrodoris iris* cautiously approaches its prey.

Photo 5

Like many other nudibranchs, *Dendrodoris iris* deposits its spawn next to its prey. The spiral spawn contains thousands of tiny eggs.

FAMILY FLABELLINIDAE

Flabellinidae and Glaucidae (pp. 702–705) and about 20 additional families all belong to the suborder Aeolidacea. All have an elongated body that tapers posteriorly into a point. Clusters or rows of appendages called cerata are arranged along the dorsum. Digestive diverticula extend into the cerata.

Aeolidaceans primarily feed upon cnidarians, but some species also consume bryozoans, various invertebrates, and spawns of fishes and other nudibranchs. When aeolidaceans feed on cnidarians they benefit through more than simple nutrition—they gain a powerful defense mechanism.

For defense and food acquisition, cnidarian tentacles have a multitude of cnidocytes, cells that hold nematocysts.

It has not yet been determined how, but a number of aeolidaceans are able to collect unfired nematocysts from cnidarians—hydroid polyps, sea anemones, and hard and soft corals—and transport them through the body to the tips of the cerata, where special pouches called cnidosacs hold the unfired nematocysts. The nematocysts are then used towards the defense of the nudibranchs.

This ability is unique among the animal kingdom. Many aeolidaceans have cerata that closely resemble the tentacles of a sea anemone. They are in fact mimics, and the presence of nematocysts lends credibility to their claim.

Their intense coloration serves as a further deterrent to predators. The cerata, the long delicate tentacles, and the rhinophores are normally brightly striped. The rhinophores may be rough or smooth and arranged into lamellae or rings. Frequently the eyes can be seen at the base of the rhinophores.

Aeolidaceans, although often very conspicuously colored, are still very difficult to identify based on external features. It is much more appropriate to differentiate species on the basis of the radula.

Dorsal appendages of the widely distributed genus *Flabellina* are arranged in parallel or U-shaped rows. The rhinophores are ringed, and the tentacles on the end of the anterior foot are long and curved posteriorly. The oral tentacles are long and mobile.

Flabellinidae, like this *Flabellina iodinea,* usually feed on cnidaria.

Flabellina affinis, Mediterranean.

Flabellina affinis (GMELIN, 1791)
Purple nudibranch

Hab.: Though found along the coast of Ghana, it is principally distributed in the Mediterranean at depths of 1 to 70 m. *F. affinis* is almost exclusively found on its prey, the hydrozoan *Eudendrium*.

Sex.: Like all nudibranchs, *F. affinis* is a hermaphrodite.

Soc.B./Assoc.: This is an appropriate tankmate for invertebrates and small fishes.

M.: Successful maintenance is contingent upon the presence of ample *Eudendrium* polyps. Temperatures in excess of 20°C are only tolerated for short periods of time.

Light: Dim light.

B./Rep.: Meandering spawns are laid on *Eudendrium* from March to October when temperatures reach 20°C. Veliger larvae hatch from the small eggs (0.08 mm) after five days.

F.: C; exclusively polyps of the hydrozoan *Eudendrium*.

S.: *F. affinis* is a feeding specialist. However, if one is successful towards colonizing the aquarium with several stalks of *Eudendrium*, it is possible to maintain this beautifully colored animal and observe its feeding habits. Although eggs may be laid, rearing is impossible. Metamorphosis of the free-swimming veliger larvae has not been observed.

T: 15°–20°C, L: 5 cm, TL: from 100 cm, WM: w, WR: b, AV: 4, D: 4

An undescribed *Flabellina* species with black rhinophores.

Flabellina iodinea, North America.

F. iodinea with its meandering spawn.

Flabellina rubrolineata, Indo -West Pacific.

Flabellina sp., Pacific.

Genus *Glaucus*

This genus is well adapted to its planktonic lifestyle: using its three long, flat lateral appendages, *Glaucus atlanticus* suspends itself in the water column. Additional buoyancy is lent by air or gas. Because the 4 cm long body is dark blue dorsally and silvery white ventrally, it is camouflaged from predators from both above and below. *Glaucus* primarily feeds on hydrozoans of the genera *Velella* (by-the-wind sailor), *Physalia* (Portuguese man-of-war), and *Porpita* encountered drifting on the water surface. Consumed nematocysts are stored in the dorsal appendages, where they produce strong stings when touched.

Glaucus atlanticus

Glaucus atlanticus occurs circumtropically. There it swims along the water surface.

Hermissenda crassicornis, eastern Pacific.

Genus *Phidiana*

Although greatly similar to members of the family Flabellinidae (see previous pages), this genus falls within the Glaucidae. Species are distinguished on the basis of morphological differences of the reproductive organs and the radula. Definite species identification, therefore, necessitates dissection. Dorsal cerata are aligned into rows, and the rhinophores—receptors for chemical and current stimuli—are smooth, ringed, or tubercular.

Hermissenda crassicornis

Phidiana sp., Pacific.

Light and dark variants of *Hermissenda crassicornis*.

Pteraeolidia ianthina, Red Sea.

Pteraeolidia ianthina, Indo-Pacific.

Genus *Pteraeolidia*

Pteraeolidia is a monotypic genus, and *Pteraeolidia ianthina* (ANGAS, 1884) is its sole member. Capable of achieving a length of 12 cm, *P. ianthina* has an elongated body covered with dark blue club-shaped cerata. The rhinophores are lamelliform. Like hermatypic (reef-forming) corals and some anemones (see Maine Atlas Vol. 1, p. 221), this nudibranch has zooxanthellae (symbiotic algae), an unusual attribute within this family. The cerata are arranged such that they do not overlap, thereby maximizing the amount of sunlight striking algae-laden tissue. The animals feed on hydrozoans, gleaning nematocysts from their prey and then transporting the weapons to the tips of the cerata. There the nematocysts serve towards the defense of the nudibranch.

Pteraeolidia ianthina

Class Scaphopoda, Tusk or Tooth Shells

Scaphopods—commonly called tusk or tooth shells—have a graduated, tubular shell which is secreted by the mantle after it encompasses the larva. Both ends of the conical shell are open, and the shell has a curved profile.
Represented worldwide by about 350 species, tusk shells live at a broad range of depths, typically in soft substrates. There they burrow into the sediment until the narrow posterior end of the shell remains just above the substrate. Because there are no gills, the mantle cavity serves as the organ of respiration. Located at the emerged posterior end of the shell, the mantle cavity absorbs oxygen over the entirety of its surface.
The bulbous foot and captaculae (capture threads) are found at the broad anterior end of the shell. Slow erratic locomotion is achieved as the foot is extended forward, then retracted.

Nutrition

Small organisms, most commonly shelled unicellular organisms called foraminiferans, are hunted from the mesopsammon (areas between the grains of sand) and captured on the sticky, clublike captaculae. Food is triturated and transported by the radula.

Reproduction

While tusk shells are dioecious, they lack differentiating external features. The numerous small eggs and sperm are released into the open water where fertilization occurs. Zygotes pass through a planktonic trochophore-like larval stage before settling onto and digging into a sandy substrate.

Species identification is based on external features of the shell. Like other molluscs, scaphopods list starfish, benthic fishes (primarily rays), and predatory slugs such as *Natica* spp. as their enemies.

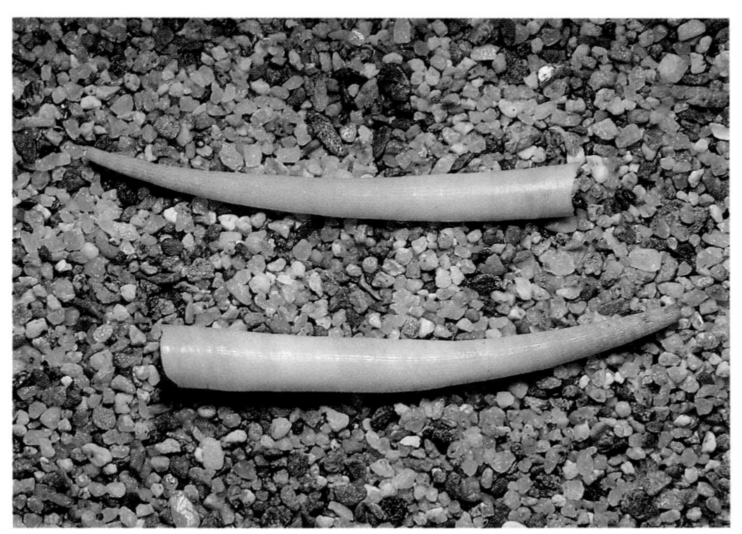

Antalis tarentinum, Mediterranean.

Antalis tarentinum (LAMARCK, 1818)
Common tusk shell

Hab.: Mediterranean. It burrows in sand substrates (see class introduction).

Sex.: Though dioecious, there are no external differentiating characteristics. Copulatory organs are lacking.

Soc.B./Assoc.: Solitary animals.

M.: In sight of its dietary preferences—foraminiferans and other benthic organisms—*A. tarentinum* is not considered a very appropriate aquarium animal. Foraminiferans and most other small sand dwellers typically have an abbreviated lifespan in aquaria because of the limited or total lack of current through the substrate. A reverse flow undergravel filter system may provide the solution, thereby allowing foraminiferans and the desired tusk shells to be successfully maintained. These problems aside, the common tusk shell requires medium to coarse sand and a strong current.

Light: Moderate light.

B./Rep.: Spawning season extends from April to May. The 0.2 mm eggs are released then fertilized in the open water. The resulting free-swimming larvae pass through a planktonic stage before settling to the substrate.

F.: Microorganisms, particularly sand-dwelling foraminiferans (unicellular organisms).

S.: None known.

T: 15°–22°C, L: 6 cm, TL: from 100 cm, WM: s, WR: b, AV: 3, D: 3–4

Index

Italics: Scientific names of genera and species.

Normal: Common names

SMALL CAPS: Taxons above genus

A

B

Index

Index

Index

Index

Index

Index

Bibliography

Abbott, D.P. (1987): Observing marine invertebrates. Stanford University Press, Stanford, California, USA.

Adijodi, K.G. & Adiyodi, R.G. (1988): Reproductive biology of invertebrates. John Wiley & Sons, Chichester, N.Y., USA.

Amsler, K. (1984): Zauberwelt der Meere. Stürtz Verlag, Würzburg, Germany.

Baensch, H.A. & Debelius, H. (1992): Meerwasser Atlas. Mergus Verlag, Melle, Germany.

Baensch, H.A. (1992): Neue Meerwasserpraxis. Tetra Verlag, Melle, Germany.

Barns, R.D. (1987): Invertebrate Zoology. Saunders College Publishing, Austin, USA.

Bauchet, P., Danrigal, F. & Huyghens, S. (1978): Living seashells. Molluscs of the English Channel and Atlantic coasts. Blandford Press, Dorset, U.K.

Baumeister, W. (1990): Meeresaquaristik. Verlag Eugen Ulmer, Stuttgart, Germany.

Bellmann, H., Hausmann, K., Janke, K. & Schneider, H. (1991): Einzeller und Wirbellose. Die farbigen Naturführer. Bertelsmann Club, Gütersloh, Germany.

Bemert, G. & Ormond, R. (1981): Red sea coral reefs. Kegan Paul International, London-Boston.

Bennet, I. (1971): The Great Barrier Reef. Landsdowne Press, Sydney, Australia.

Bennet, I. (1966): The fringe of the sea. Rigby Ltd., Adelaide, Australia.

Brusca, G.J. & Brusca, R.C. (1978): A naturalist 's seashore guide: Common marine life of the northern California coast and adjacent shores. Mad River Press, Eureca, California, USA.

Brusca, R.C. & J.G. Brusca (1970): Invertebrates. Sinauer, Sunderland.

Carefood, T. (1977): Pacific seashores. A Guide to intertidal biology. University of Washington Press, Seattle, USA.

Clark, M. & Rowe, F.W. (1971): Monograph of shallow water Indopacific echinoderms. British Museum of Natural History, London, UK.

Clark, A.M. (1977): Starfishes and related echinoderms. TFH Publications Inc., Neptune City, New Jersey, USA.

Coleman, N. (1981): What shell is that? Landsdowne Press, Sydney, Australia.

Colin, P.L. (1978): Caribbean reef invertebrates and plants. TFH Publications Inc., Neptun City, New Jersey, USA.

Czihak, G. & Dierl, W. (1961): *Pinna nobilis* L. Großes Zoologisches Praktikum 16a. Fischer Verlag, Stuttgart, Germany.

Dakin, W.J., Bennet, I. & Pope, E. (1963): Australian seashores. Angus & Robertson, Sydney, Australia.

Dance, P.(1977): Das große Buch der Meeresmuscheln. Ulmer, Stuttgart, Germany.

Devaney, D.M. (1977): Reef and shore fauna of Hawaii. Bernice P. Bishop Press, Honululu, Hawaii, USA.

Ditlev, H. (1980): A field guide to the reef-building corals of the Indo-Pacific. Scandinavian Science Press, Klampenborg, Denmark.

Doubilet, D. (1989): Light in the sea. Swan Hill Press, Oxford, N.Y., Frankfurt.

Edwards, A.J. & Head, St.M. (1987): Red Sea. Pergamon Press. Oxford, N.Y., Frankfurt.

Endean, R. (1982): Australia's Great Barrier Reef. University of Queensland Press, St. Lucia, London, UK.

Fechter, R. & Falkner, G. (1990): Weichtiere. Europäische Meeres- und Binnenmollusken. Steinbachs Naturführer. Mosaik Verlag, München.

Bibliography

Fielding, A. & Robinson, E. (1987): An underwater guide to Hawaii. University of Hawaii Press, Honululu, USA.

Fioroni, P. & Meister, G. (1974) *Loligo vulgaris* Lam. Gemeiner Kalmar. Großes Zoologisches Praktikum 16c/2. Fischer Verlag, Stuttgart, Germany.

Fischer, A. (1975): Juwelen des siebenten Kontinents. Molden Verlag, München, Germany.

Fricke, H.W. (1972): Korallenmeer. Belser Verlag, Stuttgart, Germany.

Friedrich, H. ((1965): Meeresbiologie. Borntraeger, Berlin, Germany.

Garnet, J.R. (1968): Venomous Australian animals dangerous to man. CLS Publications, Parkville.

George, D. & George, J. (1979): Marine life. An Illustrated Encyclopedia of Invertebrates in the Sea. Harrap, London, UK.

Gessner, F. (1957): Meer und Strand. VEB Deutscher Verlag der Wissenschaften, Berlin, Germany.

Gillett, K. (1976): The Australian Great Barrier Reef in Colour. A.W. Reed, Sydney, Australia.

Glynn, P.W. & Wellington, G.M. (1983): Corals and coral reefs of the Galapagos island. University of California Press, Berkeley, USA.

Göthel, H. (1992): Farbatlas Mittelmeerfauna. Niedere Tiere und Fische. Ulmer Verlag, Stuttgart.

Götting, K.- J. (1974): Malakozoologie. Grundriß der Weichtierkunde. Fischer Verlag, Stuttgart, Germany.

Gosliner, T. (1987): Nudibranchs of southern Africa. Brill, Leiden, Neatherlands.

Gosner, K.L. (1979): A field guide to the Atlantic seashore. Houghton Mifflin C., Boston, USA.

Gotshall, D.W. & Laurent, L. L. (1979): Pacific Coast Subtidal Marine Invertebrates. Sea Challengers, Monterey, California, USA.

Grzimek, B. (1971): Enzyklopädie des Tierreiches. Band 1 und 3. Weichtiere und Stachelhäuter, Kindler Verlag, Zürich, Switzerland.

Guille, A.P., Laboutte & Menou, J.L. (1986): Guide des étoiles de mer, oursins et autres échinoderms du lagon de Nouvelle-Calédonie. Édition de L'Orston, Collection Faune Tropical 25, France.

Habe, T. (1971): Shells of Japan. Hoikusha Publishing Co., Osaka, Japan.

Harmelin, J.K., Vacelet, J. & Petron, Ch. (1987): Méditerranée vivante. Glénat Grenoble, Switzerland.

Harris, V.A. (1990): Sessile animals of the seashore. Chapman and Hall, N.Y., USA.

Haywood, M. & Wells, S. (1989) : The manual of marine invertebrates. Salamander Books Limited, London, New York.

Heezen, B. C. & Hollister, C.D. (1971): The face of the deep. Oxford University Press, N.Y., USA.

Hobson, E. & Chave, E.H. (1972): Hawaiian reef animals. University of Hawaii Press, Honululu, Hawaii, USA.

Human, P. (1992): Reef creature identification. Florida, Caribbean, Bahamas. New World Publications, Jacksonville, Florida, USA.

Jaeckel, S.H. (1957): Kopffüßer Tintenfische. Neue Brehm Bücherei. Zimsen Verlag, Wittenberg, Germany.

Jamgoux, M. & Lawrence, J.M. (1989): Echinoderm studies. Balkema Publishers, Rotterdam, Neatherlands.

Bibliography

Janke, K. & Kremer, B.P. (1988): Düne, Strand und Wattenmeer. Kosmos Naturführer. Frankh Verlag, Stuttgart, Germany.

Johnson, M.E. & Snook, H.J. (1927): Seashore animals of the Pacific. Dover Publications, N.Y., USA.

Jones, O.A. & Endean, R. (1976): Biology and geology of coral reefs. Vol.1-3. Academic Press, N.Y., London, San Francisco.

Kaestner, A.: Lehrbuch der speziellen Zoologie. VEB Gustav Fischer Verlag, Jena, Germany.

Kandel, E.R.(1979): Behavioral biology of *Aplysia*. A contribution to the compara- tive study of opisthobranch molluscs. Freeman Co., USA.

Kaplan, E.H. (1982): A filed guide to coral reefs. Houghton Mifflin Company, Boston, USA.

Kerstitch, A. (1989): Sea of Cortez marine invertebrates. Sea Challenger, Monterey, California, USA.

Kipper, H.E. (1986): Das optimale Meerwasseraquarium. Aqua Documenta Verlag, Bielefeld, Germany.

Kotzloff, E.N. (1987): Marine invertebrates of the Pacific northwest. University of Washington Press, Seattle, USA.

Kühlmann, D. (1984): Das lebende Riff. Landbuch Verlag, Hannover, Germany.

Kühlmann, D., Kilias, R. & Rauschert, M. (1993): Wirbellose Tiere Europas. Neu-mann Verlag, Radebeul, Germany.

Kükenthal, W. & Matthes, E. (1960): Leitfaden für das zoologische Praktikum. Gustav Fischer Verlag, Stuttgart, Germany.

Lalli, C.M. & Gilmer, R.W. (1989): Pelagic snails. The biology of holoplanctonic gastropod mollusks. Stanford University Press, Stanford, USA.

Lange, J. & Kaiser, R. (1991): Niedere Tiere tropischer und kalter Meere. Ulmer Verlag, Stuttgart, Germany.

Lawrence, J. (1987): A functional biology of echinoderms. Croom Helm, London, UK.

Levine, J.S. & Rotman, J.L. (1985): Faszinierende Unterwasserwelten. Birkhäuder Verlag, Basel, Switzerland.

Lindner G. (1975): Muscheln und Schnecken der Weltmeere. BLV, München, Germany.

Loughurst, A.R. (1981): Analysis of marine ecosystems. Academic Press, London, UK.

Mackie, G.O. (1976): Coelenterate ecology and behavior. Plenum Press, N.Y., USA.

Margalef, R. (1985): Western Mediterranean. Pergamon Press, Oxford, UK.

Matthes, D. (1978): Tiersymbiosen und ähnliche Formen der Vergesellschaftung. Fischer Verlag, Stuttgart, Germany.

Mayland, H.J. (1979): Niedere Tiere im Aquarium. Wirbellose im Meeresaquarium. Philler Verlag, Minden, Germany.

Möhres, F.P. (1964): Welt unter Wasser. Tiere des Mittelmeeres. Belser Verlag, Stuttgart, Germany.

Moosleitner, H. & Patzner, R.A. (in Druck): Mittelmeer. Niedere Tiere. Naglschmid Verlag, Stuttgart, Germany.

Morris, P.A. (1966): A field guide to Pacific coast shells. Mifflin Co., Boston, USA.

Morris, R.H., Abbott, D.P. & Haderlie, E.C. (1980): Intertidal invertebrates of California. Stanford University Press, Stanford, California, USA.

Muscatine, L. & Lenhoff, H.M. (1974): Coelenterate biology: Reviers and new perspectives. Academic Press, N.Y., USA.

Nesis, K.N. (1987): Cephalopods of the world. Squids, cuttlefishes, octopuses, and allies. T.F.H. Publications Inc., Neptune City, New Jersey, USA.

Nichols, D. (1967): Echinoderms. Hutchinson University Library, London, UK.

Nordsieck, F. (1968): Die europäischen Meeres-Gehäuseschnecken (Prosobranchia). Vom Eismeer bis Kapverden und Mittelmeer. Fischer Verlag, Stuttgart, Germany.

Nordsieck, F. (1969): Die europäischen Meeresmuscheln (Bivalvia). Vom Eismeer bis Kapverden, Mittelmeer und Schwarzes Meer. Fischer Verlag, Stuttgart, Germany.

North, W.J. (1976): Underwater California. Berkely, University of California Press, California, USA.

Olivier P.G. (1992): Bivalved seashells of the Red Sea. Hemmen Verlag, Wiesbaden, Germany.

Patzner, R. & Debelius, H. (1984): Partnerschaft im Meer. Pfriem Verlag, Wuppertal, Germany.

Pearse,V., Pears, M. & Buchsbaum, M. (1987): Living invertebrates. Blackwell Sciences Publication, Palo Alto, California, USA.

Poppe, G.T. & Goto, Y. (1991): European seashells. Vol 1. Hemmen Verlag, Wiesbaden, Germany.

Probst, K. & Lange J. (1975): Das große Buch der Meeresaquaristik. Ulmer Verlag, Stuttgart, Germany.

Randall, R.H. & Meyers, R.F. (1983): The corals. Guide to the costal resources of Guam. University Press of Guam.

Ray, C. & Ciampi, E. (1956): The underwater guide to marine life. A. S. Barnes, San Diego, California, USA.

Remane, A., Storch, V. & Welsch, U. (1985): Kurzes Lehrbuch der Zoologie. Fi-scher Verlag, Stuttgart, Germany.

Ricketts, E.F. & Calvin, J. (1968): Between Pacific Tides. Fourth Edition. Revised by J.W. Hedgepeth. Stanford University Press, Stanford, California, USA.

Riedl, R. (1983): Fauna und Flora der Adria. Paul Parey, Hamburg, Germany.

Roper, C.F.E., Sweeney, M.J. & Nauen, C.E. ((1984): Cephalopods of the world. An annotated and illustrated catalogue of species of interest to fisheries. FAO. Fish. Synop. Vol. 125.

Rose, K.J. (1980): Classification of the animal kingdom. David Mckay Company Inc., N.Y., USA.

Ruppert, E.E. & Fox, R. S. (1988): Seashore Animals Of The Southeast. A guide to common shallow-water invertebrates of the southeastern Atlantic coast. University of South Carolina Press, South Carolina, USA.

Salvini-Plawen, L.v. (1971): Schild- und Furchenfüßer. Neue Brehm Bücherei. Zimsen Verlag, Wittenberg, Germany.

Saunders, W.B. & Landman, N.H. (Hrsg.)(1987): *Nautilus*. The biology and paleobiology of a living fossil. Plenum Press, New York, USA.

Schmekel, L. & Portmann, A. (1982): Opisthobranchia des Mittelmeeres. Nudibranchia und Saccoglossa. Springer Verlag, Heidelberg, Germany.

Schmid, P. & Paschke, D. (1987): Rotes Meer. Niedere Tiere. Naglschmid Verlag, Stuttgart, Germany.

Schuhmacher, H. (1976): Korallenriffe. BLV Verlag, München, Germany.

Sefton, N. & Webster, St. K. (1986): Caribbean reef invertebrates. Sea Challenger, Monterey, California, USA.

Sharabati, D. (1984): Red Sea Shells. Routledge & Kegan Paul, Boston, USA.

Bibliography

Shepherd, S.A. & Thomas, I.M. (1982): Marine invertebrates of southern Australia. Woolman, Government Printer, Adelaide, Australia.

Sheppard, Ch.R.C. (1983): A natural history of the coral reef. Blandford Press.

Shirai S. (1970): The story of pearls. Japan Publications, Tokyo, Japan.

Smith, R.I. & Carlton, J. (1975): Lights Manual: Intertidal invertebrates of the central California coast. University of California Press, Berkeley, USA.

Spies, G. (1989): Praxis Meerwasseraquarium. Landbuch-Verlag, Hannover, Germany.

Stephenson, T.A. & Stephenson, A. (1972): Life between tidemarks and rocky shore. W.H. Freeman, San Francisco, USA.

Sterrer, W. (1986): Marine fauna und flora of Bermuda. John Wiley & Sons, N.Y., Chichester, Brisbane, Toronto.

Sweeney, M, Roper, C.F.E., Mangold, K.M., Clarke, M.R. & Boletzky, S.v. (1992): "Larval" and juvenile cephalopods. A manual for their identification. Smithsonian Contributions to Zoology 513. Smithsonian Institution Press, Washington, USA.

Thompson, T.E. (1976): Biology of opisthobranch molluscs. Ray Society, London, UK.

Thompson, T.E. (1976): Nudibranchs. T.F.H. Publications Inc., Neptune City, New Jersey, USA.

Thompson, T.E. (1988): Molluscs: Benthic opisthobranchs. Linnean Society, London, UK.

Valentin, C. (1986): Faszinierende Unterwasserwelt des Mittelmeeres. Einblicke in die Meeresbiologie küstennaher Lebensräume. Pacini Editore, Paul Parey, Hamburg, Germany.

Veron, J.E.N.: Corals of Australia and the Indo-Pacific. Angus & Robertson Publishers, London, UK.

Vine, P. (1986): Red sea invertebrates. IMMEL Publishing, London, UK.

Von Prahl, H., Brando, A. & Erhardt, H. (1988): Arrecifes del Caribe Colombiano. Villegas Editores, Bogotá, Colombia.

Von Prahl, H. & Erhardt, H. (1985): Colombia - Corales y arrecifes coralinos. FEN-Colombia, Bogotá, Colombia.

Voss, G.L. (1976): Seashore life of Florida and the Caribbean. Seamann Publishing, Miami, Florida, USA.

Walls, J.G. (1982): Encyclopedia of marine invertebrates. T.F.H. Publications Inc., Neptune City, New Jersey, USA.

Wilkens, P. (1987): Niedere Tiere im tropischen Seewasseraquarium. Engelbert Pfriem Verlag, Wuppertal - Elberfeld, Germany.

Wilkens, P. & Birkholz, J. (1986): Niedere Tiere. Röhren-, Leder- und Hornkorallen. Pfriem Verlag, Wuppertal - Elberfeld, Germany.

Wood, E. (1989): Coral Reefs. Salamander Book, London, UK.

Zann, L.P. (1980): Living together in the sea. T.F.H. Publications Inc., Neptun City, New Jersey, USA.

Zea, S. (1987): Esponjas del Caribe Colombiano. Catálogo Científico, Bogotá, Colombia.

Zeiller, W.: (1974): Tropical marine invertebrates of southern Florida and the Bahama islands. John Wiley & Sons, N.Y., London, Toronto.

Zlatarski, V. N. & Estalella, N.M. (1982): Les scléractinaires de Cuba. Édition de l'Academie bulgare des Sciences, Sofia, Bulgaria.

Photo Credits

Charles Anderson: 513, 602, 615, 627.

Hans A. Baensch: 42, 74, 139, 144 b, 148, 178, 312, 434.

Clay Bryce: 43, 140, 177, 184, 210, 372, 550, 551, 582, 593, 595 b., 606 t, 608.

Mr. Couet: 649 b., 651 b., 659 t., 670 t., 671 b.l., 674 b.

Helmut Debelius: 640, 647 b., 652 t., 660 b., 662 (2), 673 t., 676 t., 677 b., 681 t., 686 b.l., 688 t., 689 b., 690 t., 694 t.

Dieter Eichler: 185, 294, 295.

Harry Erhardt: 23, 25 b., 26, 27, 28, 29, 30, 31 t., 32, 33, 34, 35, 36, 37, 38, 39, 40, 41, 44, 45, 46, 47, 48, 49, 50, 51, 52, 53, 54, 55 t., 56, 57, 58, 59, 61 (2), 62, 63, 64, 65, 67, 68, 69, 70, 72, 73, 75, 77 (2), 78, 79, 80, 81, 82, 83, 84, 85, 86, 87, 88, 89, 90, 91, 92, 93, 94, 96, 97, 99 (2), 100, 101, 103 (2), 104, 105, 106, 108, 109, 111, 112, 113, 114, 115, 116, 117, 118, 120, 121, 122, 123, 124 (2), 125, 126, 142, 143, 145, 146, 147, 148, 149, 150, 151, 153 (2), 157, 158, 159, 160, 161, 162, 164, 165, 166, 167, 168, 169, 170, 171, 172, 173, 174, 175, 176, 186, 187, 207, 211, 212, 213, 214, 215, 216, 218, 220, 221, 222, 223, 225, 226, 227, 229, 230, 231, 232, 233, 234, 235, 236, 237, 250, 251, 252, 253, 254, 255, 256, 257, 259, 260, 261, 262, 263, 264, 265, 266, 267, 268, 269, 270, 271, 272, 273, 274, 275, 276, 277, 278, 279, 280, 281, 282, 283, 284, 286, 287, 288, 289, 290, 291, 292, 293, 294, 296, 297, 298, 299, 300, 301, 303, 305, 309, 312, 322, 323, 324, 325, 326, 327, 328, 329, 330, 331, 332, 333, 334, 335, 363, 364, 365, 366, 367, 368, 369, 370, 371, 373, 374, 375, 376, 377, 378, 379, 380, 381, 382, 383, 384, 385, 386, 387, 388, 389, 390, 391, 392, 393, 394, 395, 396, 397, 398, 399, 400, 401, 402, 403, 404, 405, 406, 407, 408, 409, 410, 411, 412, 413, 414, 415, 416, 417, 418, 419, 420, 421, 422, 423, 424, 425, 426, 427, 428, 429, 430, 431, 432, 433, 435, 436, 437, 438, 440, 441, 442, 443, 444, 445, 446, 447, 448, 449, 450, 451, 452, 453, 454, 455, 456, 457, 458, 459, 460, 461, 462, 463, 464, 465, 466, 467, 468, 469, 470, 473, 474, 475, 476, 477, 478, 481, 482, 483, 484, 485, 486, 487, 488, 489, 490, 491, 502, 508, 509, 519, 521, 526, 527, 529, 530, 531, 532, 533, 534, 535, 536, 537, 539, 565, 573, 574, 576, 578, 581, 584, 585, 586, 587, 589 b., 597 t., 598, 603 t., 607, 610, 611, 612, 614, 617, 619, 620, 623 (2), 625, 626, 629.

Wolfgang Fiedler: 645 t., 650 b., 652 b.r., 653 (2), 655 b., 663 b., 668, 675, 687 m.r., b.l. + b.r., 689 m., 693.

Herbert Frei: 654 t., 681 b., 687 m.l., 691 t.

Helmut Göthel: 665 t.

Otto Gremblewski - Strate: 13, 55 b., 66, 71, 110, 119, 127, 144 t., 249, 308, 512, 516, 517, 575, 588, 589 t., 603 b.r., 630 (2), 631 (2), 645 b.

Großkopf: 217, 219, 228, 238, 285, 439.

Howard Hall: 637 b., 639 (2), 659 b., 703 b.l.

Photo Credits

Johann Hinterkircher: 180, 496, 507, 509, 511, 518, 545 t., 564, 571, 580 (2), 590 (2), 597 b., 601, 605, 606 b., 616 (2), 621, 628.

Reimund Hübner: 661.

Achive IKAN: 313, 554, 644, 647 t., 650 t., 666, 669 t., 671 t. + b.r., 679 b., 682 t., 686 b.r., 691 b., 696 (3), 697 (3), 699, 700, 701 m.l. + m.r. + b.l.

Scott Johnson: 701 b.r., 702 t., 703 b.r., 705 b.

Alex Kerstitch: 687 t., 701 t.

Gerhard Krämer: 514.

Rudie Kuiter: 643 t., 649 t., 672 b., 705 t.

Horst Moosleitner: 95, 155, 179, 182, 188, 190, 192, 193, 307 (2), 498, 510, 547, 562, 609, 695.

John Neuschwander: 663 t.

Arend van den Nieuwenhuizen: 197, 599, 667 b., 682 b.

Aaron Norman: 568, 592, 594, 595 t., 596 b., 600 (2), 618, 622, 660 t., 672 t.

Horst Owsjanny: 596 t.

Robert A. Patzner: 31 b., 258, 306, 310, 311, 541 (2), 545 b., 546, 548, 549, 552, 553, 563, 569, 570, 572, 577, 579, 583, 604, 613, 624, 657, 658, 684, 685, 707.

Thomas Paulus: 651 t., 667 t.

Peter Reiserer: 683.

Ingo Riepl: 655 t., 669 b.

Ed Robinson: 676 b.

Arnd Rödiger: 603 b.l.

Ingo Schulz: 209.

Manfred Sieger: 141, 154, 156, 163, 166, 167, 471, 528.

Roger C. Steene: 656 b.

Dietrich Stüber: 362 (8).

W. A. Tomey: 208.

Horst Uecker: 189, 515.

Boris Unger: 25 t., 107, 191, 472.

Herwarth Voigtmann: 656 t., 692.

Bill Wood: 673 b.

Norbert Wu: 12, 181, 183, 194 (2), 195 (2), 196 (2), 495, 497, 499, 500 (2), 501 (2), 520, 566 (2), 567 (2), 704 t.

Natalie Yonow: 637 t., 641 (2), 643 b., 652 b.l., 654 b., 665 b., 670 b., 674 t., 677 t., 679 t., 688 b., 689 t., 690 b., 694 b., 702 b., 703 t., 704 b.

The Authors

Dr. Harry Erhardt works at the Institute of Zoology and Comparative Anatomy at the University of Kassel, Germany.
Born in 1938 at Riga (Latvia) and raised in the eastern Germany, Dr. Erhardt studied mining, then began formal studies in biology after graduating from high school. He specialized in zoology at the University of Gießen, Germany. In 1969 Dr. Erhardt received a scholarship from the German Academic Exchange Service (Deutscher Akademischer Austauschdienst = DAAD) and spent several years along the Caribbean coast of Colombia studying the fertility of various economically important fishes.
For more than 30 years, Dr. Erhardt has been diving and photographing marine organisms both professionally and as a hobby. He has explored virtually all tropical and subtropical seas including the Pacific, the Indian Ocean, Australia's Great Barrier Reef and, especially, the Caribbean where he has logged many hundreds of hours. But he is most at home in the Mediterranean where he first embarked on his fascinating hobby in 1963. Marine invertebrates hold the majority of his interest, and accordingly, he has written and lectured on the subject. Since 1963 Dr. Erhardt has used the same camera and equipment: as a friend of the medium format and of ROLLEI-technology, he uses a ROLLEIFLEX 3,5 F in a ROLLEIMARIN housing in conjunction with ROLLEINAR I and II lenses. The art of his expertise is beautifully displayed in this book.

The Authors

Dr. Horst Moosleitner was born on July 30. 1936. Presently he is an elementary school principal in Hallein, Austria, and an associate at the University Salzburg.

When he was just 14 years old he began to dive, and through this pathway developed an interest in biology. He began underwater photography in 1959, traveling to Dalmatia, Greece, and Creta, and in 1961 he drove to the Red Sea. His career as an author for specialty magazines coincided with his success as an underwater photographer. While working, he began studying biology at the University of Salzburg. In 1977 he completed his dissertation on the brain of *Blennius pavo* and received his doctorate. Since then Dr. Moosleitner has written several scientific papers, about 200 articles in aquarium and diving magazines, and has lent his knowledge towards numerous books.

Dr. Moosleitner's interest is largely centered upon fishes, sea anemones, starfish, the behavior of marine animals, and the various types of symbiotic relationships. Recognizing the need for keys that allow divers and aquarists to identify organisms *in situ*, Dr. Moosleitner puts his efforts towards designing keys that eliminate the need for sampling.